D0842949

# BIOMETRIC
# INVERSE
# PROBLEMS

# BIOMETRIC INVERSE PROBLEMS

Svetlana N. Yanushkevich
University of Calgary
Alberta, Canada

Adrian Stoica
California Institute of Technology
Pasadena, California, USA

Vlad P. Shmerko
University of Calgary
Alberta, Canada

Denis V. Popel
Baker University
Baldwin City, Kansas, USA

Taylor & Francis
Taylor & Francis Group

Boca Raton  London  New York  Singapore

A CRC title, part of the Taylor & Francis imprint, a member of the
Taylor & Francis Group, the academic division of T&F Informa plc.

Published in 2005 by
CRC Press
Taylor & Francis Group
6000 Broken Sound Parkway NW, Suite 300
Boca Raton, FL 33487-2742

### Library of Congress Cataloging-in-Publication Data

Biometric inverse problems.
      p. cm.
  by Svetlana N. Yanushkevich and others.
  Includes bibliographical references and index.
  ISBN 0-8493-2899-3
  1. Biometric identification--Mathematics. 2. Inverse problems (Differential equations) I. Yanushkevich, Svetlana N.

TK7882.B56B4578 2005
006.4--dc22                                             2005041944

Taylor & Francis Group
is the Academic Division of T&F Informa plc.

Visit the Taylor & Francis Web site at
http://www.taylorandfrancis.com

and the CRC Press Web site at
http://www.crcpress.com

# CONTENTS

# Preface

This book aims to contribute to the inverse problems of biometrics, a direction relevant to the synthesis of biometric data. It provides theory, tools and examples for practitioners and academicians working in fields which require the processing of biometric information such as those of security, personal identification, etc.

Synthetic biometric data has been the focus of many previous studies. These attempts were limited in that the synthesis was either non-automated or only semi-automated. Some examples include:

*Synthetic fingerprints.* Albert Wehde was the first to "forge" fingerprints in the 1920's. Wehde "designed" and manipulated the topology of synthetic fingerprints at the physical level. The forgeries were of such "high quality" that professionals could not recognize them[*]. Today's interest in automatic fingerprint synthesis addresses the urgent problem of testing fingerprint identification systems, training security personnel, biometric database security, and protecting intellectual property.

*Synthetic signatures.* The indirect study of synthetic signatures and handwriting began in the 16th century. Signatures and handwriting were considered to contain information about some of the psychological features of a person. Some elements of the psychological state of a person were thought to be "encoded" in his handwriting and could be "decoded". For such an analysis, the text can be selected arbitrarily because the information was carried by the features of handwriting. For example, I. Morgenschtern[†] analyzed some features in the signatures and handwriting of Napoleon, Nietzsche, Bismark, Goethe, Schiller, Shakespear, Chopin, Tolstoy, and many other famous people at various periods of their life. Morgenschtern stated that by analyzing signatures or handwriting, it is possible to receive psycho-graphological information from the original writer. Current interest in signature analysis and synthesis is motivated by the development of improved devices for

---

[*]S. A. Cole, *Suspect Identities – A History of Fingerprinting and Criminal Identification*, Harvard University Press, 2001

[†]I. Morgenschtern, *Psycho-Graphology*, Vayerman Publishing House, Saint-Petersburg, 1903

human-computer interaction which enable input of handwriting and signatures. The focus of this study is the formal modeling of this interaction.

*Synthetic irises.* Prior to computerized iris acquisition and processing, ocularists used natural techniques to synthesize the iris. The ocularist's approach to iris synthesis was based on the composition of painted primitives, and utilized layering semi-transparent textures built from topological and optic models. These methods are widely used by today's ocularists: vanity contact lenses are available with fake iris patterns printed onto them (designed for people who want to change eye colors). Notice that colored lenses, i.e., synthetic irises, cause trouble for current identification systems based on iris recognition. Additional efforts and synthesis tools are needed to overcome this problem.

*Synthetic speech and singing voices* have evolved considerably since the first experiments in the 1960s. Intelligibility seems to have reached nowadays a sufficiently high level, and new challenges have emerged. New targets in speech synthesis include improving the audio quality and the naturalness of speech, developing techniques for emotional "coloring", and combining it with other technologies, for example, facial expressions and lip movement. Synthetic voice should carry information about age, gender, emotion, personality, physical fitness, and social upbringing. The synthesis of an individual's voice will be possible too, the imitation based upon the actual physiology of the person. A closely related but more complicated problem is generating a synthetic singing voice. This problem has been approached from various directions for training singers, studying the famous singers' styles, and designing synthetic user-defined styles combining voice with synthetic music.

*Synthetic music.* Since the 1960s, efforts have been made at developing music synthesis algorithms based on the musical acoustics of live instruments. These results are claimed by the music community to improve the quality of music instruments. They can also be used to design automated systems for imitating the acoustic characteristics of various instruments. Synthetic music is widely used in training musicians and singers. There are many studies on the influence of different musical styles, played to humans, animals, and plants. To do this, systems for producing synthetic music with user-defined characteristics are useful.

*Synthetic emotions and expressions* are more sophisticated real world examples of synthesis. People often use their smile to mask sorrow, or mask gladness with a neutral facial expression. Such facial expressions can be thought of as artificial or synthetic in a social sense. In contrast to synthetic fingerprints and irises, the carrier of this synthetic facial information is a person's physical face rather than an image on the computer. The carrier of information can be thought of as facial topologies, indicative of emotions. To investigate the above problems, techniques to model facial expressions, i.e., the generation of synthetic emotions, must be developed. These results can be used, in particular, in a new generation of lie detectors. A related problem is how music or an instrument expresses emotions. To examine whether music produces emotions, a measuring methodology might be developed. Many ideas and algorithms have been proposed to solve these problems.

*Humanoid robots* are artificial intelligence machines which include challenging direct and inverse biometrics: language technologies, such as voice recognition and synthesis, speech-to-text and text-to-speech; face and gesture recognition of the "moods" of instructor, following of cues; dialog and logical reasoning; vision, hearing, olfaction, tactile, and other sensing (artificial retinas, e-nose, e-tongue)[‡].

This book focuses on the automated synthesis of biometric data (signatures, fingerprints, etc.). By understanding the problem of biometric data synthesis (the reverse problem), insights may be found in biometric data analysis (the direct problem).

## Why synthetic biometrics?

The primary focus of biometric technology is for personal verification and identification based on naturally possessed biological properties. Over the last decade, biometric technology has grown steadily and became a multi-billion dollar market for personal security and identification, access control and automated document processing. Currently deployed biometric systems use comprehensive methods and algorithms (such as pattern recognition, decision making, database searching, etc.) to *analyze* biometric data collected from individuals. We will consider the inverse task, *synthesis*, the creation of artificial biometric data to create biologically meaningful data for existing biometric systems useful, for example, to test biometric tools.

Inverse problems arise in many fields, where one tries to find a model that phenomenologically approximates observational data. In these problems, some direct model typically predicts a set of data that one is interested in.

There are different kinds of texts covering biometric systems: fingerprints for crime analysis, signatures for document automation, etc. This book addresses problems that are important for both analysis and synthesis of biometric data. The book emphasizes and contributes to the generation of synthetic biometric data which can also be thought of as *biometric data forgeries*. We argue that synthetic biometric data can improve the performance of existing identification systems by automatically generating biometric forgeries, and by modeling strategies and tactics of forgery ranging from so called *random forgeries* to *skilled forgeries*. An excellent example is a

---

[‡]A. Stoica, Humanoids for urban operations, *White paper, Humanoid Project, NASA, Jet Propulsion Laboratory,* Pasadena, 2005, http://ehw.jpl.nasa.gov/humanoid.

system by Cappelli, SFinGe, which generates synthetic fingerprints and uses them to test fingerprint identification systems.

Synthetic biometric data will play an important role in enhancing the security of identification and verification of biometric systems. Traditionally, security strategies (security levels, tools, etc.) are designed based on a *hypothetical* robber or forger. Properly created artificial biometric data, however, provides another approach to enhancing security through the detailed and controlled modeling of a wide range of training skills, strategies and tactics.

We show that some synthetic biometric data can be created using existing tools and modeling procedures. Connections are drawn to contemporary biometric tool design approaches based on traditional biometric information analysis techniques. We build on these traditional biometric techniques, and focus on the functional requirements of existing and emerging biometric devices.

## The key features and the scope of the book

The book is written in textbook style. There are about

- ▶ 200 examples throughout the text,
- ▶ 200 figures,
- ▶ 60 problems and
- ▶ 50 Matlab code fragments.

Existing biometric techniques and biometric tools have not previously considered the synthesis of biometric data, except in a few particular cases. The focus of the book is on developing design paradigms that can help discover novel algorithms, and perform functional synthesis, as well as to describe modelling, analysis, design, and optimization techniques for biometric data synthesis. Pursuing the idea of making the field of inverse problems of biometrics accessible to engineers, students and independent readers, the goals of the book are as follows:

(a) To introduce *data structures* to represent synthetic biometric information;

(b) To introduce *models* for the synthesis of biometric data with a focus on visual synthetic biometric information;

(c) To introduce *techniques* and *tools* for generating synthetic biometric data.

To achieve these goals, the authors reviewed existing methods and developed novel methods and algorithms in the area of contemporary digital image and signal processing, computational geometry, modeling, and the controlled distortion of biometric information. We assume that our gentle reader possesses fundamental knowledge of signal and systems, digital image and signal processing, probability theory and random processes.

We point to the following features that distinguish this book from others:

▶ A systematic study of the inverse problems of biometrics, including automated techniques for generating synthetic biometric information.

▶ The first attempt to discuss the impact of studying the inverse problems of biometrics on improving biometric systems for identification, and to discuss other practical applications such as training systems, security systems, and intellectual property protection.

Broadly, this book addresses the following:

Topological models as the central element of new biometric data synthesis systems. These utilize interpolation and approximation techniques, controlled distortions, controlled noise wrapping, and the assembly of synthetic biometric objects.

Techniques for advanced digital image processing (inverse transformations) used to synthesize specific biometric characteristics.

Innovative ideas and solutions inspired by recent advances in biometrics, such as various techniques for generating and using synthetic biometric information.

The book consists of eight self-contained chapters. To strengthen the intuitive approach, each chapter covers a specific biological object, describing its characteristic, and presenting techniques to synthesize data representative of it. This gives an unambiguous representation of the methods and tools for synthesis of each particular type of biometric. Each chapter contains a list of problems that can be used for self-study and self-evaluation, a list of bibliographical sources and interesting facts for further reading that allows the reader to investigate other methods and tools, and a list of references. The problems in each set range from simple applications of procedures developed in the book to challenging solutions of more complex problems inspiring suggestions, algorithms, or hypotheses, which may extend the ideas presented in this book. If you, ambitious reader, work out any of these ideas, we would be interested in seeing your results.

The ethical and social aspects of studying the inverse problems of biometrics are also discussed.

We introduce some areas of particular interest: testing of biometric systems, security, training of personnel, and intellectual property protection. In particular, we argue that in the field of image processing, image synthesis serves as a source for many innovative technologies, for example, virtual image simulation. Similarly, solutions to the inverse problems of biometrics are fostering pioneering applications.

More specifically, each self contained chapter covers the following:

*Chapter 1, Introduction*, gently introduces the directions and methodology used to solve the inverse problems of biometrics. We start by investigating forgeries, because these can be considered as a particular case of synthetic biometric data (forged signatures, faked fingerprints, etc.) The chapter describes how to use results from synthetic biometrics to increase the security of the original biometric data. Although biometric data tends by its nature to be unique, it lacks secrecy and can be recovered and used by an attacker, for example, a person leaves a record of their fingerprint when they touch things.

Also, in this chapter, the application of synthetic biometric information to solve security problems is discussed. We refer to several practical applications of synthetic biometric data: (a) intellectual property protection using the watermarking technique (the watermarking technique involves embedding information into the host data itself); (b) encryption of biometric templates (encoded samples of biometric data such as fingerprint, face etc.); (c) the technique known as steganography that involves hiding information in an unsuspected carrier data.

*Chapter 2, Basics of synthetic biometric data structure design,* lays the technology groundwork for developing automated synthetic biometrics tools. The chapter starts by defining invertible systems, and describing how they are represented by models and simulated. To introduce the concept of synthetic biometric data, various related inverse problems are introduced (direct and inverse Fourier transform, convolution and deconvolution, for example). Also the common problem of reversing or equalizing the distortion in a non-ideal system is considered from the viewpoint of synthetic data generation. Then, methods of generating random biometric attributes are considered. Some renowned solutions are introduced for assembling topological structures from synthetic primitives. Finally, techniques for distorting the topology of biometric data using various advanced computational geometry techniques (affine transform, approximation and interpolation techniques, fuzzy logic, etc.) are presented.

Chapters 3-6 introduce techniques to design synthetic signatures, fingerprints, faces, and irises. Each technique is different because of the specific nature of each biometric information source. However, there are many generic features. For example, distortion and assembling operators and data warping by noise are used for each type. These techniques can be implemented with automated tools, i.e., software tools for computational geometry, image processing and pattern recognition, and optimization based on artificial intelligence paradigms.

*Chapter 3, Synthetic signatures,* describes the synthesis of signatures based on various geometrical models. In contrast to iris and fingerprints, which do not change significantly over time, signatures evolve over time and vary each time they are produced. More precisely, the signature is a behavioral biometric. Statistical models for analysis and synthesis of behavioral information are imperfect. Though the chapter focuses on statistical approaches to synthesize signatures, other models are outlined. Signatures are easier to forge than, for example, irises or fingerprints. However, the automatic generation of synthetic signatures requires specific methods. Techniques used for signature synthesis depend on how the signature will be verified: off-line (the signature is on a document) and on-line (the signature is recorded in real-time on a special device). In on-line verification, writing dynamics such as pen pressure, line orientation, and pen speed are also captured, making it difficult to forge. We discuss an approach for generating synthetic signatures based on assembling topological primitives such as Bezier curves and other splines.

*Chapter 4, Synthetic fingerprints*, discusses generating synthetic fingerprints. The aim of automatic fingerprint synthesis is to construct a realistic image of a fingerprint without taking a real fingerprint. Thus, this new image is unique and not based on an actual person's fingerprint.

The synthesis of realistic fingerprint images is a multi-phase design process. In this Chapter, the different aspects of fingerprint image processing are discussed. A master fingerprint, which is an intermediate fingerprint-like topological structure in the design cycle, is discussed in detail with various MATLAB examples. Realistic fingerprints are achieved by adding stochastic information to the master fingerprint.

*Chapter 5, Synthetic faces*, discusses the automated synthesis of human faces and facial expressions. Synthetic facial attributes are needed for various applications, in particular: testing systems for face identification and recognition, including testing for animation techniques, training systems, medicine and psychology. Facial expressions carry information about affective state (fear, anger, enjoyment, etc.); cognitive activity (e.g., perplexity); temperament and personality (e.g., hostility, sociability); truthfulness (e.g., leakage of concealed emotions), and psychopathology (e.g., diagnostic information relevant to depression and other states).

*Chapter 6, Synthetic irises*, describes the technique for iris image generation. The human eye converts information in the form of light energy to nerve actions (electrical spikes). The *iris* collects and focuses the light onto the *retina* where it is detected. Both the iris and the retina are unique to an individual and contain useful biometric information. Several features make the eye a good carrier of unique individual information, in particular: (a) the spatial patterns that are apparent in the iris and retina are unique to each individual; (b) compared with other biometrics (such as face, voiceprints, etc.), the iris and retina are more stable and reliable for identification; (c) iris-based and retina-based personal identification systems are noninvasive because the iris and retinas are overt bodies. These are the motivations for studying the generation of synthetic irises and retinas for various practical applications, and for improving the characteristics of existing systems.

The eye carries short-term (behavioral) and long-term (topological structure) information. A synthesized image is created by assembling components from a library of primitives and basic colors. Many examples of synthetic iris generation are presented with MATLAB sources.

*Chapter 7, Biometric data structure representation by on Voronoi diagrams*, focuses on the properties of Voronoi diagrams useful for generating visual biometric data. Using the incremental properties of Voronoi diagrams, it is possible to manipulate primitives from a library to design biometric images and to color metric space via Voronoi regions. Voronoi transforms can be used during different phases of image analysis and synthesis. Various useful image effects can be achieved by using Voronoi transforms after or before other types of transforms, in particular, *distance* transforms. Details of the application of Voronoi diagrams to analysis and synthesis of signatures, facial topology and expressions, irises, and fingerprints are demonstrated with several examples.

*Chapter 8, Synthetic DNA*, contributes to the development of a new generation of biometric devices and systems. Current biometric technologies such as fingerprints, iris scans, face scans, voice patterns, signatures and combinations of those suffer from various disturbances (both environmental and physical) which lower the accuracy of personal identification systems as well as performance characteristics. DNA should be considered as a future biometric having no such limitations. DNA signatures used in forensics have shown that people can be identified with high accuracy. The main problem with having DNA as a biometric is the enormous time and resources required for sequencing and processing (alignment, matching, storing, etc.). Future DNA based biometric technology will ensure real time DNA sequencing and processing. One problem is a privacy concern during development and testing which can be dealt with by using synthetic DNA instead of actual sequences taken from humans. In this chapter, methods for reading and generating DNA sequences based on sequence analysis and synthesis techniques are introduced. Current genomic data is accessible through databases which contain billions of nucleotides. Both the meaning of DNA sequences and their statistical characteristics have been considered to ensure variability and separatability for artificially generated DNA.

## Targeted audience

This book is written for various audiences:

*Customers of biometric devices and systems.* Many organizations employ biometric technology. These include investigation agencies, corporations, and medical facilities that monitor people's fingerprints, retinas, and other physical traits to verify their identity. Customers of biometric devices and systems may find this book beneficial for the following reasons:

▶ The book contains practical examples for certain inverse problems in biometrics. The programs can be directly reused by customers of biometric devices and systems. Data generated would be useful for testing and comparing the quality of different manufactured systems.

▶ The book is structured in a way to cover as many physical biometrics devices as possible, assuring that all methods and algorithms have immediate implementation. At the end of each chapter, there is a section surveying existing biometric systems and presenting possible ways to integrate these algorithms into existing systems.

▶ This book also has a great educational component for those who want to explore biometric devices and systems.

*Researchers in the field of biometrics.* Research in biometrics is growing at an exponential pace, evolving from an immature set of techniques to characterize the physical features of a person to a multi-million dollar research market supported by industrial establishments and government institutions. The lack of publications in the general area of biometrics is due to the absence of a critical mass of research generalizing existing techniques. This book naturally targets researchers and provides the following:

▶ The book introduces the field of inverse problems of biometrics and provides a conceptual comparison with traditional techniques. All addressed problems are formalized, and the methods and algorithms used to solve them are described.

▶ Practical implementations of methods and algorithms are supplied with necessary technical justification. Ready-to-go algorithmic implementation samples and various scenarios for experimentation facilitate reading and understanding.

▶ Where well-known research approaches in particular areas of inverse biometrics are presented, constructive criticism is offered to motivate other researchers to contribute to the field.

*Vendors of biometric devices and systems.* Biometrics is a relatively new discipline and a novel commercial area. Most companies working to produce biometric hardware and to design processing software are relatively small, with an average of twenty employees each. There is a need to educate entrepreneurs, keeping them up-to-date with recent trends in biometrics. Our book provides the following:

▶ Practical recommendations aim to establish widely recognized standards and benchmarks for manufacturer product testing, tuning and implementation.

▶ The book makes recommendations on how to better view existing biometric systems by understanding the inverse problems and their solutions. Also, it can be used as a handbook for possible applications and existing configurations of biometric systems.

▶ The book is a self-education guide for those who are considering adoption of our methods and algorithms.

*Students.* The book has adoption possibilities for classes in biometrics, security, simulation, pattern recognition, and image analysis, and can be used as

▶ A textbook for graduate students,

▶ A special topics textbook for undergraduate students,

▶ A collection of problems for inter-department team projects (electrical engineering, computer science, psychology, medicine, mechanical engineering), and

▶ A textbook for continuing education.

There are three reasons to adopt this book as a textbook:

*The first reason.* Authors use many features of the textbook style, like detailed explanation, illustrations, and examples. Every chapter is equipped with a list of recommendations for further reading.

*The second reason.* The set of coding examples in the text itself can serve as a laboratory manual for various modeling and synthesis processes in biometrics.

*The third reason.* The book embraces a balanced approach to covering existing research topics and introducing new methods and algorithms. The book has a flexible structure of self-contained chapters. Readers can pick a set of these chapters while excluding the rest. This book is based on a number of the authors' lectures given in academia, and to government and commercial organizations, focusing on various aspects of biometrics. There are sources we recommend for further reading[§].

---

[§]Bolle R, Connell J, Pankanti S, Ratha N and Senior A. *Guide to Biometrics*, Springer, 2004. Jain A, Bolle R, and Pankanti S, Eds. *Biometrics, Personal Identification in Networked Society*, Kluwer Academic Publishers, 1999. Maltoni D, Maio D, Jain AK, and Prabhakar S, Eds. *Handbook of Fingerprint Recognition*, Kluwer Academic Publishers, 2002, Springer, 2003. Chirillo J and Blaul S. *Implementing Biometric Security*, John Wiley and Sons, 2003.

*Inquisitive persons.* There are many specialists from other fields with links to biometrics. We made a special effort to attract the attention of this category of readers. In particular, specialists from the following fields are likely to have vested interests in biometrics: medical imaging and video, security and authentication, robotics, etc.

This book does not attempt to be theoretically comprehensive: rather it is a first attempt to present the basics of synthetic biometrics, and should be treated as an introduction to the subject. By combining a focus on basic biometric principles with a description of current technology and future trends, this book aims to inspire the next generation of engineers to continue to develop the theory and practical tools of biometric technologies.

*Svetlana N. Yanushkevich*
BIOMETRIC TECHNOLOGIES: MODELING AND SIMULATION
Laboratory
University of Calgary, Canada

*Adrian Stoica*
BIOLOGICALLY INSPIRED TECHNOLOGY AND SYSTEMS Group
California Institute of Technology, CA, U.S.A.

*Vlad P. Shmerko*
BIOMETRIC TECHNOLOGIES: MODELING AND SIMULATION
Laboratory
University of Calgary, Canada

*Denis V. Popel*
Research Group on BIOMETRICS, Baker University, KS, U.S.A.

*Calgary*
*California*
*Kansas*

# Acknowledgments

We would like to acknowledge several people for their useful suggestions and discussion:

Dr. S. V. Ablameyko, Laboratory of *Pattern Recognition and Image Processing* Institute of Cybernetics, National Academy of Sciences, Minsk, Republic of Belarus.

Dr. L. Bruton, Department of Electrical and Computer Engineering, University of Calgary, Canada.

Dr. R. Fraga, Computer Science Department, Baker University, U.S.A.

Dr. R. Cappelli, The *Biometric Systems Laboratory* at Dipartimento di Elettronica, Informatica e Sistematistica, University of Bologna, Italy

Dr. O. Johnsen, Ecole d'Ingenieurs and Architekts de Fribourg, Switzerland.

Mr. Phil Phillips, Group on *Information Security, Office for National Statistics*, U.K.

Mr. W. Rogers, Publisher of *Biometric Digest*, U.S.A.

Dr. S. N. Srihari, Department of Computer Science and Engineering, State University of New York at Buffalo, Director of CEDAR, *Center of Excellence for Document Analysis and Recognition* U.S.A.

Mr. R. Walton, the *Biometric Technologies: Modeling and Simulation Laboratory*, Department Electrical and Computer Engineering, University of Calgary, Canada.

We would like to thank our graduate students and research assistants *Oleg Boulanov, Sergey Kiselev* and *Penlong Wang*, as well as *Ian Pollock* from from University of Calgary (Canada), and students *Antoine Widmer* and *Olivier Mulhauser* from the Ecole d'Ingenieurs and Architekts de Fribourg (Switzerland) for their valuable suggestions and assistance that were helpful to us in ensuring the coherence of topics and material delivery "synergy".

This work is the result of collaboration of:

Laboratory of *Biometric Technology: Modeling and Simulation*, University of Calgary, Canada,

*Biologically Inspired Technology and Systems Group*, California Institute of Technology, CA, U.S.A., and

Research Group on *Biometrics*, Baker University, KS, U.S.A.

In experimental study the following software tools have been used:

*SFinGe,* software for generation of synthetic fingerprints, University of Bologna, Cesena, Italy.

*FaceGen,* software for face modeling, Singular Inversions Inc., British Columbia, Canada.

*Comnetix,* software and Live-Scan station and software for fingerprint acquisition and identification, Ontario, Canada.

*MATLAB* environment and the software modules for image preprocessing, synthesis, warping etc., Biometric Technologies Laboratory, University of Calgary, Canada.

*signClone and DNAClone,* software packages for synthesis of signatures and biological sequences, Neotropy LLC, KS, U.S.A.

*Silicon Graphics* facilities at the University of Calgary, Canada.

Special thanks to Ms. *Nora Konopka*, Ms. *Rachael Panthier*, and Ms. *Linda Manis*, Taylor and Francis LLC for providing a friendly atmosphere at all phases of our joint work on the manuscript.

# 1

## Introduction to the Inverse Problems of Biometrics

This chapter focuses on the role of synthetic biometric information in state-of-the-art biometric systems. Contemporary techniques and achievements in biometrics are considered from two sides: analysis of biometric information (direct problems) and synthesis of biometric information (inverse problems) (Figure 1.1).

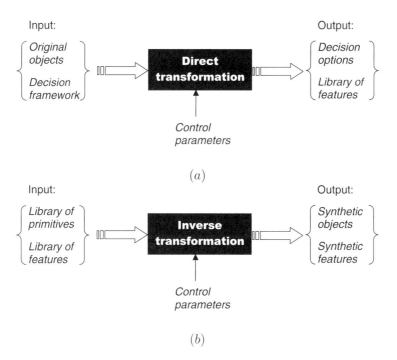

**FIGURE 1.1**
Biometrics include: direct problems (a) and inverse problems (b).

In this chapter, the methodological unity of the direct and inverse problems of biometrics is considered: analysis of biometric information results in the set of features used for recognition and design, and synthesis of biometric information results in synthetic biometric data designed by a set of rules.

This chapter starts with an explanation of possible attacks on a biometric system. This is motivated by one of the possible applications of synthetic biometric information: biometric systems using an imitation of a forger's attacks. In this, a forger is a natural generator of synthetic biometrics. Thus, an automated modeling of the forger's strategy and style might be the main principle of the generator of synthetic biometric information. The latter have controlled parameters associated with various forgeries.

In this book, an intruder (hacker, forger) is considered with respect to the possibility of concerting and coordinating recurrent attacks. Under this assumption, a biometric system is vulnerable. To protect the system, a possible scenario of the attacks, strategies and tactics of forgers must be analyzed and modeled. This is one of the reasons for developing inverse problems of biometrics aimed at designing synthetic biometric data with controlled characteristics. This means that synthetic data can be generated with respect to various control parameters, including extreme situations. The parameters are chosen

▶ To make it difficult to attack a biometric system, and

▶ Maximally resemble the skills, strategy, and tactics of an attacker.

This chapter also introduces the potential benefits of the solution to the inverse problems of biometrics through real-world examples. A brief introduction to Computer Aided Design (CAD) tools provides an understanding of the architecture and requirements of the components of direct problems, as well as an extension of the system to solving inverse problems.

The structure of this chapter is as follows. A possible scenario of attacks to a biometric system is introduced in Section 1.1. A forger is considered as a generator of synthetic biometric information. In Section 1.2, a brief overview of typical direct and inverse problems is introduced (the pair of Fourier transforms, convolution and deconvolution, etc.). It is shown that direct and inverse data manipulation can be considered as a complete solution to the various problems. From this position, direct and inverse problems of biometrics are considered in Section 1.3. An overview of biometric data structures and types is given in Section 1.4. Approaches to generating synthetic biometric data are represented in Section 1.5. Section 1.6 introduces the basic ideas of the conversating of biometric information used in biometric interfaces. Section 1.7 describes design tools for biometrics systems. Applications of devices that generate synthetic biometric information are studied in Section 1.8. Section 1.9 discusses the ethical and moral problems of inverse problems. After the summary (Section 1.10), a sample of problems for team projects are given (Section 1.11). Section 1.12 concludes the chapter with recommendations for further reading.

## 1.1 Attacks on biometric systems

Inverse problems of biometrics are aimed at enhancing the security of information or physical access to systems, as well as biometric systems themselves. Study of forgeries of biometric data addresses the problem of training a biometric system on a set of forged data.

### 1.1.1 Forgery of biometric data

A forger is understood as a highly-professional specialist who can produce high-quality copies of original biometric data. His/her knowledge and skills include:

▶ Understanding the weaknesses of biometric devices and systems (threshold of sensitivity, vulnerability of information flows, etc.),

▶ Awareness of the scheme of biometric data transmission in a biometric system,

▶ Skills to produce synthetic biometric information and use this information in a biometric system,

▶ Ability to choose a scenario of an effective, active attack on a biometric system.

Hence, a forgery of biometric data is defined as synthetic data that carries information (features) that maximally approximates original biometric data. There are two classes of synthetic data: (a) synthetic data that is an acceptable copy of original data, and (b) synthetic data that is hypothetical.

> **Example 1.1** *A biometric system must be carefully tested with respect to:*
>
> (a) *Thresholds of sensitivity (acceptance and rejection rate) of identification and recognition algorithms,*
>
> (b) *Access to data flows transmission in various phases of data processing,*
>
> (c) *System response to a "forger" attack. The "forger" can be imitated by automatically generated synthetic biometric data.*

The study of a scenario of attacks provides the possibility to improve a biometric system. Attacks should be modeled with parameters corresponding to real-life conditions.

The most significant role of the inverse methods of biometrics is:

▶ To advance in modeling of an attack,
▶ To automatically generate a possible critical situation for a biometric system,
▶ To study its behavior, and
▶ To develop the corresponding tools to prevent the potential for unwanted behavior of biometric systems in the future.

### 1.1.2   Scenario of attacks

Attacks on a biometric system can happen at the sensor level, at the level of the data communication stream, and at the component level of a system (the template database, etc.). Attacks on computer systems and networks can be divided into

*Active,* that involve the altering of a data stream or the creation of a fraudulent stream, and

*Passive,* that are inherently eavesdropping on, or snooping on, transmission. The goal of the attacker is to access information that is being transmitted.

Study of attacks includes:

▶ Modeling of a *scenario* of attack (skills of a forger, his strategy and tactics), and
▶ Modeling of a *defense* is divided into *passive defense* and *active defense.* This classification resembles the above definition of active and passive attacks.

In Figure 1.2 various attacks on a biometric system are illustrated.

> **Example 1.2** *Below examples of attacks are given:*
>
> (a) *Attacks involving presentation of biometric data at the sensor are given in Figure 1.2.*
> (b) *A hacker could replace the biometric library on the computer with a library that always declares a true match for this user.*
> (c) *Attack on the database could result in authorization of a fraudulent individual, or removal of a person from a screening list.*

Finally, a new scenario of attacks should be considered based on the solutions of the inverse problems of biometrics. We indicate only two of them:

▶ *Massive attacks* involve the emission of a large amount of randomly generated synthetic biometric information that can be incorporated at various levels of the biometric system.

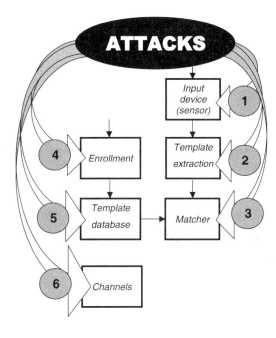

Scenario 1. The true biometric is presented but in some unauthorized manner.

Scenario 2. After the features have been extracted from the input signal they are replaced with another feature set.

Scenario 3. The matcher is attacked directly to produce a high or low match score resulting in a change of decision making strategy.

Scenario 4. Authentication protocol.

Scenario 5. Unauthorized modification of data in the database.

Scenario 6. Attacks on the channels. For example, an attack on the channel between the sensor and biometric system could consist of a resubmission of a previously stored biometric signal.

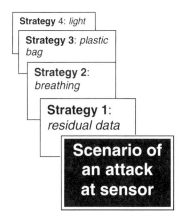

A scenario of a sensor attack can be modeled based on the following strategies:

Scenario 1. Using residual biometric data on the sensor surface

Scenario 2. Breathing on the sensor's surface,

Scenario 3. Placing a thin-walled water-filled plastic bag on the sensor surface

Scenario 4. Using a halogen lamp to create a kind of snow blindness in the sensor

**FIGURE 1.2**

Possible scenarios of attack on a biometric system and a scenario of sensor attacks on a biometric system consists of attacks with respect to some different strategies and tactics of forgers (Example 1.2).

▶ *Increasing the probability of unauthorized accesses* by generating synthetic information that resembles biometric data very closely.

From a technical viewpoint, it is possible to develop effective strategies, algorithms and tools for defense.

### 1.1.3    Examples of forgeries

Here, two classes of forgeries, signature and fingerprints, which have been well-studied, are introduced. A signature carries the *behavioral* characteristics of a person: a signature can be changed at any time, the psychological state of a person influences a signature, too (Figure 1.3). A fingerprint carries *physical* characteristics: it cannot be changed during the person's life. The known models of these forgeries provide possibilities for the automated generation of synthetic signatures and fingerprints.

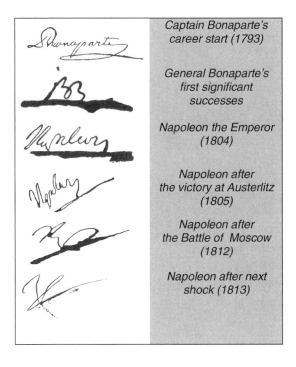

Captain Bonaparte's career start (1793)

General Bonaparte's first significant successes

Napoleon the Emperor (1804)

Napoleon after the victory at Austerlitz (1805)

Napoleon after the Battle of Moscow (1812)

Napoleon after next shock (1813)

A signature is a means of personal identity authentication and verification. A signature does not have characteristics of permanence in that a person can change his or her signature at any time. Stability and style variation are the main characteristics in handwriting analysis, recognition, and synthesis. A handwritten signature is the result of a rapid movement. The shape of the signature remains relatively the same over time when the signature is written down on a pre-established frame (context). This physical constraint contributes to the relative time-invariance of the signatures.

**FIGURE 1.3**
A signature carries the behavioral characteristics of a person.

**Forgeries of signatures.** Traditionally, in a scenario of attack on a system for signature and handwriting identification and verification, two kinds of forgers are assumed (Figure 1.4):

*Random* forgers, whose activity is recognized as *zero-effort* forgeries (random scribble), *home-improved* forgeries, and

*Skilled* (professional) forgers, that produce high quality forgeries based on skills in handwriting techniques.

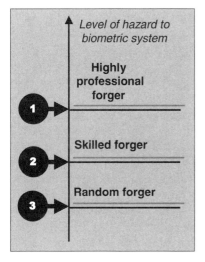

*Highly professional forgers:* Level of danger is 1:
Behavior of biometric system - unpredictable:

*Skilled forgers:* Level of danger is 2:
High-quality forgeries
Behavior of biometric system - inprecise

*Random forgers:* Level of danger is 3
Forgery is classified as:
(*a*) a random scribble of a signature,
(*b*) a first attempt to forge the signature of another person,
(*c*) a home-improved forgery,
Behavior of a biometric system - regular

**FIGURE 1.4**
Classification of forgers of signatures with respect to level of danger and the behavior of a biometric system.

In this book, *highly professional* forgers are considered as forgers who can *produce* and *manipulate* a synthetic signature with respect to possibilities of advanced automated techniques of identification and verification. These kinds of forgers employ knowledge and skills as follows:

▶ There are a lot of cases when it is difficult to discriminate a genuine signature from a forged one, i.e., probability of correct decision making is not satisfactory.

▶ *Static* characteristics of a signature carry information about its topological structure.

▶ *Dynamic* characteristics include not only the topological features of a
   signature but also features like acceleration, velocity, and trajectory
   profiles.

**Forgeries of fingerprints.** In the case of a *registered* finger, the intruder can
force the legitimate user to press his/her finger against the fingerprint sensor
under duress. Also, the intruder can give the legitimate user a sleeping drug,
in order to use either the finger directly against the sensor, or by making a
mold of the finger.

The defense against attacks with respect to the first scenario depends on
many factors (type of biometrics, type of sensor system, etc.).

Finally, a synthetic fingerprint is a fingerprint made to imitate a real (living)
fingerprint. It can be made of gelatin, silicone, play-doh, clay, or other
materials.

### 1.1.4   Trends

In response to new requirements for the reliability of biometric devices and
systems, heightened security measures (such as personnel training for extreme
situations) and the need to combine the efforts of researchers from various
disciplines to tackle the problem, the demand for novel biometric methods
and equipment becomes obvious.

Biometric systems and devices inevitably become more complex, requiring
a wide range of technologies and techniques to be used in their creation. In
order to expedite the evolution and testing of biometric systems, new tools
based on the inverse biometrics paradigm are required.

One can observe that biometric study often leads to a set of experimental
results whose value is limited, because they reveal only a limited insight into
the problem, or fail when repeated for similar, but not identical, situations.

> **Example 1.3** *The performance and characteristics reported
> by vendors often fail to be confirmed in practice. To check the
> vendor's data, extensive tests with various synthetic biometric
> information must be performed.*

Another trend in biometrics is related to reconstructive surgery that aims at
rebuilding damaged faces. At the first phase, a face model is constructed. At
the second phase, a physical prototype from computer generated data is built,
traditionally, by the stereolithography approach (layer by layer from a pool of
liquid resin). In this way it is possible to create unique, anatomically-accurate
"bio models" with respect to individual patient's data.

> **Example 1.4** *There are several approaches to recognizing a
> synthetic face - for example, a thermography image.*

Visible light, as a form of electromagnetic energy, has wavelengths ranging
from about 400 $\mu$m for violet light to about 700 $\mu$m for red light. It is

well documented that the human body generates waves over a wide range of the electromagnetic spectrum, from $10^{-1}$ Hz (resonance radiation) to $10^9$ Hz (human's organs) (Figure 1.5). The human body radiates nonvisible infrared light (waves of 3-12 $\mu$m long) in proportion to its temperature.

The Stefan-Boltzmann law states that the total radiation emitted by an object is directly proportional to the object's area and emissivity and the fourth power of its absolute temperature. All objects with a temperature above absolute zero emit infrared radiation.

Brain waves can be visualized as electroencephalograms and have a frequency range of about 1-60 Hz. Less known mechanisms are also present, for example, body cells generate waves in the range of about 1.5 MHz to 9.5 MHz.

The temperature of an area of the body is a product of cell metabolism and local blood flow. With some processes (disease, emotion, etc.) there is a reduction or change in blood flow to affected tissues. This will result to alterations surface temperature.

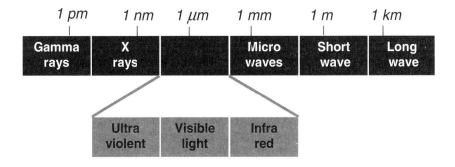

**FIGURE 1.5**
Electromagnetic waves provide information used in biometric technologies; mainly visible and infrared light is used.

The human face and body emit infrared light both in the mid-infrared (3-5 $\mu$m) and the far-infrared (8-12 $\mu$m). Most light is emitted in the longer-wavelength, lower energy far-infrared band. Infrared thermal cameras can produce images of temperature variations in the face at a distance. Infrared light in the band (1-3 $\mu$m) is reflected in the way that visible light is rather than emitted by warm body. It does not give a measure of temperature but can still provide other useful information.

**Example 1.5** *Consider the upper-band of the near-infrared (1.4–2.4 mμ) image.*

(a) *Human skin and hair have low and high reflectance respectively and appear as dark and bright areas in a thermal image accordingly.*

(b) *Artificial hair has lower reflectivity than natural human hair and results in a darker image. However, it is difficult to detect surgical face alterations in this band.*

It should be noted that facial thermal images are correlated with effects of specific drugs and alcohol. Therefore, these images can be used to remotely monitor a crowd.

There are many phenomena that can be useful in developing non-contact techniques for human emotion identification, but this area is not yet well studied.*

## 1.2    Classical direct and inverse problems

In this section, typical direct and inverse problems are introduced. The aim is to show that solution of the problem is based on an understanding of direct (analyzing) and inverse (synthesis) modeling.

Inverse problems (also called *model inversion*) arise in many fields, where one tries to find a model that phenomenologically approximates observational data. In these problems, some direct model (also called *forward* model) typically predicts a set of data that one is interested in. The problem is that a large number of parameters remains undetermined in the model.

**Example 1.6** *Often in the above cases one has to resign oneself to a manual trial and error process. If a linear relationship holds between the parameters in the search space and those obtained by the model, this might be a straightforward method.*

However, if nonlinear behavior is observed, the modeling becomes a rather difficult process. In this case, guided search methods are used to optimize the process in nonlinear environments.

---

*The field of bioenergy science that studies a bodies radiation is called *bioelectromagnetics*. In particular, so-called *Kirlian effect* or *aura* images carry useful information about emotions. Historically, S. Kirlian was the first who demonstrated body radiation by photography and electrofield imaging. W. Kilner used filters to analyze body radiation in the visible spectrum. In today laboratory systems, this information can be measured using various methods, in particular, by devices based on sub-harmonic analysis in the range of 100 to 800 MHz.

**Example 1.7** *Evolutionary algorithms use error information from previous trials in the feedback loop in order to choose the next model parameters to be tested. The adaptation of the model to a given environment is thereby transformed into an optimization task. A set of model parameters is sought that results in the best possible performance of the model. In cases where the forward model is not exactly invertible, one looks for the best possible approximation, as often occurs in real world applications where models are, at best, good approximations.*

Given a function $f$, the relation $g$ is called the inverse function of $f$ if and only if $g$ is a function, and $g$ is obtained by interchanging the components of each of the ordered pairs in $f$.

**Example 1.8** *Let $f = (x, y)|y = 2x - 3$, then the inverse function is specified as $f = (x, y)|x = 2y - 3$*

**Signal and systems.** There are a lot of techniques in signal processing that are based on inverse paradigm, in particular, inverse filtering, inverse Fourier, Laplace, $Z$, Walsh, and Haar transforms, inverse Euler formulas, invertible system, and inverse operators. They are relevant to two aspects of the Fourier theory: analysis and synthesis. Given a signal $x[n]$ (discrete time domain), calculation of coefficients $X[k]$ (frequency domain) is called *Fourier analysis*. The reverse process that generates $x[n]$ from $X[k]$ is called *Fourier synthesis* (Figure 1.6).

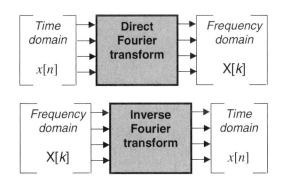

Direct Fourier transform:

$$x[n] = \sum_{k=0}^{N-1} X[k]e^{jk\Omega_0 n}$$

Inverse Fourier transform:

$$X[k] = \frac{1}{N} \sum_{n=0}^{N-1} x[n]e^{-jk\Omega_0 n},$$

$x[n]$ is a periodic signal with fundamental period $N$ and fundamental frequency $\Omega_0 = 2\pi/N$,
$X[k]$ are the coefficients of the signal $x[n]$.

**FIGURE 1.6**
Direct and inverse Fourier transforms.

A *system* is a set of related entities, *components* or *elements*. A system is *invertible* if the input to the system can be recovered from the output except for a constant scale factor. This requirement implies the existence of an inverse system that takes the output of the original system as its input and produces the input of the original system (Figure 1.7).

> **Example 1.9** *In a linear time invariant system, the process of recovering $x(t)$ from $h(t) * x(t)$ is termed* deconvolution, *since it corresponds to reversing the convolution operator. Thus, an inverse system performs deconvolution. A common problem is that of reversing or* equalizing *the distortion introduced by a nonideal system. In many equalization applications, an exact inverse system may be difficult to find or implement. An approximate solution is often sufficient in such cases.*

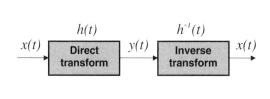

An inverse system performs deconvolution

$$x(t) = y(t) * h^{-1}(t)$$
$$= [x(t) * h(t)] * h^{-1}(t)$$
$$= x(t) * [h(t) * h^{-1}(t)]$$
$$= x(t)$$

where $h(t) * h^{-1}(t) = \delta(t)$

**FIGURE 1.7**
A linear time invariant system (direct transform) and inverse system (inverse transform).

**Synthetic benchmarks.** Traditionally, benchmark sets are used for testing new algorithms. The problem arises if testing conditions are not adequate to real conditions. A possible solution is to develop *generators of synthetic data*, so that circuit parameters can be controlled. There are several benefits from synthetic benchmark generation, in particular:

▶ The specified parameters of a constrained synthetic data can be taken either from the default profile or chosen by the user.

▶ Realism of the obtained circuit can be achieved by combining the above strategies, i.e., the advantages of parameterized random data generation (model), actual characteristics, combining local and global automated correction and control, and the user's decision making.

▶ It is possible to generate data that are not available in standard benchmark sets. For example, one parameter can vary while others are fixed or scaled appropriately. This way, a set of benchmarks can be generated by variations of the chosen parameters.

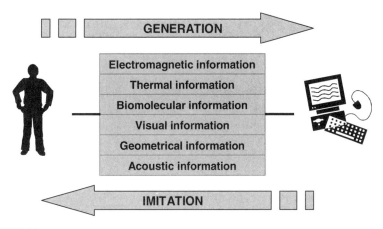

**FIGURE 1.8**
A biological object as a source of different types of information. Imitation of biometric information is a solution to the inverse problem.

Interaction between the analysis of circuits and the generation tools is of fundamental importance.

## 1.3 Direct and inverse problems of biometrics

As stated earlier, the *direct* and *inverse* problems of biometrics correspond to the analysis and synthesis of biometric information, respectively. The inverse problems (also called *model inversion*) can be formulated as finding a model that phenomenologically approximates observational data. In these problems, some direct model (also called *forward* model) typically predicts a set of data that one is interested in.

### 1.3.1 Automated analysis of biometric information

The direct problems can be used to analyze biometric information. For solving inverse problems of biometrics, methods for modeling and simulation of biometric information are needed. For example, speech can be synthesized directly from text. However, if a genuine voice is given, we define its simulation by a device as voice (speech) imitation. The direct problem (analysis) is a state-of-the-art concept of biometric technology (Figure 1.8), whereas the inverse problem is less studied. Synthetic biometric data can be represented in various forms (electrical, electromagnetic, visual, etc.) For example, synthetic signatures, hand-geometry and fingerprints are topological structures.

### 1.3.2   Automated synthesis of biometric information

Automated synthesis of biometric information is the inverse problem to the analysis of biometric information. For example, in the field of image processing, image synthesis serves as a source for many innovative technologies, such as virtual image simulation. Similarly, solutions to inverse problems in biometrics are now fostering pioneering applications, such as unique biometric imitators that reflect both psychological and physical characteristics.

The solution of the inverse problems can help to

- ▶ Discover novel data structures to represent biometric information;
- ▶ Discover novel architectures for biometric devices and systems;
- ▶ Perform functional synthesis of biometric information, and carry out its analysis and modeling;
- ▶ Extend methods of computer aided design of biometric systems;
- ▶ Perform optimization of biometric information.
- ▶ Study different attacks on biometric systems.

In Table 1.1, the direct and inverse problems associated with various biometric objects is given.

The efficiency of algorithms used in biometrics largely depends on our understanding of inverse problems. A reasonably efficient representation of artificial biometric information is of fundamental importance for the quality of biometric devices and systems. One criteria for choosing an artificial biometric data structure is its potential for analysis. The concept of likeness of biometric information is theoretically bounded by different methods from different fields of biometrics, and it is the base for testing biometric devices and systems. These methods allow us to analyze the behavior of a system at different levels of abstraction and offer powerful methods and algorithms for manipulating biometric data structures.

## 1.4   Basic notion of biometric data

In this section, the basic notion of biometric data used in automated systems is introduced. There are several pieces of terminology that are used in the inverse problems of biometrics:

- ▶ Biometric data *classification* and *representation*,
- ▶ Biometric data *meaningfulness*, and
- ▶ Biometric data *design*.

**TABLE 1.1**
Direct and inverse problems of biometric technology.

| DIRECT PROBLEM | INVERSE PROBLEM |
|---|---|
| Signature identification | Signature *forgery* (signature imitation) |
| Handwriting character recognition and keystroke dynamics identification | Handwritten text *forgery* (handwritten imitation) |
| Face recognition | Face *reconstruction* and mimics animation |
| Voice identification and speech recognition | Voice and speech *synthesis/imitation* |
| Iris and retina identification | Iris and retina *image synthesis* |
| Fingerprint identification | Fingerprint *reconstruction/imitation* |
| Hand geometry identification | Hand geometry *reconstruction/imitation* |
| Infrared identification | Infrared image *reconstruction* |
| Ear identification | Ear geometry *reconstruction/imitation* |

## 1.4.1 Representation of biometric data

In automated synthesis of biometric data, the following basic definitions are used:

*Synthetic* or *artificial* biometric data is data obtained by the rules taken from original biometric data. For example, given the set of rules and appropriate approximation technique it is possible to generate a set of synthetic signatures that can be considered as forgeries of given biometric data. Synthetic biometric data is generated by a *generator* or *synthesizer*.

*Primitives* are data that carry small portions of information about a given biometric object. *Topological* primitives are the results of topological decomposition of biometric data. They can be identified and recognized in topological Cartesian space and frequency domain. A collection of primitives that are used in design is called a *library*.

*Master-data* is a representation of biometric data after only initial synthesis steps have been performed. Master-data carries features of original biometric data but still does not possess the important characteristics of real images such as noise, sensor parameters etc.

## 1.4.2   Description of biometric data

It is reasonable to describe synthetic data in terms of qualitative information because synthetic data is the source of information:

▶ Particular information on a given type of biometric is incorporated into some carrier of information.

▶ This particular information becomes available for synthesis after analysis and categorization of information.

The term "qualitative information" is a synonym of the term "data". However, the term "quantitative information" is used to a numerical evaluation of the information carried by data. A quantitative evaluation of information is needed, for example, in evolutionary algorithms and artificial neural networks for identification, recognition, and generation of synthetic biometric data.

The following definitions are used in automated biometric information generation:

*Recoverable* and *unrecoverable* data are related to the problem of direct and inverse transforms. For some classes of transform, it is impossible to recover initial data because the inverse transform does not exist.

*Acceptable and non-acceptable synthetic data.* One of the possible definitions of acceptable synthetic data is as follows: this is biometric data that can be tested by experts and automated identification systems and can not be recognized as synthetic data. For example, an expert and fingerprint identification system and can not distinguish a synthetic and original fingerprint. In contrast, *non-acceptable synthetic data* is the intermediate results of the generation of biometric information and defined as data that can be recognized as synthetic biometric data through testing.

*Hidden* biometric information is defined as parameters or features that are incorporated into a hidden carrier of information and/or parameters or features composed with other data. An example of hidden biometric information is the watermarking technique. Hidden biometric information can be also defined as synthetic biometric information created from original biometric information by distortion transforms. An inverse distortion transform allows for recovery of original biometric information.

**Example 1.10** *To prevent theft of information from a database of fingerprints, distorted copies of original information can be stored in a database.*

*Data* fusion is the process of deriving a single consistent knowledge base from multiple knowledge bases. This process is important in many cognitive tasks such as decision making, planning, design, and specification, that can involve collecting information from a number of potentially conflicting perspectives, sources, or participants.

**Example 1.11** *Various levels of abstraction are used in information fusion:*

(a) *In biometrics, fusion is known as a multiple biometric that uses various sources of biometric information, for example, facial, fingerprint, and hand geometry.*

(b) *Given biometric data, for example, a fingerprint; a fusion is an extraction of information from the primitives that composed this fingerprint.*

In the inverse problems of biometrics, an approach that is *inverse to data fusion* is considered:

$$< \texttt{Single knowledge base} >$$
$$\Rightarrow \quad < \texttt{Multiple component knowledge bases} > .$$

### 1.4.3 Design of synthetic biometric data

Design of synthetic biometric data is based on several paradigms. The most important are defined as follows:

The *Scenario of synthesis* of biometric data closely relates to the scenario of real world conditions of collection and analysis of biometric data, for example, on attack on a biometric system, skills of a forger, his strategy, tactics, etc.

*Modeling* is the generation of synthetic biometric data.

*Decomposition* is the *partitioning* of data into topological primitives. In image processing, this procedure is called the *segmentation* of an image into regions that can be represented in a form suitable for further processing. In analysis, decomposition reduces complexity and simplifies the description, for example, reducing boundary complexity by convex hull computing. In synthesis, the decomposed biometric data is represented by a set of topological primitives that are used, for example, in assembling synthetic data.

*Composition* is the *assembling* of biometric data from primitives. *Assembly* means the construction of more complex systems from the components provided, in particular, with features identical to the components which began the process. An assembling algorithm is based on *design rules*, a set of rules for assembly of a fingerprint image from primitives (macroprimitives).

> **Example 1.12** *Two synthetic primitives of a fingerprint can combined to create a larger primitive. The assembly procedure must satisfy various general topological criteria (joint points, directions, etc.) and particular criteria of a fingerprint (not all primitives can be used for assembling, directions of connections can not be arbitrarily chosen, etc.).*

A *design technique* aims to synthesize acceptable synthetic data, for example, with minimal errors, reasonable control parameters (local and global topological regulations), local (small size) and global (large size) application, etc. A design technique includes data structures, models, and algorithms for data manipulation.

*Transforms* are considered in *topological space* and the *spectral domain* related to the representation of biometric data as a topological structure (image) or a set of spectral coefficients. A *Topological transform* converts any initial topological structure into another topological configuration.

> **Example 1.13** *Examples of topological transforms are distance transforms, affine transforms, and coordinate transforms. The deformation or warping technique is based on various topological transforms.*

*Local and global synthesis.* Merging several primitives is a local design of biometric data. Global correction can be applied to synthetic biometric data after non-satisfactory results of testing.

*Topological compatibility.* One of the possible definitions of topological compatibility is as follows: this is the characteristic of processing algorithms that indicates that data do not require preprocessing when using different algorithms for manipulation of topological data.

*Hidden biometric information* is synthetic biometric data that is created from original biometric data by distortion transforms. Inverse distortion transforms allow recovery of original biometric data.

*Controlled distortion* is a modification of original biometric information; this process is under control, i.e., modification is made with respect to a set of parameters and/or characteristics.

> **Example 1.14** *The fingerprint attribute is generated by wrapping some additional details randomly (with given distribution) generated (points, lines, etc.). The result is an attribute that is similar to an attribute in a real fingerprint.*

### 1.4.4 Artificial intelligence paradigm in synthetic biometric data design

Artificial intelligence techniques are a natural way to synthesize biometric data.

A *learning paradigm* is a technique that uses artificial intelligence methods and evolutionary strategies in some of its phases.

> **Example 1.15** *The evolutionary strategy can be used for assembling synthetic fingerprints and testing the results with respect to specific biometric criteria.*

*Self-assembly* is the process of construction of a unity from components acting under forces/motives internal or local to the components themselves, and arising through their interaction with the environment. Self-assembling structures create their own representations of the information they receive. That is, assembly according to a distributed plan or goal implicit in the structure and behavior of the parts, not explicitly stated or held by a central controlling entity. Components in self-assembling systems are unable to plan but respond to their surroundings. The stimuli to which a component is capable of responding are dictated by that component's physical composition and properties, for example, minimum energy states. This is quite different from traditional programming methodology, which requires all data to be explicitly specified by the programmer. Self-assembling structures are well suited for problems for which it is either difficult or impossible to define an explicit model, program or rules for obtaining the solution.

> **Example 1.16** *A design system based on the evolutionary strategy is the typical self-assembling system. This system is able to process noisy, distorted, incomplete or imprecise data, and it has higher-level properties that cannot be observed at the level of the elements (primitives), and that can be seen as a product of their interactions.*

Neural networks perform parallel information processing functions; their design was originally inspired by biological neural systems. A neural network consists of two general elements:

▶ The processing units, called *neurons*, and
▶ The connections between the neurons, the *weights*.

*In direct processing*, weights are the main information storage locations in the network. In order to determine the weights of a network, one has to give a "training algorithm" that extracts the weights from a set of training data. However, training algorithms have to deal with the danger of convergence to local optima, especially in large search spaces and for large training data sets. One often observes that the algorithm gets stuck in a local optima before a satisfactory solution is found. One way to circumvent this problem is a time-consuming trial and error process. An alternative is the use of a *global search algorithm*. This searches global optima instead of local ones and the danger of premature convergence to false solutions becomes small.

*In the inverse optimization problem,* the global search algorithm alters the weights of the network and receives feedback on how well the network performs. To cast this into inverse modeling, the algorithm can be seen as being in the feedback loop of an iterative process with the neural network as mathematical model to adapt to some given data set (Figure 1.9).

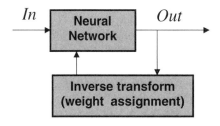

*Inverse optimization problem: An optimization algorithm in the feedback loop of an iterative process within the neural network*

**FIGURE 1.9**
Inverse modeling in a neural network.

## 1.5    Examples of synthetic biometric data

This section is a brief introduction to the problem of synthetic data generation. Various types of synthetic biometric data are considered. The solution of inverse problems requires particular methods and algorithms for each type of biometric - there is no general approach. This chapter introduces the practical aspects of the problem of the relationship of various types of biometric information.

**Synthetic signature.** Generating s fictitious signature is the simplest example of an inverse problem in biometrics. The task of verification of handwriting objects is to determine whether a given signature or handwriting is genuine or forged. The forged handwriting object has the proper shape, but differs from the genuine object in the quality of the strokes. Free-hand forgeries, on the other hand, can differ from a genuine signature with respect to the values of various size and shape features.

Like a signature imitation, an imitation of handwriting is a typical inverse problem of graphology. In contrast to signature imitation, experts have more statistical data. The relevant imitation problem of keystroke dynamics is similarly characterized.

**Synthetic voice and speech.** Speech characteristics contain prosody (pitch structure, stress structure, and tempo information), phoneme and vocal quality. There are two aspects to the problem:

(*a*) Voice synthesis (that satisfies given characteristics), and

(*b*) Voice imitation (given a voice, imitate another voice).

There are also available techniques and tools for detecting emotions in speech. For example, the level of emotion can be identified by speech as neutral, angry, sad, and joyful. Solutions of the inverse problem, speech synthesis with given characteristics, are well known.

**Synthetic face.** Face recognition systems detect patterns, shapes, and shadows in the face. Face reconstruction is a classic problem of criminology. However, due to the complexity of the problem, computer-aided methods for *face reconstruction* are still far from satisfying quality requirements. This inverse problem is relevant to the direct problem of face recognition, which is also a complex one.

> **Example 1.17** *Many devices are confused when identifying the same person smiling or laughing. Devices can be improved by training on a set of synthetic facial expressions generated from real face image (Figure 1.10). Another problem is identifying speech that was spoken with a different voice. The inverse problem is mimicry synthesis (animation) given a text to be spoken.*

> **Example 1.18** *Behavioral biometric information can be used in evaluation of truth in answers of questions, or the truth of a person speaking (Figure 1.11).*

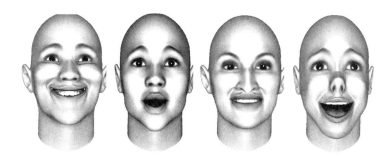

**FIGURE 1.10**
Synthetic faces (Example 1.17).

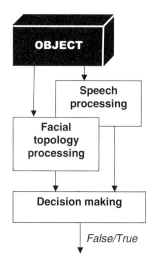

Behavioral biometric information:
- *Speech, and*
- *Facial topology.*

*Processing:*
- *Non-controlled variation of acoustic signal,*
- *Non-controlled variation of facial topology.*

*Decision making: threshold of variation both non-controlled acoustic signal and facial topology.*

**FIGURE 1.11**
Lie detector based on processing speech and facial topology (Example 1.18).

**Synthetic retina and iris.** Retina recognition systems scan the surface of the retina and compare nerve patterns, blood vessels and such features. iris recognition systems scan the surface of the iris to compare patterns. To the best of our knowledge, automated methods of *iris* and *retina image reconstruction*, or *synthesis* have not been developed yet.

**Synthetic fingerprints.** Traditionally, two possibilities of fingerprint imitation are discussed with respect to obtaining unauthorized access to a system: (i) the authorized user provides his fingerprint for making a copy, and (ii) a fingerprint is taken without the authorized user's consent, for example, from a glass surface (a classical example of spy-work) by forensic procedures.

One of the methods for fingerprint synthesis is based on the continuous growing for initial random set (Figure 1.12). The method itself can be used to design fingerprint benchmarks with rather complex structural features.

**FIGURE 1.12**
Synthetic fingerprint assembling (growing).

**Synthetic palmprints.** Recognition of larger parts of the skin on the hand called *palmprints*, makes a comparison of patterns in the skin (similar to fingerprint recognition systems). The difference between a fingerprint recognition system and a palmprint recognition system lies mostly in the size of the scanner and the resolution of the scanning array. Hence, methods of synthetic fingerprint design can be applied to synthetic palmprint generation.

**Synthetic hand topology.** Hand topology is useful for person identification. Generation of synthetic hand topologies is a trivial problem.

**Synthetic vein pattern topology.** Vein pattern recognition systems detect veins in the surface of the hand. These patterns are considered to be as unique as fingerprints, but have the advantage of not being as easily copied or stolen as fingerprints are.

**Synthetic infrared images.** Human thermograms are affected by changes in ambient temperature. Physiological conditions, in particular, include:

▶ Ingestion of substances which are vasodilators or vasoconstrictors,
▶ Sinus problems,

▶ Inflammation,

▶ Arterial blockades,

▶ Incipient strokes,

▶ Soft tissue injuries.

Infrared video cameras are passive, emitting no energy or other radiation on the subject, but merely collecting and focusing the thermal radiation spontaneously continuously emitted from the surface of the human body (Figure 1.13). Facial thermography is a robust biometric, meeting the dual requirements of uniqueness and repeatability.

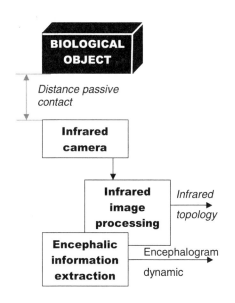

*Infrared camera*

▶ *Operating in the mid (3-5 micron) or long (8-12 micron) infrared bands.*

▶ *Produce images of patterns caused by superficial blood vessels which lay up to 4 cm below the skin surface.*

*The human body is bilaterally symmetrical. The assumption of symmetry facilitates the assigning of face axes. The reality of minor local asymmetries in each person's face facilitates alignment of images for comparison, classification and identification of unique thermogram for each person.*

**FIGURE 1.13**

Infrared information provides a unique topological structure and encephalogram dynamics.

> **Example 1.19** *Thermography, which is the use of cameras sensitive in the infrared spectrum, provides highly secure, rapid, noncontact positive identification of human faces or other parts of the body, even with no cooperation from the subject being identified.*

**Synthetic encephalogram dynamics.** Encephalic information is a mesure of the electrical state of the brain, and well understood for various physical and psychophysical states. Using encephalograms, various psychological states can be recognized and detected, in particular

▶ Deep depression,
▶ Extreme excitement,
▶ Extreme irritability,
▶ Anomalous, and
▶ Sleeping.

Encephalic information can be obtained by *contact*, and *non-contact* methods. Non-contact methods are preferable but they require additional tools for signal processing. Radiometric infrared camera systems can produce a non-contact encephalogram by exhibiting local temperature fluctuations associated with the cycle of heart beats (Figure 1.13). Synthetic encephalogram dynamics can be generated by signal processing, for example, via inverse Fourier transforms (master data) with further controlled generation of particular features.

## 1.6 Conversating of biometric information

In this section, the problem of the relationships between various types of biometric information and between biometric and non-biometric information are discussed. This problem is discussed from an engineering point of view. However, this is a phenomenological problem that is not well understood yet. The focus of this section is the role of synthetic biometric information in converging technologies, i.e., in interfacing various types of information.

> **Example 1.20** *The following interfaces utilize properties of interaction and conversion information:*
>
> (*a*) *Brain-To-Brain,*
> (*b*) *Brain-To-Machine,*
> (*c*) *Speech-To-Vision,*
> (*d*) *Speech-To-Machine.*

### 1.6.1 Types of conversions

There are two cases of information conversions:

▶ Conversions between non-biometric information and biometric information
   $<$ `Non-biometric information` $>$ $\Leftrightarrow$ $<$ `Biometric information` $>$ .
   For example, conversion of information from text (non-biometric) and information carried by lip movement (facial biometric information).

▶ Biometric information can be converted to another type of biometric information

 `< Biometric information >` ⇔ `< Biometric information >`.

 Typical examples are the conversion of voice to facial emotion, lip movement to speech, and facial emotion to encephalogram.

## 1.6.2  Brain-machine interface

A brain-machine interface can be defined as a communication system in which:

(a) Messages are sent by an individuals brain in the form of electrophysiological signals,

(b) Messages do not pass through the normal output (peripheral nerves and muscles), and

(c) Messages are mapped into an environment.

It follows from this definition that:

▶ A brain-machine interface can be considered as an alternative approach for acting,

▶ The subject's mental state in the form of electrophysiological signals can be mapped into actions,

▶ Nerves and muscles are replaced with electrophysiological signals, and the system translates these signals into actions.

> **Example 1.21** *The simplest functions of the brain-machine interface are motor actions such as right or left hand movement or foot movement, and cursor control.*

The electroencephalogram (EEG) is the most popular technique for measuring brain activity. Brain-machine interfaces use a correlation between

 EEG signals and actual or imagined movements, and
 EEG signals and mental tasks.

A brain-machine interface is based on the assumption that a person's intent can be identified from measurement taken from an internal carrier such as blood volume, electrophysiological signals, etc. Determining the reliability of information is the most important task when attempting to estimate a person's intent. The reliability depends on:

▶ The quality of extracted electrophysiological signals (the vast number of electrically active neuronal elements generate mutually correlated signals), and

▶ Spatial topology of the brain and head.

Three ways for improving the reliability of electrophysiological signals by using the following information sources:

(*a*) *Mental* information,

(*b*) *Facial* information, and

(*c*) A *learning* paradigm.

**Example 1.22** *Figure 1.14 illustrates three levels for improving reliability of electrophysiological signals.*

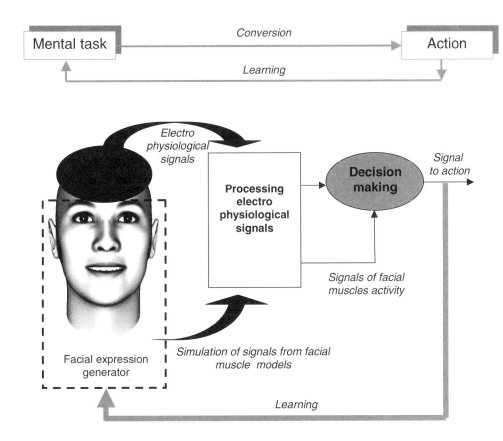

**FIGURE 1.14**

Modeling the brain-machine interface: synthetic facial expressions combined with electrical activity recorded from the brain (Example 1.22).

In Figure 1.14, facial muscle activation and eye movement can contribute to electrical activity recorded from the scalp. Facial muscle signals are easily detected and support decision making based to determine a person's intent. Finally, the output signal can be used for various purposes. A facial expression generator forms the controlled muscle signals into a wide range of facial expressions using facial muscles models. This modeling technique is related to the development of *learning* brain-machine interfaces. A learning paradigm is one in which mutually adaptive interactions occur between the user and the interface and involves two aspects:

▶ *User adaptation* to brain-machine interface via training, i.e., development of specific skills for controlling electrophysiological signals, and

▶ *Interface adaptation* to user via training, i.e., recognition of specific, individual components in processing the electrophysiological signals (electrophysiological portraits of intentions). This includes the creation of an individual library of intentions, the identification of unwanted or dangerous intensions, etc.

A learning paradigm can provide significant improvement of decision making reliability.

> **Example 1.23** *Users learn over a series of sessions to control cursor movement and develop their own style, for example, imagination of hand movement or without thinking about the details of performance. These data are collected in the interface and analyzed with the goal to find an effective strategy for interaction.*

### 1.6.3   Implementation

Conversions of biometric information are useful in various practical applications, in particular, robotics, communication, and devices for people with disabilities. In conversion, the processing of the original and synthetic information is synchronized. In known conversion-based systems, behavioral biometric information is used (facial expressions, voice, and electroencephalogram dynamic).

> **Example 1.24** *Figure    1.15    illustrates    some    useful conversions of biometric information.*

## 1.7   Design of biometric devices and systems

In this section, a design methods for biometric devices and systems is considered. This methods includes tools for multibiometric system design and generators of synthetic biometric data.

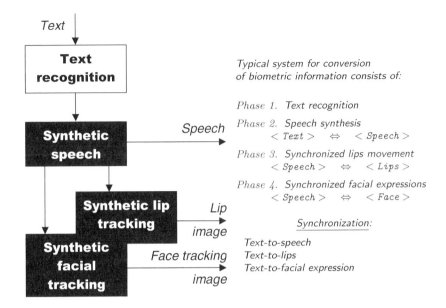

**FIGURE 1.15**
Converting biometric information (Example 1.24).

### 1.7.1 Functions

There are several reasons to combine tools for the design of biometric devices for analysis and synthesis into biometric systems:

*Reason 1.* Most biometric devices can be designed using the same tools.

*Reason 2.* Information fusion can be implemented in various levels of design and various biometrics.

*Reason 3.* Reliability of biometric devices and systems can be improved by combining several biometrics.

*Reason 4.* Application of biometric devices can be extended, for example, to robotics, reconstructive surgery, intelligent decision support, and lie detectors.

CAD of biometric devices and systems provides the possibility for design of biometric tools for analysis and synthesis of *physical* (fingerprint, facial, vein pattern, retina, iris, infrared, ear topology, etc.) and *behavioral* (voice, signature, keystroke dynamics, EEG dynamics, etc.) biometric information.

**Example 1.25**  *Characteristics of voice are pitch, tone, and frequency. Signatures are characterized by a pen pressure and frequency of writing. Characteristics of keystroke dynamics are time between keystrokes, word choices, word combinations, general speed of typing etc.*

### 1.7.2  Design styles for various applications

Below, the most important requirements for generators of synthetic data with respect to applications are listed:

- In *testing*, generation of many synthetic benchmarks is required. Each benchmark must carry user-defined features. These features reflect various aspects of using biometric devices.

- In *database security*, all original biometric data must be transformed to a synthetic copy. A database of biometric data is defined after these transformations as collections of synthetic biometric data.

- In *robotics*, facial expressions must be generated with synchronized synthetic speech and content of interactive conversation;

- In *artificial intelligence decision support*, the level of automated support and the contribution of automated tools in the decision making process are important factors in the development of artificial intelligence tools and human-machine interfaces. For example, in a new generation of lie-detectors, the artificial intelligence tools are used in various levels: from local optimization problems to global problems like generating questions based on analysis of a person's behavior.

- In *reconstructive surgery*, topological information about the patient's face scanning, for example, by tomograph, must be modified for the purpose of physical prototyping.

- In *watermarking techniques*, unique synthetic biometric data must be incorporated (embedded) into the host carrier.

- In *training*, synthetic biometric data must be generated according to a scenario of training. For example, real-time facial expressions and speech must be generated in training immigration and police officers.

### 1.7.3  Architecture

Biometric system architecture contains a flexible relationship (configuration) of computing tools and depends on several factors, in particular, levels of design, the role of artificial intelligence tools and human decision making.

**Configuration.** Configuration of biometric system depends on many factors, in particular, the most important requirements of the biometric system include:

▶ The type of biometric and the type of direct problem, (verification or identification),

▶ The type of inverse problem (modeling of scenario of attacks, design of extremely hard benchmarks, testing biometric systems, etc.),

▶ The concept of library of primitives of biometric information,

▶ Metrics, including measures of topological structures and information theoretical measures,

▶ Decision making (availability of artificial intelligence support, the role of human-based interface, etc.).

For example, in generation of synthetic biometric data, there are several criteria for testing biometric data at each step of automated design. However, satisfying the criteria of acceptability at a local level does not guarantee the same at a global level. There are several criteria for testing synthetic data at a global level, but the role of an expert is still important.

Optimization of the role of an expert reflects the problem of the human-machine interface. An example below demonstrates this problem in an interactive training system.

> **Example 1.26** *The task of a banking officer is to recognize signature imitations/forgeries. Human processing can be supported by an automated system. Hence, there is a need for both human and machine-based approaches.*

**Artificial intelligence support** is based on techniques that can analyze and collect knowledge, and return it into knowledge of a different forms[†]. For this, specific data structures are used. Artificial intelligence tools are classified as

(a) Artificial neural networks,

(b) Algorithms that implement the evolutionary strategies, and

---

[†]A knowledge-based system is software or hardware that acquires, represents, and uses knowledge for a specific purpose. The basic components of knowledge-based systems are a knowledge base which stores knowledge and an inference engine which makes inference using the knowledge.

Artificial neural networks are based on model of biological neurons and solve the problems by self-learning and self-organization. They derive their intelligence from the collective behavior of simple models of individual neurons.

Evolutionary-based artificial intelligence tools explore principles of biological evolution and can manipulate a much richer set of possibilities in the design space that are beyond the scope of traditional methods.

(c) Expert systems.

There are several levels of application of artificial intelligence tools, in particular:

▶ Local optimization problems,

▶ Global optimization problems, and

▶ Joint human-machine decision making.

> **Example 1.27** *In applying a library of primitives to the design of synthetic biometric information, the key idea is to adapt computing technology that enables us to navigate, explore, and discover a set of primitives representing a concept. The essence of learning through the concept of biometric primitives lies on the formalism that information coded in primitives can be explicitly represented as*
>
> > *(a) A mathematical component that could be simply sequences such as a geometric series, splines, systems of differential equations, or complex systems for the description of fuzzy data.*
> >
> > *(b) A topological component that could be various kinds of simple plots, or surface contour plots,*

### 1.7.4   Databases

Biometric information can be decomposed with respect to various criteria (the storage memory, topological complexity, uncertainty of information, quality, dimensions, etc.). A particular case is the topological form of biometric information. Decomposition with respect to topological complexity results in the so-called $0 - level$ of complexity or *topological primitives*. A set of topological primitives is called a *library of primitives*, which is the base for assembling of the process of analysis.

Primitives are often defined as patterns and may refer to regularities exhibited in biometric data, or their existence in an entity of particular interest.

**Decision making.** In this book, decision making is considered in:

▶ Synthetic data design and

▶ Automated support of expert decision making.

Decision making addresses the statistical problems[‡]. In decision making, four rates are calculated (Figure 1.16):

---

[‡]Confidence intervals on the FRR and FAR, and errors in the FRR and FAR are introduced in [8].

*Success rate,*
*False rejection rate* (FRR),
*False acceptance rate* (FAR), and
*Equal error rate* (EER).

> **Example 1.28** *The FRR often called a false negative identification error and can be interpreted as follows: the search of biometric data fails to find the previously enrolled biometric data, presenting the risk of a fraudulent application.*
> *The FAR is called a false positive identification error, that is, given biometric data, erroneously matches that of another biometric data already enrolled, requiring further checks against the possibility of a fraudulent application.*
> *Figure 1.16 provide details of the FRR, FAR, and EER.*

It follows from this example that security requires a low probability of false positive identification, while usability requires a low probability of false negative identification.[§]

## 1.8 Applications of inverse biometrics

In section 1.7, various applications of synthetic data have been introduced with respect to CAD of devices for generation of synthetic biometric data. In this section, detailed aspects of several applications are discussed.

### 1.8.1 Training

A new generation of biometric devices and simulators can be used for educational and training purposes (immigration control, banking service, police, justice, etc.). They will emphasize decision-making skills in non-standard and extreme situations.

> **Example 1.29** *A simulator for immigration control officer training must account for the following scenario: the imitator generates biometric information about a customer: face image, voice, passport data, and other documents. This information is the system's input data. The system then analyzes the current information (collates the passport photo with present image, verifies signatures, analyzes handwriting, etc.).*

---

[§]If each identification involves a comparison against the entire set of prior enrolments, error rates for this one-to-many identification search will depend on the size of the database.

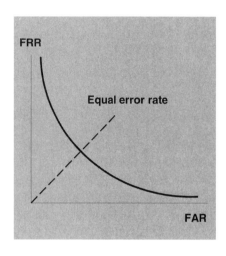

*Success rate:*
> successful      verifications      or
> identifications are made compared
> to the total number of trials

*False rejection rate:*
> the    system    falsely    rejects    a
> registered  user  compared  to  the
> total number of trials

*False acceptance rate:*
> the      system      falsely      accepts
> a    nonregistered    (or    another
> registered) user as a registered one
> compared  to  the  total  number  of
> trials

*Equal error rate:*
> The common value:

$$FAR = FRR$$

- *A low EER value indicates a high level of accuracy in the system and better security*
- *A big FRR often means a low FAR*
- *A big FAR often means a low FRR*

**FIGURE 1.16**
The relationship between FRR, FAR, and EER (Example 1.28).

With some simplification, the evaluation of training persons is given in Table 1.2. Levels A, B, C and D characterize the skills are to identify biometric data. The difference between A and D levels is that machine-based identification has failed, and human-based decision making has no support, i.e., the role of a human decision is determinative. Level B is critical for automatic identification because of a critical amount of uncertainty: results of identification are not stable. Hence, the results of a machine-based identification are contradictory and deceptive, and the role of human-based decision making is of prime importance.

Notice that such a training system is closely related to the discovery learning paradigm, which is a component of an intelligent tutoring system.

## 1.8.2   Intellectual property privacy protection

While biometric data provides uniqueness, it does not provide secrecy.

**TABLE 1.2**
Training in biometric technology.

| LEVEL | QUALIFICATION | EXAMPLE |
|---|---|---|
| A | Human based decision making, biometric systems fail | Surgical operation changes (face, mimics, voice, fingerprint) |
| B | Human based decision making, support of biometric system is unreliable | Difficult conditions for identification, combination of false information |
| C | Recognition of simplest forgeries is supported by biometric system | Hair style, glasses, beard, whiskers |
| D | Recognition of significant differences (forgeries) is supported by biometric system | Photo, passport, and a person |

**Example 1.30** *Signatures are on every cheque a person signs and fingerprints are on every surface a person touches. There are various sources of attacks that are possible in a generic biometric system, in particular: a fake biometric, resubmission of digitally stored biometric data, use of a feature detector to produce feature values chosen by the attacker, and features extracted using the data obtained from the sensor are replaced with a synthetic feature set. If the matcher component is attacked, high or low matching scores can be produced.*

Techniques for increasing security or protection of biometric data include:

▶ *Encryption* or *ciphering* consists in transforming a message into a non-accessible message. A ciphering can be applied to biometric templates: during authentication, encrypted templates can be decrypted and used for generating a matching result with the online biometric data.

▶ *Watermarking* involves embedding information into the host data itself; this means that watermarking can be implemented by hiding an arbitrary mark using a key in public data.

▶ *Steganography* involves hiding message that can be public, into another one. The term is derived from the Greek language and means secret communication involving hiding critical information in an unsuspected carrier data.

Cryptography focuses on methods to make encrypted information meaningless to unauthorized parties. Steganography is based on concealing the information itself. Watermarking is a form of steganography where the goal is to reduce the chances of illegal modification of the biometric data. Image watermarking methods can be divided into two groups according to the domain of application: spatial methods and spectral domain methods.

Watermarking involves embedding information into the host data itself. It can provide security even after decryption. Some of the different types of watermarks are:

▶ Visible watermarks,

▶ Fragile, semi-fragile and robust watermarks,

▶ Spatial watermarks,

▶ Image-adaptive watermarks, and

▶ Blind watermarking techniques.

## 1.9    Ethical and social aspects of inverse biometrics

Ethical and social aspects of inverse biometrics include several problems, in particular, the prevention of undesirable side-effects, and targeting of areas of social concern in biometrics.

### 1.9.1    Roles for practical ethics linked to social sciences

The focus of *practical* ethics is on collaboration among practitioners to solve problems that have ethical components. There are a number of roles for practical ethics linked to social problems, in particular: (a) prevention of undesirable side effects of synthetic biometric data, (b) facilitation of quality research in biometrics by social science, and (c) targeting of social concern in biometrics. In addition, ethics must be incorporated into science education.

### 1.9.2    Prevention of undesirable side effects

Prevention of undesirable side effects aims to study the potential negative impacts of biometrics, as far as important segments of society are concerned, and how can these be prevented. Undesirable ethical and social effects of solutions of inverse biometric have not been studied yet. However, it is possible to predict some of them. The particular examples are given below.

**Example 1.31** *At least three negative ethical and technical problems are related to inverse problems of biometrics:*

(a) *Synthetic biometric information can be used not only for improving the characteristics of biometric devices and systems but also can be used by forgers to discover new strategies of attack.*

(b) *Synthetic biometric information can be used for generating multiple copies of original biometric information.*

(c) *Synthetic biometric information can be used for "improving" a state-of-the-art forger's technique.*

### 1.9.3 Targeting of areas of social concern in biometrics

Practical ethics and social sciences should not be limited to anticipating and preventing problems. Both can play an important role in facilitating the development of biometrics by encouraging reflective practice. An important goal of this reflection is to eliminate the compartmentalization between the technical and social that is so predominant in science engineering. Practical ethicists can work with engineers and scientists to identify interesting and worthy social concerns to which the latest developments in biometrics could be applied. Philosophers and social scientists cannot simply dictate which problems practitioners should try to solve, because not all social problems will benefit from the application of biometrics, and not all future generations of biometric devices and systems are equally likely.

Directing biometrics toward social problems does not eliminate the possibility of undesirable side effects, and equally biometric techniques designed to produce harm may have beneficial spin-offs.

## 1.10 Summary

Biometric technology is one of the possible ways to secure human society and to extend knowledge about human nature and behavior. Humans rely heavily on the natural senses (touch, sight, sound, smell) to keep out of danger. Today, biometric technologies utilize some of the unique features of human existence, i.e., intellectual (thinking, deciding, learning, remembering, etc.), physical (moving, reaching, lifting, etc.), and emotional (feeling, expressions, etc.). Biometric technologies utilize these features in personal identification, protection, and for the prediction of human behavior.

Typical problems in biometrics can be classified as direct and inverse ones. Direct problems are associated with analysis of information and are aimed at identification, verification and recognition of biometrical data.

Inverse problems are associated with synthesis of information and are aimed at designing new, synthetic data with properties that satisfies the user's requirements. This book concentrates on some inverse problems of biometrics.

1. Inverse problems of biometrics are related to security issues of biometric systems: modeling possible attacks and corresponding defence, modeling behavioral responses in critical situations, and modeling specific environments in which biometric systems operate. The results of the modeling have an impact on advancing biometric systems by improving testing techniques.

2. Some inverse problems are solved by appropriate models which rely on approximation of original data. These models become the *generators* of synthetic biometrical data, if reasonable control strategies are developed, and control procedures are defined. This approach is known in modeling as *analysis-by-synthesis*.

3. Generation of synthetic biometric data is considered with respect to

   ▶ The user's role in decision making,
   ▶ The data abstraction level in decision making, and
   ▶ The design strategy.

4. Synthetic representations are used in the following fields:

   *Random data and stochastic process generators* are classical examples of inverse problems. Usually the input of these generators is a random process with a uniform distribution or white noise. The second input is a set of controlled parameters that is associated with the desired characteristics of the output of random data or a stochastic process.

   *Evolutionary algorithms* based on natural evaluation and selection techniques are typical examples of generators of synthetic information. For example, evolutionary strategies have proven their effectiveness in image synthesis, fingerprint recognition, topological design, logic circuit design, and generation of chemical, physical and biological objects.

   *Converging technologies* for improving human performance are a field where synthetic data plays a crucial role, in particular in human-machine interfaces, for example, $<$ `Brain-To-Brain` $>$ and $<$ `Brain-To-Machine` $>$. For example, systems based on the conversation of biometric information,

   $$< \texttt{Text} > \quad \Leftrightarrow \quad < \texttt{Speech} > \text{ and } < \texttt{Speech} > \quad \Leftrightarrow \quad < \texttt{Lips} >,$$

   include generators of synthetic information (voice, speech, lip movement). Synthetic voice and speech are solutions to the inverse problems of voice and speech analysis. By analogy, lip movement is generated via inverse models from the analysis of lip expressions.

5. In general, there are two main phases in synthetic data generation:

    (a) Generate data with characteristics similar to the original data, and
    (b) Make decisions on the quality of generated data aimed at testing synthetic data by several acceptability criteria.

6. Studying inverse problems of biometrics address ethical, legal, moral, and social concerns throughout research, development, and implementation. This will require new mechanisms to ensure representation of public interest, and to incorporate ethical and social-scientific education in the training of engineers and personnel.

## 1.11 Problems

**Problem 1.1** In [60], the suitability of biometrics for a given application is evaluated in terms of its ability to address a deployer's specific authentication needs

▶ The *scope* is rated on a scale from 1 (narrow) to 10 (broad); biometrics can be used in a limited set of problems (narrow scope) or to address problems encountered by a large number of users (broad scope).

▶ *Urgency* rated on a scale from 1 (low) to 10 (high);

▶ *Effectiveness* rated on a scale from 1 (low) to 10 (high);

▶ *Exclusivity* rated on a scale from 1 (low) to 10 (high); biometrics might be the only solution to a problem or one of many potential solutions.

▶ *Receptiveness* rated on a scale from 1 (low) to 10 (high); biometrics might be welcomed as a necessary authentication solution. For example, the public has been generally receptive to the use of biometrics, but only to solve specific problems.

In Figure 1.17a, the criminal identification solution matrix is defined based on the primary role of biometrics to identify an individual in order to conduct law enforcement functions. A citizen identification solution matrix (Figure 1.17b) is the use of biometrics to identify or verify the identity of individuals in their interaction with government agencies (issuing cards, voting, immigration, social services, etc.). Surveillance is the use of biometrics to identify or verify the identity of individuals present in a given space or area (Figure 1.17c). An alternative solution is, for example, manual monitoring of cameras.

Evaluate from this position the following situations: (a) citizen identification, (b) criminal identification, (c) network access, and (d) E-commerce.

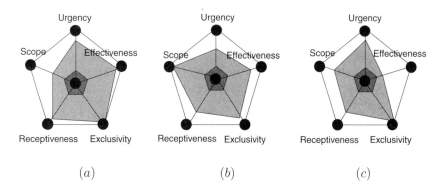

**FIGURE 1.17**
Biometric solution matrix: criminal (a), citizen (b), and surveillance (c) (Problem 1.1).

**Problem 1.2** In many systems, the false match rate, referred to also as the *false acceptance rate* (FAR), is the most critical accuracy metric. A false match does not always mean that the biometric device or system has failed in the following real-world situations: (a) casino: identification of known card counters; (b) finger-scan system: false match rate is 1/10,000 (the likelihood of templates from two distinguished person in single match attempt).

**Problem 1.3** In Section 1.8, the ethical and social aspects of inverse problems of biometrics were introduced. In Section 1.1, the scenario of attacks on biometric systems is discussed, including a new scenario where massive attacks using synthetic biometric information are introduced.
(a) Propose a technical idea to prevent new types of attacks.
(b) Give your understanding of ethical and social problems of inverse problems of biometrics.
(c) Compare positive and negative ethical and social aspects of biometrics with respect to inverse problems.

## 1.12 Further reading

**Biometrics.** Below, we refer to the literature on various aspects of biometrics.

> ***Fundamentals.*** An excellent introduction to biometrics can be found in papers by Jain et. al. [35] and collection of papers edited by Jain et al. [36, 38]. The selected papers edited by Polonnikov and Korotkov [67] introduce the trends and vision of a new generation of biometric

systems. State-of-the-art trends in biometrics are discussed in paper by Shmerko et al. [33] and the book by Zhang [105]. The *Guide to Biometrics* by Bolle et al. [8] provides the reader with practical aspects and recommendation on development of biometric devices and systems. The authors of [100] introduced their vision on the problem of identity assurance in biometrics. Chalmond [16] introduces the fundamentals of direct and inverse problems in image analysis.

Data fusion is aimed at improvement of the reliability of biometric devices and systems and is of special interest in today's biometrics. Clark and Yuille [20] introduced techniques for data fusion for sensory systems. Prabhakar and Jain [68] proposed an algorithm for decision-level fusion in fingerprint verification.

Fundamentals of image processing can be found in [30]. Practical MATLAB implementation useful in synthesis of biometric data can be found in [31].

*Artificial intelligence support of decision making in biometric systems.* In collection of papers [38], the biometric techniques based on artificial intelligence paradigm are introduced.

*Security.* Chirillo and Blaul [19] introduced the fundamentals of implementation of biometric security. Jain and Uludag [39] introduced the concept of hiding biometric data. Ratha et al. [71] introduced the technique for enhancing security and privacy in biometrics-based authentication systems based on the concept of cancellable biometrics data.

*Testing, standards, and synthetic benchmarks.* Testing of biometric devices and systems is considered in [30, 50, 51]. Matsumoto et al. [31] studied the impact of gummy fingers on fingerprint systems. Tilton [89] discussed various approaches to biometric standards development. Wayman [94] introduces the Federal biometric technology legislation.

*Signature and keystroke dynamic analysis and synthesis.* Modeling of skilled forgers was the focus of much study. In particular, in 1977 Nagel and Rosenfeld [58] proposed an algorithm for detection of freehand forgeries. Ammar [2] analyzed the progress in this area. Brault and Plamondon [4] studied a modelling of dynamic signature forgery. Rhee et al. [29] proposed an algorithm for on-line signature verification and skilled forgeries modeling. A comprehensive survey on modeling, on-line, and off-line handwriting recognition can be found in a paper by Plamondon and Srihari [28]. Fundamentals of handwriting identification can be found in the book by Huber and Headrick [34]. Comprehensive references to various aspects of signature-based identification system design are included in [47, 59]. Oliveira et al. [22] demonstrated an algorithm for generation of signatures based on a deformation technique.

Bergadano et al. [4] reported results on application of keystroke dynamics technique. A keystroke-based authentication algorithm has been described by Obaidat and Sadoun [62].

*Iris and retina.* Various aspects of iris-based identification have been developed, in particular, in [21, 23, 95, 96]. A paper by Hill [14] is useful for design of retina-based identification devises.

*Fingerprint analysis and synthesis.* A paper by Bery [5] is an excellent tour of the history of fingerprinting. Cappelli et al. [12, 13, 14, 15] developed an effective and robust algorithm for generation of synthetic fingerprints. Various techniques of fingerprint identification have been studied in [29, 46, 56, 57, 70, 93].

*Facial analysis and synthesis.* Blanz and Vetter [7] have developed a morphable model for the synthesis of 3D faces. Plenty of techniques that are useful in synthesis and analysis of facial expressions have been proposed in [10, 17, 26, 43, 77, 90, 104].

*Ear.* An introduction to this type of biometrics can be found in the overview by Burge and Burger [11].

*Speech* synthesis is, perhaps, the oldest among biometric automated synthesis techniques, and has a vast variety of techniques and implementations. Current trends in design of speech-based biometric systems are discussed by Reynolds [72].

*Hand geometry* is defined as a surface area of the hand or fingers and corresponded measures (length, width, and thickness). Zunkel [107] introduced an algorithm for hand geometry based authentication. This direction is studied by Zhang [105], Sanchez et al. [75], and Kumar et al. [44].

*Palmprint.* The inner surface of the hand from the wrist to the root of fingers is called the *palm*. Palmprint is represented on the surface by topological combination of three types of lines: flexure lines, papillary ridges, and tension lines. Automatic palmprint authentication techniques have been developed by Zhang [96], Han et al. [32], Duta et al. [25], Kumar et al. [44], and Shu et al. [81]. The concept of an eigenspace was used by Zhang [96] in modeling of palprints using eigenpalms (a palprint in the training set was represented by an eigenvector).

*Intellectual property protection.* Chen and Leung [18] studied the effects of chaotic spreading codes in the development of image watermarking. Pankanti and Yeung [65] proposed an algorithm for verification watermarks based on fingerprint recognition technique. In addition, Wolfgang and Delp [97, 98] analyzed image watermarking with biometric data and trends in this direction. Techniques of data hiding based on synthetic biometric information have been reported in [101].

*Banking and financial applications.* Using biometric technologies in various levels of banking systems and financial communications is discussed in collection of papers edited by Shmerko and Yanushkevich [34].

*Humanoid robots.* Synthetic biometrics are largely employed in research aimed at endowing robots with instincts and reflexive behaviors, and, eventually, with the capability to "dream" (imagine or simulate potential solutions), "play" (test or validate solutions in real-world), and "imitate" (follow guidance of a human). For example, a key characteristic of the approach proposed in [86] is an emphasis on teaching and fostering, a component overlooked by other approaches, yet at least as necessary as learning. Teaching is performed initially by humans, who select appropriate means, lessons, and examples; consequently, the robots themselves exercise teaching, as a proof of their learning and "understanding" what was learned, proving this awareness of self and others, exercising internal models, and showing imagination and adaptation to new situations. This development involves language technologies, including voice recognition, speech-to-text and voice synthesis, which are sufficiently developed to allow simple interaction with the robot; face and gesture recognition, which is sufficiently developed to allow robots to read the "moods" of the instructor, follow cues, etc.; knowledge base, dialog and logical reasoning; improved vision, hearing, olfaction, tactile, and other sensory capacities, developed to a certain extent and incorporated in various commercial devices (artificial retinas, e-nose, e-tongue). Some of the capabilities and technologies still need to be developed to include efficient and human-friendly means to transfer cognitive and motor skills to robots, cognition and self-awareness, perception from big sensory arrays (e.g. skin) and an integrated platform that combines available technologies. This is important in achieving human-friendly means for cognitive and motor skill transfer and interaction through dialog in natural language, perceiving the world with multiple sensors, development of perceptual maps and schemes, neural and fuzzy neural techniques [76], etc.

*Training.* Lecture notes by Stoica [84], Yanushkevich and Shmerko [54] include various aspects of biometric system design for training users and developers.

In addition, useful ideas for synthetic biometric data design can be taken from [55, 83].

**Convergent technologies** refers to the synergy of various rapidly progressing directions of science and technology, in particular, nanoscience and nanotechnology, biotechnology and biomedicine, and biometric technology. Convergent technologies are fields where synthetic data play a crucial role, particularly, in human-machine interfaces. The idea is to measure and simulate processes observed at the neuron level. The brain-machine interfaces provide the interaction with the neural system, that is a non-muscular communication and control channel for conveying messages and commands to an external world.

There are several effects which can be measured to evaluate neuron activity, in particular, cardiovascular and electrophysiological effects. A brain-machine technology might be based on monitoring brain activity using the following measurement techniques:

▶ EEG,
▶ Magnetoencephalography,
▶ Position emission tomography,
▶ Functional magnetic resonance imaging, and
▶ Video imaging.

These techniques can be used in the brain-machine interfaces. However, there are several constraints to the above techniques. For example, because the basic phenomena measured by position emission tomography, functional magnetic resonance imaging, and visual-based methods is blood flow change, it is difficult to achieve real-time communication. Review on brain-machine technology can be found in Wolpaw et al. paper [99]. Details of experimental studies can be found, in particular, in the papers by Millán [53, 54], and Kostov and Polak [42].

Converging technologies for improving human performance have been discussed in [74]. Oliver et al. [64] developed an algorithm for facial expression recognition based on face and lip movement. Sproat [82] proposed a multilingual text-to-speech synthesis algorithm.

**Cancellable techniques** are defined as intentional, repeatable distortions of a biometric data. To distort biometric data, an appropriate transform should be chosen. Distortion transform can be invertible and noninvertible. In most applications, the distortion transform is selected to be noninvertible, i.e., even if the exact transform is known and so the resulting transformed data, the original data cannot be recovered. Cancellable technique is very flexible in applications and opportunities to secure biometric data. For example, cancellable transforms can be applied in spatial domain, to features, or in spectral domain. There are several ways: the data can be transformed directly after acquisition, or the extracted features can be transformed. Various aspects of cancellable biometrics are discussed in [8, 68].

**Error rates of biometric systems.** Today biometric systems can achieve the following false match error rates:

▶ Fingerprint identification system: of 1 in $10^5$ using a single fingerprint,
▶ Iris identification system: of 1 in $10^6$ using a single iris,
▶ Facial identification system: of 1 in $10^3$ using a single face appearance.

In order to uniquely identify one person in a population, for example, of 50 million, a fingerprint system should use at least four fingers per person and an iris system should use both eyes. Facial biometrics could not provide a sufficient accuracy of identification of this population. However, facial identification system can be used in one-to-one comparison as an aid to identity checking, for example, for passport holders. For more details, see [8, 19, 23, 35, 17, 46].

**Attacking biometric identifiers.** The term *spoofing* is often used to indicate a security attack where an attacker (even a legitimate user of a system), runs a program with a false login screen and the unsuspecting user provides password and ID to an attacker. Several scenarios are possible in this situation, which depend on the attacker's strategy.

**Thermography** which is the use of cameras sensitive in the infrared spectrum, provides highly secure, rapid, and noncontact identification. This identification, including faces or other parts of the body, can be achieved with no cooperation from the identifying person. This fact is used in the development of a new generation of biometric systems based on distant, noncontact principles.

Infrared devices can detect hidden micro parameters which lie below the skin surface. These micro parameters cannot be forged. The infrared image cannot be changed by applique or surgical intervention. This is because the temperature distribution across artificial facial items (hair, skin, plumpers, or other appliques and dental reconstruction) is different from normal. Reconstruction, for example plastic surgery, may:

▶ Redistribute fat,

▶ Add or subtract skin tissue,

▶ Add inert materials,

▶ Create or remove scars,

▶ Resurface the skin via laser or chemicals,

▶ Apply permanent color by tattooing eyelids and lips,

▶ Remove tattoos and birthmarks by laser, and

▶ Implant or dipilate hair

Any the above procedures can confuse automatic identification and recognition systems because algorithms that are used in this systems can not detect these nonnatural distortions and improvement of the algorithm is impossible. Infrared devices provide additional information that can be efficiently used in the above scenarios.

Bharadwaj and Carin [6] studied methods of comparing two infrared images using hidden Markov model. Infrared thermography has been introduced by Gaussorgues [28]. Fujimasa et al. [27] studied the problem of converting infrared images to other data. Of use for development of infrared system is a paper by Anbar et al. [2]. Details of thermal image processing can be found in [27, 28, 41, 69].

Brain responses can be recognized in terms of local changes in blood volume and oxygen saturation around the active brain regions. Since hemoglobin is a natural and strong optical absorber, changes in this molecule can be monitored with near infrared detection [48, 92].

Fundamentals of Markov random fields and application in image analysis and synthesis were introduced by Li [25].

**Chernoff faces** are a method of visualizing multidimensional data developed by statistician Herman Chernoff. Chernoff facial expressions are described by ten facial characteristic parameters: head eccentricity, eye eccentricity, pupil size, eyebrow slant, nose size, mouth shape, eye spacing, eye size, mouth length and degree of mouth opening. Each parameter is represented by a number between 0 and 1. For example, facial expressions can be represented by the vectors [0000000000], or continuously animated transitions between randomly chosen facial expressions.

**Inverse problems** including image recovery, and tomography are discussed in [61]. This collection of 15 research papers edited by Nashed and Scherzer, can be recommended for detailed study of inverse problems. Inverse problems in [61] are defined as determining for a system an input that produces an observed or desired output. Formally, if normed spaces $X$ and $Y$ correspond to the input and output, and the system is modeled as an operator $T$, then the problem is formulated as solving the equation $T(x) = y$ to find $x$ for a given $y \in Y$. This is the simplest example of canonical formulation and solution of an inverse problem. In most cases, inverse problems are *ill-posed*, for example, tomography. In theses cases, techniques based on the regularization of linear and nonlinear ill-posed problems, bounded variation regularization, fractal and wavelet sets, etc. should be used. In addition, see results by Terzopolous [88].

**An additional source** of ideas for synthesis biometric information can be found in another area: circuit design. Synthetic benchmark circuits have been discussed, in particular, in [55, 83]. From these papers, we observe that the philosophy of generating logical circuits given parameters is based on utilization of uncertainty. The uncertainty is the source of flexibility in generating a set of circuits. In the next chapters of the book, this philosophy is utilized for generating the synthetic biometric topologies. In particular, the equivalent of a profile circuit, or master-circuit, is a master fingerprint, master-iris, etc.

**Evolutionary algorithms** in synthetic information generation. The principle of evolution is the primary concept of biology, linking every organism together in a historical chain of events. For example, every creature (circuit) in the chain is the product of a series of events (subcircuits) that have been sorted out thoroughly under selective pressure (correct or noncorrect subcircuits) from the environment (design technique). Over many generations, random variation and natural selection shape the behavior of individuals (circuits) to fit the demands of their surroundings (principles of circuit design). For example, using some from the above principles, Sekanina and Ružička [76] generated a class of image filters.

Plenty of useful information on computational techniques inspired by nature can be found in the *IEEE Transactions on Evolutionary Computing* and *Proceedings of the NASA/DoD Conference on Evolvable Hardware*¶.

In particular, the collection of papers by Dasgupta covers various computational aspects of the immune system [22]. The natural immune system is a complex adaptive pattern-recognition system that defends the body from foreign pathogens. Rather than rely on any central control, it has a distributed task force that has the intelligence to take action from a local and also global perspective using its network of messengers for communication. The immune system has evolved innate (nonspecific) immunity and adaptive (specific) immunity. From the computation point of view, the natural immune system is a parallel and distributed adaptive system. The immune system uses learning, memory, and associative retrieval to solve recognition and classification tasks. Specifically, it learns to recognize relevant patterns, remember patterns that have been seen previously, and use combinatorics to construct pattern detectors efficiently.

**Watermarking** is defined as a mechanism for embedding specific data into host data. This specific data must satisfy a number of requirements, in particular: author (source, the used tools, technique, etc.) identification, and being difficult to detect and remove. The goal of watermarking is to protect the intellectual property rights of the data. For example, watermarking is used for multimedia data authentication. Watermarking techniques can be classified with respect to various criteria, in particular:

▶ Visible watermarks which are designed to be easily perceived and identified,

▶ Fragile (can be distorted by the slightest changes to the image), semi-fragile (can be distorted by all changes that exceed a user-specified threshold), or robust (withstand moderate to aggressive attacks) watermarks.

▶ Spatial and spectral watermarks are embedded into the image in the spatial and frequency domains respectively.

---

¶For example, in the collection of papers [85] various aspects of applications of evolutionary algorithms in synthesis of systems are introduced, including embrionics and bio-inspired approaches.

There are several techniques that explore various particular effects and properties of image and watermarks. For example, image-adaptive watermarks which locally adapt the strength of the watermark to the image content through perceptual models for human vision. These models are implemented in image compression. In particular, a reference to watermarking technique is the paper by Kahng et al. [40] where the natural properties of host data have been utilized. It is reasonable in some cases to use biometric data as watermarks.

In [39], the security of the biometric data based on encryption have been analyzed. The idea was to apply an encryption to the biometric templates. Uludag et al. [91] have discussed so-called *biometric cryptosystems* which are traditional cryptosystems with biometric components. This paper can be recommended as an introduction to biometric cryptosystems. Lach et al. [45] proposed an watermarking algorithm using fingerprints for field programmable gate arrays (FPGA). Yu et al. [103] investigated a robust watermarking technique for 3D facial models based on a triangular mesh. In this approach, watermark information is embedded into a 3D facial image by perturbing the distance between the vertices of the model to the center of the model. Bas et al. [1] developed a so-called *content* based method of watermarking and demonstrated useful properties in design of robust watermarking algorithms. Shieh et al. [78], used an evolutionary strategy (a genetic algorithm) to optimize a watermarked image in the spectral domain with respect to two criteria: quality and robustness.∥

## 1.13   References

[1] Ammar M. Progress in verification of skillfully simulated handwritten signatures. *Pattern Recognition and Artificial Intelligence,* 5:337–351, 1993.

[2] Anbar M, Gratt BM, and Hong D. Thermology and facial telethermology. Part I. History and technical review. *Dentomaxillofacial Radiology,* 27(2):61–67, 1998.

[3] Bas P, Chassery J-M, and Macq B. Image watermarking: an evolution to content based approaches. *Pattern Recognition,* 35:545–561, 2002.

[4] Bergadano F, Gunetti D, and Picardi C. User authentication through

---

∥Robust watermarks should survive various attacks, for example, attacks in the spectral domain (low-pass and/or median filtering, adding white noise, etc.), or attacks in the spatial domain by topological distortions (affine transforms, re-meshing, local deformations, cropping, etc.).

keystroke dynamics. *ACM Transactions on Information and System Security*, 5(4):367–397, 2002.

[5] Bery J. The history and development of fingerprinting. In Lee HC and Gaensslen RE, Eds. *Advances in Fingerprint Technology*, pp. 1–38, CRC Press, Boca Raton, FL, 1994.

[6] Bharadwaj P and Carin L. Infrared-image classification using hidden Markov trees. *IEEE Transactions on Pattern Analysis and Machine Intelligance*, 24(10):1394–1398, 2002.

[7] Blanz V and Vetter T. A morphable model for the synthesis of 3D faces, In Rockwood A, Ed., *Computer Graphics Proceedings*, pp. 187–194, Addison Wesley Longman, Boston, 1999.

[8] Bolle R, Connell J, Pankanti S, Ratha N, and Senior A. *Guide to Biometrics*. Springer, Heidelberg, 2004.

[9] Brault JJ and Plamondon R. A complexity measure of handwritten curves: modelling of dynamic signature forgery. *IEEE Transactions on Systems, Man and Cybernetics*, 23:400–413, 1993.

[10] Brunelli R and Poggio T. Face recognition: features versus templates. *IEEE Transactions on Pattern Analysis and Machine Intelligence*, 15(10):1042–1052, 1993.

[11] Burge M and Burger W. Ear biometrics. In Jain A, Bolle R, and Pankanti S, Eds., *Biometrics, Personal Identification in Networked Society*, pp. 273–285, Kluwer, Dordrecht, 1999.

[12] Cappelli R, Lumini A, Maio D, and Maltoni D. Fingerprint classification by directional image partitioning. *IEEE Transactions on Pattern Analysis and Machine Intelligence*, 21(5):402–421, 1997.

[13] Cappelli R. SFinGe: an approach to synthetic fingerprint generation. In *Proceedings of the International Workshop on Biometric Technologies*, University of Calgary, Canada, pp. 147–154, 2004.

[14] Cappelli R. Synthetic fingerprint generation. In Maltoni D, Maio D, Jain AK, Prabhakar S, Eds., *Handbook of Fingerprint Recognition*, pp. 203–232, Springer, Heidelberg, 2003.

[15] Cappelli R, Maio D, and Maltoni D. Synthetic fingerprint-database generation. In *Proceedings of the 16th International Conference on Pattern Recognition,* vol. 3, pp. 744–747, Canada, 2002.

[16] Chalmond B. Modeling and inverse problems in image analysis, *Applied Mathematical Sciences*, vol. 155, Springer, Heidelberg, 2003.

[17] Chellappa R, Wilson CL, and Sirohey S. Human and machine recognition of faces: a survey. *Proceedings of the IEEE*, 83(5):705–740, 1995.

[18] Chen S and Leung H. Application of chaotic spreading codes to image watermarking. In *IEEE International Symposium on Intelligent Signal Processing and Communication*, pp. 256–260, Nashville, Tennessee, 2001.

[19] Chirillo J and Blaul S. *Implementing Biometric Security*, John Wiley & Sons, New York, 2003.

[20] Clark J and Yuille A. *Data Fusion for Sensory Information Processing Systems*. Kluwer, Dordrecht, 1990.

[21] Cui J, Wang Y, Huang J, Tan T, Sun Z, and Ma L. An iris image synthesis method based on PCA and super-resolution, In *Proceedings of the International Conference on Pattern Recognition*, 2004.

[22] Dasgupta D. *Artificial Immune Systems and Their Applications*. Springer, Heidelberg, 1999.

[23] Daugman J. Recognizing persons by their iris pattern. In Jain AK, Bolle RM, and Pankanti S, Eds., *Biometrics: Personal Identification in Networked Society*, pp. 103–122. Kluwer Academic Publishers, Boston, MA, 1999.

[24] Duc B, Fisher S, and Bigun J. Face authentication with Gabor information on deformable graphs. *IEEE Transactions on Image Processing*, 8(4):504–516, 1999.

[25] Duta N, Jain AK, and Mardia KV. Matching of palmprints, *Pattern Recognition Letters*, 23(4):477–485, 2001.

[26] Fua P and Miccio C. Animated heads from ordinary images: a least-squares approach, *Computer Vision and Image Understanding*, 75(3):247–259, 1999.

[27] Fujimasa I, Chinzei T, and Saito I. Converting far infrared image information to other physiological data. *IEEE Engineering in Medicine and Biology Magazine*, 19:71–76, May/June 2000.

[28] Gaussorgues G. *Infrared Thermography*. In Microwave Technology Series 5, Chapman and Hall, London, 1994.

[29] Germain RS, Califano A, and Coville S. Fingerprint matching using transformation parameter clustering. *IEEE Computational Science and Engineering*, pp. 42–49, Oct.-Dec. 1997.

[30] Gonzalez RC, Woods RE, and Eddins SL. *Digital Image Processing Using MATLAB*. Pearson, Prentice Hall, 2004.

[31] Hamamoto Y. A Gabor filter-based method for identification. In Jain LC, Halici U, Hayashi I, Lee SB, and Tsutsui S, Eds., *Intelligent Biometric Techniques in Fingerprint and Face Recognition*, pp. 137–151, CRC Press, Boca Raton, FL, 1999.

[32] Han CC, Chen HL, Lin CL, and Fan KC. Personal authentication using palm-print features. *Pattern Recognition*, 36(2):371–381, 2003.

[33] Hill R. Retina identification. In Jain AK, Bolle RM, and Pankanti S, Eds., *Biometrics: Personal Identification in Networked Society*, pp. 123–142, Kluwer, Dordrecht, 1999.

[34] Huber RA and Headrick A. *Handwriting Identification: Facts and Fundamentals*. CRC Press, Boca Raton, FL, 1999.

[35] Jain AK, Ross A, and Prabhakar S. An introduction to biometric recognition, *IEEE Transactions on Circuit and Systems for Video Technology*, 14(1):4–20, 2004.

[36] Jain A, Pankanti S, and Bolle R, Eds., *Biometrics: Personal Identification in Networked Society*, Kluwer, Dordrecht, 1999.

[37] Jain AK, Hong L, and Bolle RM. On-line fingerprint verfication. *IEEE Transactions on Pattern Analysis and Machine Intelligence*, 19(04):302–313, 1997.

[38] Jain LC, Halici U, Hayashi I, Lee SB, and Tsutsui S, Eds., *Intelligent Biometric Techniques in Fingerprint and Face Recognition*, CRC Press, Boca Raton, FL, 1999.

[39] Jain AK and Uludag U. Hiding biometric data. *IEEE Transactions on Pattern Analysis and Machine Intelligence*, 25(11):1494–1498, 2003.

[40] Kahng AB, Lach J, Mangione-Smith WH, Mantic S, Markov IL, Potkonjak M, Tucker P, Wang H, and Wolf G. Constraint-based watermarking technique for design IP protection. *IEEE Transactions on Computer-Aided Design of Integrated Circuit and Systems*, 20(10):1236–1252, 2001.

[41] Keyserlingk J, Ahlgren P, Yu E, Belliveau N, and Yassa M. Functional infrared imaging of the breast. *IEEE Engineering in Medicine and Biology Magazine*, 19:30-41, May/June 2000.

[42] Kostov A and Polak M. Parallel man-mashine training in development of EEG-based cursor control. *IEEE Transactions Rehabilitation Engineering*, 8:203–204, 2000.

[43] Kumar VP, Oren M, Osuna E, and Poggio T. Real time analysis and tracking of mouths for expression recognition. In *Proceedings DARPA98*, pp. 151–155, 1998.

[44] Kumar A, Wong DCM, Shen HC, and Jain A. Personal verification using palmprint and hand geometry biometric. *Lecture Notes in Computer Science*, vol. 2688, pp. 668–678, 2003.

[45] Lach J, Mangione-Smith WH, and Potkonjak M. Fingerprinting techniques for field-programmable gate array intellectual property

protection. *IEEE Transactions on Computer Aided Design of Integrated Circuits and Systems*, 20(10):1253–1261, 2001.

[46] Lee HC and Gaensslen RE, Eds., *Advances in Fingerprint Technology.* CRC Press, Boca Raton, FL, 1994.

[47] Lee LL, Berger T, and Aviczer E. Reliable on-line human signature verification systems. *IEEE Transactions on Pattern Analysis and Machine Intelligence*, 18(6):643–647, 1996.

[48] Luo Q, Nioka S, and Chance B. Functional near-infrared image. In Chance B and Alfano R, Eds., *Optical Tomography and Spectroscopy of Tissue: Theory, Instrumentation, Model, and Human Studies,* Proceedings SPIE, 2979:84–93, 1997.

[49] Maio D, Maltoni D, Cappelli R, Wayman JL, and Jain AK. FVC2000: fingerprint verification competition. *IEEE Transactions on Pattern Analysis and Machine Intelligence*, 24(3):402–412, 2002.

[50] Mansfield T, Kelly G, Chandler D, and Kane G. Biometric product testing final report. Issue 1.0. *National Physical Laboratory of UK*, 2001.

[51] Mansfield A and Wayman J. Best practice standards for testing and reporting on biometric device performance. *National Physical Laboratory of UK*, 2002.

[52] Matsumoto H, Yamada K, and Hoshino S. Impact of artificial gummy fingers on fingerprint systems, In *Proceedings SPIE, Optical Security and Counterfeit Deterrence Techniques* IV, 4677:275–289, 2002.

[53] Millán JdelR. Adaptive brain interfaces. *Communication of the ACM*, 46:74–80, 2003.

[54] Millán JdelR. Brain-computer interfaces. In Arbib MA, Ed., *Handbook of Brain Theory and Neural Networks*, pp. 178–181, MIT Press, Cambridge, Massachusetts, 2002.

[55] Hutton MD, Rose J, Grossman JP, and Corneil DG. Characterization and parameterized generation of synthetic combinational benchmark circuits. *IEEE Transactions on Computer-Aided Design of Integrated Circuits and Systems*, 1999.

[56] Moayer B and Fu KS. A synatactic approach to fingerprint pattern recognition. *Pattern Recognition*, vol. 7, pp. 1–23, 1975.

[57] Moayer B and Fu KS. A tree system approach for fingerprint pattern recognition. *IEEE Transactions on Computers*, 25(3):262–274, 1976.

[58] Nagel RN and Rosenfeld A. Computer detection of freehand forgeries. *IEEE Transations on Computers*, 26(9):895–905, 1977.

[59] Nalwa VS. Automatic on-line signature verification. *Proceedings of the IEEE*, 85(2):215–239, 1997.

[60] Nanavati S, Thieme M, and Nanavati R. *Biometrics. Identity Verification in a Networked World.* John Wiley & Sons, New York, 2002.

[61] Nashed MZ and Scherzer O, Eds., *Inverse Problems, Image Analysis, and Medical Imaging.* American Mathematical Society, Providence, Rhode Island, 2002.

[62] Obaidat MS and Sadoun B. Keystroke dynamics based authentication. In Jain AK, Bolle RM, and Pankanti S, Eds., *Biometrics: Personal Identification in Networked Society*, pp. 218–229, Kluwer Academic Press, Boston, MA, 1999.

[63] Oliveira C, Kaestner C, Bortolozzi F, and Sabourin R. Generation of signatures by deformation, In *Proceedings of the BSDIA97*, pp. 283–298, Curitiba, Brazil, 1997.

[64] Oliver N, Pentland AP, and Berard F. LAFTER: a real-time face and lips tracker with facial expression recognition, *Pattern Recognition*, 33(8):1369–1382, 2000.

[65] Pankanti S and Yeung MM. Verification watermarks on fingerprint recognition and retrieval, In *Proceedings SPIE* 3657:66–78, 1999.

[66] Plamondon R and Srihari SN. On-line and off-line handwriting recognition: a comprehensive survey. *IEEE Transactions on Pattern Analysis and Machine Intelligence*, 22:63–84, 2000.

[67] Polonnikov R and Korotkov K, Eds., *Biometric Informatics and Eniology*, "OLGA" Publishing House, St. Petersburg, 1995.

[68] Prabhakar S and Jain AK. Decision-level fusion in fingerprint verification. *Pattern Recognition*, 35(4):861–874, 2002.

[69] Prokoski FJ and Riedel RB. Infrared identification of faces and body parts. In Jain A, Bolle R, and Pankanti S, Eds., *Biometrics: Personal Identification in Networked Society*, Kluwer, Dordrecht, 1999, pp. 191–212.

[70] Ratha NK, Karu K, Chen S, and Jain AK. A real-time matching system for large fingerprint database. *IEEE Transactions on Pattern Analysis and Machine Intelligence*, 18(8):799–813, 1996.

[71] Ratha N, Connell J, and Bolle R. Enhancing security and privacy in biometrics-based authentication systems. *IBM Systems Journal*, 40(3):614–634, 2001.

[72] Reynolds DA. Automatic speaker recognition: current approaches and future trends. In *Proceedings of the IEEE AutoID2002*, pp. 103–108, Tarrytown, NY, 2002.

[73] Rhee TH, Cho SJ, and Kim JH. On-line signature verification using model-guided segmentation and discriminative feature selection for skilled forgeries. In *Proceedings of the 6th International Conference on Document Analysis and Recognition*, pp. 645–649, 2001.

[74] Roco MC and Bainbridge WS, Eds., *Converging Technologies for Improving Human Performance: Nanotechnology, Biotechnology, Information Technology and Cognitive Science*. Kluwer, Dordrecht, 2003.

[75] Sanchez-Reillo R, Sanchez-Avilla C, and Gonzalez-Marcos A. Biometric identification through hand geometry measurements. *IEEE Transactions on Pattern Analysis and Machine Intelligence*, 22(10):1168–1171, 2000.

[76] Sekanina L and Ružička R, Easily testable image operators: the class of circuits where evolution beats engineers. In Lohn J, Zebulum R, Steincamp J, Keymeulen D, Stoica A. and Ferguson MI, Eds. *Proceedings of the 2003 NASA/DoD Conference on Evolvable Hardware, Chicago, IL*, pp. 135–144, IEEE Computer Press, 2003.

[77] Senior A. A combintation fingerprint classifier. *IEEE Transactions on Pattern Analysis and Machine Intelligence*, 23(10):1165–1174, 2001.

[78] Shieh CS, Huang HC, Wang FH, and Pan JS. Genetic watermarking based on transform-domain techniques. *Pattern Recognition*, 37:555–565, 2004.

[79] Shmerko V and Yanushkevich S, Eds., *Banking Information Systems*. Academic Publishers, Technical University of Szczecin, Poland, 1997.

[80] Shmerko V, Perkowski M, Rogers W, Dueck G, and Yanushkevich S. Bio-technologies in computing: the promises and the reality. In *Proceedings of the International Conference on Computational Intelligence Multimedia Applications*, pp. 396–409, Australia, 1998.

[81] Shu W, Rong G, and Bian Z. Automatic palmprint verification, *International Journal of Image and Graphics*, 1(1):135–151, 2001.

[82] Sproat RW. *Multilingual Text-to-Speech Synthesis: The Bell Labs Approach*, Kluwer, Dordrecht, 1997.

[83] Stroobandt D, Verplaetse P, and Van Campenhout J. Towards synthetic benchmark circuits for evaluating timing-driven CAD tools. In *Proceedings of the International Symposium on Physical Design*, Monterey, CA, pp.60–66, 1999.

[84] Stoica A. Biometrics. *Lecture notes.* Private communication. 2005.

[85] Stoica A. Lohn J, Katz R, Keymeulen D, and Zebulum RS, Eds. *Proceedings of the 2002 NASA/DoD Conference on Evolvable Hardware, Alexandria, VA,* IEEE Computer Press, 2002.

[86] Stoica A. Robot fostering techniques for sensory-motor development of humanoid robots. *Journal of Robotics and Autonomous Systems,* Special Issue on Humanoid Robots, 37:127–143, 2001.

[87] Stoica A. Learning eye-arm coordination using neural and fuzzy neural techniques. In Teodorescu HN, Kandel A, and Jain L, Eds., *Soft Computing in Human-Related Sciences,* pp. 31–61, CRC Press, Boca Raton, FL, 1999.

[88] Terzopolous DT. Regularization of inverse visual problems involving discontinuities. *IEEE Transactions on Pattern Analysis and Machine Intelligence,* 8(4):413–424, 1986.

[89] Tilton CJ. An emergin biometric standards. *IEEE Computer Magazine.* Special Issue on Biometrics, 1:130–135, 2001.

[90] Turk M and Pentland A. Eigenfaces for recognition. *Journal of Cognitive Neuro Science,* 3(1):71–86, 1991.

[91] Uludag U, Pankanti S, Prabhakar SB, and Jain AK. Biometric cryptosystems: issues and challenges *Proceedings of the IEEE,* 92(6):948–960, 2004.

[92] Villringer A and Chance B. Noninvasive optical spectroscopy and imaging of human brain functions. *Trends in Neuroscience,* 20:435–442, 1997.

[93] Wang R, Hua TJ, Wang J, and Fan YJ. Combining of Fourier transform and wavelet transform for fingerprint recognition. In *Proceedings SPIE,* vol. 2242, Wavelet Applications, pp. 260–270, 1994.

[94] Wayman JL. Federal biometric technology legislation. *IEEE Computer* 33(2):76–80, 2000.

[95] Wildes RP, Asmuth JC, Green GL, Hsu SC, Kolczynski RJ, Matey JR, and McBride SE. A machine-vision system for iris recognition. *Machine Vision and Applications,* 9:1–8, 1996.

[96] Wildes RP. Iris recognition: an emerging biometric technology. *Proceedings of the IEEE,* 85(9):1348–1363, 1997.

[97] Wolfgang RB and Delp EJ. A watermark for digital images. In *Proceedings of the IEEE International Conference on Image Processing,* pp. 219–222, Lausanne, Switzerland, 1996.

[98] Wolfgang RB and Delp EJ. A Watermarking technique for digital imagery: further studies. In *Proceedings International Conference on Imaging Science, Systems, and Technology*, pp. 279–287, Las Vegas, 1997.

[99] Wolpaw JR, Birbaumer N, McFarland DJ, Pfurtscheller G, and Vaughan TM. Brain-computer interfaces for communication and control. *Clinical Neurophysiology*, 113:767–791, 2002.

[100] Woodward DJrJ, Orlands NM, and Higgins PT. *Biometrics: Identity Assurance in the Information Age*. McGraw-Hill/Osborne, Emeryville, CA, 2003.

[101] Yanushkevich SN., Supervisor. Data hiding based on synthetic biometric information. *Technical Report, Biometric Technologies: Modeling and Simulation Laboratory*, University of Calgary, Canada, 2005.

[102] Yanushkevich SN and Shmerko VP. Biometrics for Electrical Engineers: design and testing. *Lecture notes*. Private communication. 2005.

[103] Yu Z, Ip HHS, and Kwok LF. A robust watermarking scheme for 3D triangular mesh models. *Pattern Recognition*, 36:2603–2614, 2003.

[104] Zhang C and Cohen FS. 3-D face structure extraction and recognition from images using 3-D morphing and distance mapping. *IEEE Transactions on Image Processing*, 11(11):1249–1259, 2002.

[105] Zhang D. *Automated Biometrics: Technologies and Systems*. Kluwer, Dordrecht, 2000.

[106] Zhang D. *Palmprint Authentication*. Kluwer, Dordrecht, 2004.

[107] Zunkel R. Hand geometry based authentication. In Jain AK, Bolle RM, and Pankanti S, Eds., *Biometrics: Personal Identification in Networked Society*, pp. 87–102, Kluwer, Dordrecht, 1999.

# 2

## *Basics of Synthetic Biometric Data Structure Design*

In this chapter, the basics of synthetic biometric data structure design are introduced. The focus is on the fundamental concepts and some advanced techniques of image processing and pattern recognition:

▶ Modeling of images,

▶ Analysis-by-synthesis paradigm,

▶ Quality of modeling and processing, acceptable and unacceptable data as the result of satisfactory and non-satisfactory modeling, and

▶ Particular approaches in the synthesis of images (controlled degradation, random distortion, and warping).

The chapter begins with basic concepts of synthetic biometric data structure design (Section 2.1): model and components, static and dynamic model behavior, solutions from the model, and testing a model as reexamining the formulation, and checking on the validity of a model. The focus is the modeling of biometric data given by topological configurations. Section 2.1 provides the key to the solution of the inverse problems of biometrics. In development of computer models, the problem is decomposed into subproblems that yield a representation of a model by a set of submodels. This approach follows the assembly strategy in design. The solutions are then *merged* together, yielding a solution to the original problem. Finally, the obtained solution and model are verified.

The rest of this chapter is organized as follows. Section 2.3 focuses on the information content of biometric information. Interpretation of biometric data in terms of information (global and local, topological and spectral, recoverable and nonrecoverable, etc.) can be useful in classification, identification, and synthesis. Techniques for generating random attributes used in modeling are introduced in Section 2.4. Image processing techniques used in the synthesis of biometric data are introduced in Section 2.5. In Section 2.6, the controlling ability of warping in synthetic biometric data design is discussed. The focus of Section 2.7 is the generation of synthetic information by deformation, restoration, and interpolation methods. In Section 2.8, the basic technique of extracting structural information from synthetic biometric objects is revised. Finally, after the summary (Section 2.9) and problems (Section 2.10), recommendations for further reading are given in Section 2.11.

---

## 2.1   Basic concepts of synthetic biometric data structure design

In this section, the generic principles of synthetic biometric data structure design are introduced. The focus is the concept of the model and decision making based on the criteria of data acceptance or rejection, and strategies to design synthetic biometric data in the form of topological configurations.

### 2.1.1   Topology

In this book, biometric data is considered as a *topological configuration*. Topology is defined as a sample of properties of a geometrical figure. The notion of topology is used at local (as topological primitives) and global levels of biometric data representation and manipulation.

> **Example 2.1** *Examples of topological properties is the number of connected (merged, composed, assembled) topological primitives.*

### 2.1.2   Model

Building a model to make predictions concerning the behavior of the system under study is the first step in designing a system. A model is an abstraction of original data that can be used to obtain predictions and formulate control strategies (Figure 2.1).

In order to be useful, a model must necessarily incorporate elements of various (sometime conflicting) attributes, in general:

▶ *Realism*, or a *reasonable level of abstraction* and *a reasonably close approximation* to the real system, and

▶ *Simplicity*.

A *system* is a set of related entities, called *components*, *elements* or *primitives*. *An invertible* system is defined as follows: input to the system can be recovered from the output except for a constant scale factor. This requirement implies the existence of an inverse system that takes the output of the original system as its input and produces the input of the original system.

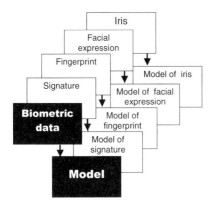

*Requirements of a model:*

- *Realism (reasonably close approximation),*
- *Simplicity, and*

*Classification of models:*

- *Static*
- *Dynamic*
- *Stochastic*
- *Deterministic*

**FIGURE 2.1**
A model is an abstraction of biometric data.

> **Example 2.2** *In a linear time invariant system (Figure 2.2), an inverse system performs deconvolution. The input $x(t)$ is not equal to the output $x(t)$ (nonideal system). The equalization can be interpreted as the simplest generation of synthetic copies of $x(t)$. This is one of the phenomena that can be used in synthetic biometric data design.*

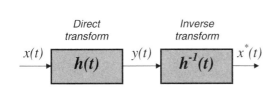

*$h(t)$ is a impulse response of a system*
*$h^{-1}(t)$ is an impulse response of an inverse system*
*The output of the cascade connection is*

$$x(t) * [h(t) * h^{-1}(t)] = x(t)$$

*The impulse response of a system $h(t)$ and inverse system $h^{-1}(t)$ must satisfy the condition*

$$h(t) * h^{-1}(t)] = \delta(t)$$

**FIGURE 2.2**
The simplest generator of synthetic data (Example 2.2).

Biometric simulation models can be classified in several ways, in particular:

*Static versus dynamic models* are those that do not evolve in time, and therefore do not represent the passage of time. In contrast, dynamic models represent systems that evolve over time.

*Deterministic versus stochastic models.* If a simulation contains *only* deterministic components, these models are called *deterministic*. In a deterministic model, all mathematical and logical relationships between the elements (variables) are fixed in advance and are not subject to uncertainty. In contrast, a model with at least one random input variable is called a *stochastic* model. Most biometric systems are modeled stochastically.

*Discrete-event dynamic models* are discrete models that represent dynamic and stochastic characteristics of data.

Having constructed a model for the problem under consideration, the next step is to derive a solution from this model. In general, there are two possibilities for this:

*Analytical* (algebraic) methods; the solution is usually obtained directly from its mathematical representation by formulas.

*Numerical* methods; the solution is generally an approximation obtained via a suitable approximation procedure.

> **Example 2.3** *In Figure 2.3, a "raw" iris primitive is represented by the equation $r = \sin 2\theta$ in a polar coordinate system. To make this model realistic, additional heuristic techniques are required.*

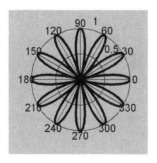

A "raw" iris topological primitive can be generated using trigonometric precedures:

▶ Equation $r = A \sin n\theta$, $A = 1$, $n = 2$ catches only the ideal shapes of an iris primitive;

▶ The generated topology is not colored;

▶ The generated topology does not carry any noise effects.

**FIGURE 2.3**

An algebraic model of an iris primitive (Example 2.3).

It follows from the above example that an algebraic solution can be considered as an intermediate solution to the problem. Typical problems in design of synthetic biometric data are as follows:

*Complexity.* Biometric data is complex and thus a formulation in terms of simple mathematical equations may be impossible.

*Limitation of analytical techniques.* Even if a mathematical model can be formulated that captures the behavior of some data structure of interest, it maybe not be possible to obtain a solution to the model by straightforward analytical techniques.

Modeling methodology in the simulation of biometrical data is called *designing*, or *generating*, synthetic biometric data structures with user-defined parameters. Some of its shortcomings and various caveats are:

▶ Generation of synthetic biometric data provides *statistical estimates* rather than the *exact* characteristics and performance measures of the model. Uncertainty is the crucial factor that is used in synthetic biometric data design.

▶ Generation of synthetic biometric data as a simulation procedure is typically time-consuming, and consequently, expensive in terms of designer time.

▶ Generation of synthetic biometric data, no matter how realistic, accurate, and impressive it is, provides consistently useful information about the data *only* if the model is a "valid" representation of the data under study.

### 2.1.3 Analysis-by-synthesis modeling

The role of the inverse problems of biometrics is closely related to the problem of modeling biometric data. Modeling methods can be distinguished as follows:

▶ *Straightforward* methods, which are based on two types (sources) of information: (a) priori knowledge and (b) knowledge obtained from analysis of data.

▶ *Analysis-by-synthesis* methods, which, in addition to the sources of information of straightforward methods, utilize a third source - the deviation between the model and original data.

The reliability of straightforward methods depends on quite precise expectations about the type of biometric data to be encountered and on the reliability and robustness of the algorithms for biometric data from two sources of information:

$$< \text{Reliability} > = < \overbrace{\texttt{Priori} \quad \texttt{knowledge}}^{Source\ 1} > + < \overbrace{\text{Knowledge} \quad \text{from} \quad \text{analysis}}^{Source\ 2} > .$$

If the modeling premises are inadequate, the modeling procedure can fail because there is no feedback to correct the model.

The reliability of analysis-by-synthesis methods depends on the reliability provided by straightforward methods and the reliability of the third source of data, that is, the deviation between the modeled and original data from three sources of information (Figure 2.4):

$$< \text{Reliability} > \; = \; < \overbrace{\text{Priori} \quad \text{knowledge}}^{\textit{Source 1}} >$$

$$+ \; < \overbrace{\text{Knowledge} \quad \text{from} \quad \text{analysis}}^{\textit{Source 2}} >$$

$$+ \; < \overbrace{\text{Deviation}}^{\textit{Source 3}} >$$

The third source of information consists of an iterative loop that successively approximates the original data. This loop is a *generator* of data: inputs are the model parameters which are yielded by analysis and outputs are recovered data.

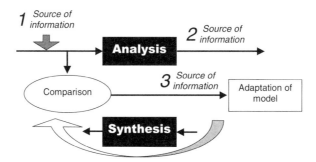

**FIGURE 2.4**

The analysis-by-synthesis approach in biometric data identification and recognition: the third source of information provided by the loop is the inverse solution of the analysis.

## 2.1.4    Acceptable and unacceptable data

Systems for synthetic biometric data structure design are often *iterative* systems. In each iteration, a pattern of data is produced. The simplest pattern includes two joint primitives (in general, randomly chosen from the library of primitives). Combining these two primitives can be done in different ways that produce different patterns of data. These patterns can be *acceptable* and *unacceptable*.

**Example 2.4** *Figure 2.5 shows three types of unacceptable topological structures.*

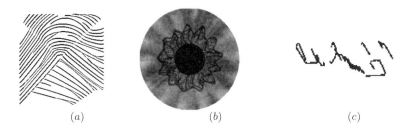

$(a)$ $(b)$ $(c)$

**FIGURE 2.5**
A sample of unacceptable biometric patterns: (a) fingerprint, (b) iris, and (c) signature (Example 2.4).

## 2.2 Synthesis strategies

One can observe from the above example that the topological patterns in Figure 2.5 cannot be classified as fingerprint-like topology. There are several strategies in the design of synthetic biometric data structure, for example (Figure 2.6):

▶ *Assembling*; this is the design of synthetic biometric data using a library of primitives and a set of assembly rules (Figure 2.6a).

▶ *Modification* of biometric data; this strategy assumes that the original biometric and new, synthetic data can be created by distortion accordingly to the set of rules (Figure 2.6b).

▶ *Functional assembling*; this is design based on combining the above two strategies: given one or several topological patterns of original biometric data, create biometric-like topology using a given library of primitives and assembling rules, and

▶ *Physical* modeling.

### 2.2.1 Assembling

A synthesis can be created by combining a library of primitives according to certain rules.

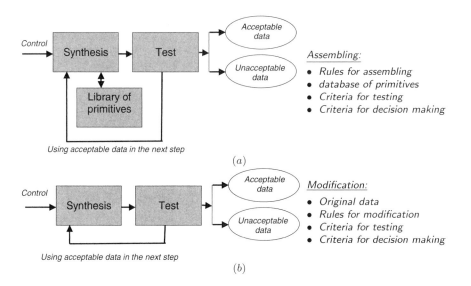

**FIGURE 2.6**
Strategies in the design of synthetic biometric data: (a) assembling, and (b) modification of biometric data.

The effectiveness of the design can be measured by the ratio of the number of acceptable synthesized results to the number of unacceptable ones:

$$< \texttt{Effectiveness} > \; = \; \frac{< \texttt{Number of acceptable copies} >}{< \texttt{Number of unacceptable copies} >}.$$

> **Example 2.5** *An evolutionary algorithm can be applied to the design, for example, of synthetic fingerprints. The topological primitives of a fingerprint are merged, that is they can be reproduced and mutate. The composed image is evaluated using the fitness function according to the evolutionary strategy. The fitness function is a set of topological features that specify "acceptable" fingerprint topology. The results at each generation are topological configurations. The number of generators depends on the specified parameters and requirements for the quality of the synthetic fingerprint.*

The evolutionary strategy is similar to the well-studied evolutionary algorithms for electronic circuit design over the library of gates (circuit primitives) that result in false (called "garbage") circuits (Figure 2.5) and correct circuits. The effectiveness of the algorithm is measured by the proportion of false and correct circuits.

The elements of a model possess certain characteristics, or *attributes*, that take on logical or numerical values. Typically, the activities of individual components interact in time. These activities cause changes in the system's state.

> **Example 2.6** *The pattern of automated design of fingerprint-like topology and iris-like topology is illustrated by Figure 2.7a and Figure 2.7b. A signature may be considered as a system, with topological primitives as elements (Figure 2.7c):*
>
> (a) *An attribute of a signature might be the number of topological primitives, their types, levels of connections, and so on.*
>
> (b) *Signatures and handwriting vary greatly, even for a single individual.*
>
> (c) *The state of a signature might be described by the time interval. When an individual is under stress, their signature is changed, and the system jumps to a new state.*

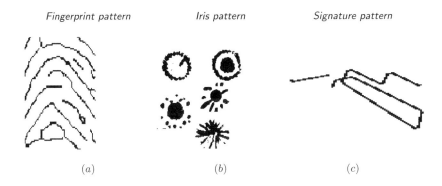

Fingerprint pattern     Iris pattern     Signature pattern

(a)     (b)     (c)

**FIGURE 2.7**
Assembling topological primitives in a synthetic fingerprint (a), synthetic iris (b), and synthetic signature design (c) (Example 2.6).

## 2.2.2   Design based on modification of biometric data

A design method based on the modification of biometric data is described by

(a) The original biometric data and

(b) The rules for its modification, or deformation.

The original biometric data (facial, fingerprint, iris, voice, etc.) is transformed to data with modified characteristics. The result is synthetic biometric data.

**Example 2.7** *Examples of designing a synthetic biometric topology based on the modification of original biometric data are as follows:*

(*a*)  *Given:  face topology, facial transformations produce synthetic face images (with another eye, nose, and face configuration).*

(*b*)  *Given: a fingerprint, a synthetic fingerprint pattern can be obtained by replacing topology configurations.*

(*c*)  *Given: an iris, a distribution of shapes (freckles, coronas, etc.) and colors can be changed that results in a synthetic iris pattern.*

(*d*)  *Given:  a voice, a synthetic voice can be designed by warping the acoustic features.*

(*e*)  *Given a hand geometry, changing its dimensions produces a synthetic hand geometry pattern.*

## 2.2.3   Functional assembling and physical modeling

The problem of functional assembling is formulated as follows: given a pattern of biometric data, design synthetic data using a library of primitives and assembling rules. Methods of functional assembling sometimes demonstrate better characteristics (in forms of the amount of acceptable synthetic data produced) compared with design using a library of arbitrary created primitives primitives.

In physical modeling, biometric data are associated with the physical nature of the carrier such as fingerprint, palm, iris, etc. The forming of this biometric data occur as a process of natural development.

**Example 2.8** *Iris topology is developed during the first months of life. This time interval can be divided into several phases. Each phase can be described in terms of physical processes and associated topological configurations. Fingerprint modeling based on anatomical concepts of skin. The upper layer of skin, called the epidermis, is formed within a few weeks of embryonic life.*

**Example 2.9** *Fingerprint development starts from several initial points.  Around these points fingerprint topology is formed, i.e., topological information is stored.  By analogy, natural development of iris topology is started from several points. The coordinates of these points are stable parameters. The physical process of topology formation around each point results in a topological primitive.*

To develop a physical model of certain biometric data, knowledge of the physical processes of formation of biometric information carriers (part of the human body) is useful. Notice that the evolution of the carrier of biometric information and the philosophy of evolutionary algorithms have several useful relationships for modeling. For example, a natural strategy in the development of irises and fingerprints, probably, can be accepted as optimal in evolutionary based synthesis of irises and fingerprints.

### 2.2.4 Testing

Testing a model addresses the following problems:

*Reexamining* the formulation of the problem and uncovering possible flaws.

*Checking* on the validity of a model to ascertain that all mathematical expressions are dimensionally consistent.

*Reconstructing* the past using additional information data, and then determining how well the resulting solution would have performed if it had been used.

*Comparing* the effectiveness of hypothetical performance with what actually happened indicates how well the model predicts "reality".

Figure 2.6 illustrates the testing of data with respect to the criterion of acceptability. Testing is applied to all stages of design of synthetic biometric data. At each step, testing is aimed at distinguishing *acceptable* and *unacceptable* synthetic biometric data.

## 2.3 Information carried by biometric data

In this section, the information carried by biometric data is categorized with respect to data representation, data quality, redundancy, and robustness.

### 2.3.1 Categories of information

Through analysis and synthesis of biometric data, the following categories of information are distinguished:

▶ *Global* and *local* information. Depending on the phase of processing and using a model of biometric data, various representations of information are required: from detailed (local) to generic (global).

▶ *Topological* and *spectral* information related to topological and spectral transformations that is aimed at extraction of features from a biometric image. For example, in some cases, processing in the frequency (spectral) domain is preferable compared with manipulation in topological space.

▶ *Recoverable* and *unrecoverable* information.

▶ *Uncertain* information is defined as data that cannot allow a decision and additional knowledge or transforms are required. For example, uncertain information corresponds to biometric data intensively "infected" by noise. Zero data is a boundary case of uncertainty.

▶ *Redundant* and *insufficient* information is relevant to the problem of optimization of space of biometric data features (minimize the number of significant features) and decision making (in the case of insufficient data).

### 2.3.2    Measuring of information

Information-theory describes how information can be measured.

> **Example 2.10** *The image redundancy problem is defined as the quantity of common information in two images.   To measure information, these images are considered as the transmitter and the receiver connected by a virtual channel.*

In this formulation, Shannon's information theory can be applied to evaluate images in terms of information (see details in the "Further Reading" Section).

> **Example 2.11** *In analysis, the entropy decreases through data processing.  In synthesis, the entropy increases through data processing (Figure 2.8).*

Analysis of biometric data is aimed at the removal of redundant information, extraction of what is useful for identification, recognition of features, and decision making. Hence, uncertainty, or entropy, decreases through analysis. The goal of synthesis is to construct synthetic biometric data that satisfies the criterion of "being as real as possible." Synthesis starts with some templates that correspond to zero entropy. In general, the design of synthetic biometric data is a stochastic process because of the random composition of topological primitives.   Moreover, additive and multiplicative noise is incorporated in models through phases of design.   Hence, the entropy increases through synthesis.

## 2.4    Generation of random biometric attributes

In contrast to the problem of denoising that is associated with the analysis of biometric data, in this section the problem of synthesizing a noisy topological structure is considered.   Various models of random biometric attributes are discussed:

$\mapsto$ *Entropy decreases through biometric data analysis*

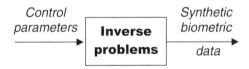

$\mapsto$ *Entropy increases through biometric data synthesis*

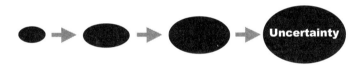

**FIGURE 2.8**
Entropy of data during analysis and synthesis (Example 2.11).

▶ The simplest noise model,

▶ A noise model with a given distribution,

▶ Periodic noise model,

▶ Random sequences* and vectors, and

▶ Random number generation with an arbitrary probability distribution.

The type of distribution and "quality" of random numbers that are used in a model can be critical in some cases to synthetic biometric data structure design. Generation of random numbers given a probability distribution is

---

*More precisely, random sequences are pseudo-random sequences because they are generated by mathematical manipulation. Random sequences are produced by devices with physical sources of noise.

based on inverse transforms. These transforms are an excellent example of how useful the solutions to inverse problems are in practice.

## 2.4.1   The simplest noise models

Spatial noise values are random numbers, characterized by a probability density function $f$ or, equivalently, by the corresponding cumulative distribution function $F$. Random number generation for the types of distributions in which we are interested follows some rules adopted from probability theory. The basic random number generator can produce: random numbers with uniform probability density function in the interval (0,1), or with normal probability density function and zero mean and unit variance.

**Noise model with given distribution.** Let $w$ be a uniformly distributed random variable in the interval (0,1). Then, a random variable $z$ with a specified cumulative distribution function $F_z$ can be obtained from the equation $z = F_z^{-1}(w)$.

**Periodic noise model.** Periodic noise in biometric data arises from electrical and/or electromechanical interference.

> **Example 2.12** *In the time domain, the model of a periodic noise source is defined by equation*
>
> $$r(x,y) = A \cdot \sin \left[ \frac{2 \cdot \pi \cdot u_0 \cdot (x + B_x)}{M} + \frac{2 \cdot \pi \cdot v_0 \cdot (y + B_y)}{N} \right],$$
>
> *where parameter $A$ is the amplitude, $u_0$ and $v_0$ determine the sinusoidal frequencies with respect to the x- and y-axis respectively, and $b_x$ and $B_y$ are phase displacements with respect to the origin. In a frequency domain, the discrete Fourier transform of $r(x,y)$ yields*
>
> $$
> \begin{aligned}
> R(u,w) = & j\frac{A}{2} \cdot \exp \frac{j2 \cdot \pi \cdot u_0 \cdot B_x)}{M} \cdot \delta \overbrace{(u + u_0, \ v + v_0)}^{\text{Impulse location}} \\
> & + j\frac{A}{2} \cdot \exp \frac{j2 \cdot \pi \cdot v_0 \cdot B_y)}{M} \cdot \delta \underbrace{(u - u_0, \ v - v_0)}_{\text{Impulse location}}
> \end{aligned}
> $$
>
> *The pair of complex conjugate impulses located at $(u + u_0, \ v + v_0)$ and $(u - u_0, \ v - v_0)$, respectively.*

**Random sequences.** In a typical stochastic simulation, randomness is introduced into simulation models via sequences $U_1, U_2, ...,$ of independent random variables with uniform distribution on the interval (0,1). The most common methods for generating random sequences use so-called *linear congruential generators*. These generate a deterministic sequence of numbers by means of a recursive formula. This formula is based on calculating the modulo-$m$ residues of a linear transformation: for some integer $m$, $X_i + 1 = aX_i + c \quad (mod \ m)$, where $X_o$ is the initial value, $a$ and $m$ are positive integers, and $c$ is a nonnegative integer.[†]

It follows from the above that:

▶ The sequence $U_i$ is completely deterministic with control parameters $X_0, a, c$, and $m$.

▶ The sequence $U_i$ can be accepted as a random sequence if it appears to be uniformly distributed and statistically independent.

▶ An arbitrary choice of $X_0, a, c$ and $m$ will not lead to a random sequence with satisfactory statistical properties, only a few combinations of these produce satisfactory results.

▶ The sequence $X_i$ will repeat itself in, at most, $m$ steps and will therefore be periodic with a period not exceeding $m$. In computer implementations, $m$ is normally selected as a largest prime number that can be accommodated by the computer word size.

> **Example 2.13** *Let $a = c = X_o = 3$ and $m = 5$. The sequence obtained from the recursive formula $X_{i+1} = 3X_i + 3 \ (mod \ 5)$ is $X_i = \{3, 2, 4, 0, 3\}$. The period of the generator is 4, while $m = 5$.*

**Random vector generation.** The multidimensional version of a random generator is

$$\mathbf{X}_i + 1 = \mathbf{AX}_i \quad (mod \ \mathbf{M}),$$
$$\mathbf{U}_i + 1 = \mathbf{M}^{-1}\mathbf{X}_i,$$

respectively, where $\mathbf{A}$ is a nonsingular $n \times n$ matrix, $\mathbf{M}$ is a vector of constants of the same dimensionality as the vector $\mathbf{X}$, and $\mathbf{M}^{-1}\mathbf{X}_i$ is an $n$-dimensional vector with components $M_1^{-1}X_1, ...., M_n^{-1}X_n$.

## 2.4.2 Generation of random numbers with prescribed distributions

Let $X$ be a variable with cumulative distribution function $F(x)$. Since $F(x)$ is a nondecreasing function, the inverse function $F^{-1}(y)$ is defined as

$$F^{-1}(y) = \inf\{x : F(x) \geq y\}, \ 0 \leq y \leq 1.$$

---

[†]Applying the modulo-$m$ operator means that $aX_i + c$ is divided by $m$, and the remainder is taken as the value of $X_{i+1}$. Each $X_i$ can only assume a value from the set $\{0,1,...,m-1\}$ and random numbers $U_i = \frac{x_i}{m}$ constitute approximations of the uniform random variables.

If $U$ is uniformly distributed over the interval $[0,1]$, then (Figure 2.9)

$$X = F^{-1}(U)$$

has the cumulative distribution function $F(x)$. In view of the fact that $F$ is invertible and $P(U \leq u) = u$, we readily obtain

$$P(X \leq x) = P[F^{-1}(U) \leq x] = P[U \leq F(x)] = F(x).$$

Thus, to generate a value, say $x$, of a random variable $X$ with a cumulative distribution function $F(x)$, first sample a value, say $u$, of a uniform distribution, $U$, compute $F^{-1}(u)$, and set it equal to $x$.

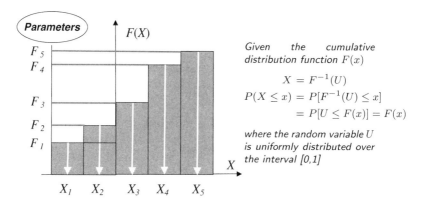

**FIGURE 2.9**
Generation of random variable $X$ from the given cumulative distribution function $F(x)$.

**Generation of a random variable with an arbitrary probability distribution.** The algorithm for generating a random variable from an empirical distribution $F(x)$ can be written as:

*Step 1.* Generate a uniform distribution over the interval $[0, 1]$, $U \in [0, 1]$.

*Step 2.* Find the smallest positive integer $k(0 \leq k \leq n - 1)$, such that $U \leq F(x_k)$, and return

$$X = F^{-1}(U) = a_k + [U - F(a_k)](a_{k+1} - a_k)/[F(a_{a+1}) - F(a_k)].$$

**Example 2.14** *Figure 2.9 illustrates*

(a) *The generation of a random variable $X$ from a given probability density function: first generate a variable $U$ from $U(0,1)$, and then take its square root.*

(b) *The generation of random variable $X$ with an arbitrary probability distribution.*

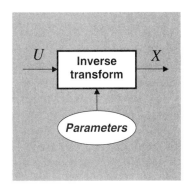

*Given:* a probability density function
$$f(x) = \begin{cases} 2x, & 0 \le x \le 1 \\ 0, & otherwise \end{cases}$$
*Generate:* a random variable with this probability density function.
The cumulative distribution function is

$$F(x) = \begin{cases} 0, & x < 0 \\ \int_0^x 2x\,dx = x^2, & 0 \le x \le 1 \\ 1, & x > 1. \end{cases}$$

$$X = F^{-1}(U) = U^{1/2}, \ 0 \le u \le 1.$$
*Given:*

$$p(x_i) = P(X = x_i), \ i = 1, 2, ..., r,$$

$$F(x_k) = \sum_{x_i \le x_k} p(x_i), \ k = 1, 2, ..., s,$$

*Generate:* a random variable with this probability distribution function.
*Algorithm*

1. *Generate $U \in [0, 1]$.*

2. *Find the smallest positive integer, $k$, such that $U \le F(x_k)$ and return $X = x_k$.*

**FIGURE 2.10**
Generation of a random variable $X$ from a given probability density function (Example 2.14).

This approach states that the cumulative distribution function, $F(x)$, exist in a form for which the corresponding inverse function $F^{-1}(\cdot)$ can be found analytically or algorithmically. Unfortunately, for many probability distributions, it is either impossible or difficult to find the inverse transform even in cases where $F^{-1}$ exists in an explicit form.

**Generation of a random vector of variables with a given probability distribution.** In generation of a random vector $\mathbf{X} = (X_1, ..., X_n)$ from a given cumulative distribution function $F(\mathbf{x})$, two cases can be distinguished:

(a) The variables $X_1, ..., X_n$ are *independent*, and

(*b*) The variables $X_1, ..., X_n$ are *dependent*.

For independent variables $X_1, ..., X_n$, the joint probability density function, $f(\mathbf{x})$, is $f(\mathbf{x}) = f(x_1, ..., x_n) = f_1(x_1)f_2(x_2), ..., f_n(x_n)$. where $f_i(x_i)$ is the marginal cumulative distribution function of the variable $X_i$. Generation of the random vector $\mathbf{x} = (X_1, ...X_n)$ from the cumulative distribution function $F(\mathbf{x})$ is defined by the equation $X_i = F^{-1}(U_i)$, $i = 1, ..., n$.

## 2.5   Degradation model in synthesis of biometric data

In this section, the phenomena of the degradation of an image widely used in analysis is considered for the purpose of synthesis of biometric data.

Image restoration is to recover a corrected image from a degraded image, as well as some information about the degradation. This information is usually called the degradation model. The model is normally either analytical or empirical. Analytic models arise from a mathematical description of the degradation process. Many degradations can be modelled as point-spread functions. Empirical models result from direct measurement of the degradations. Geometric degradations can be modelled empirically by imaging an object of known geometry and then measuring the distorted image.

### 2.5.1   Direct and inverse models of image degradation

In the synthesis of biometric data, several techniques of image processing are used, in particular:

▶ Restoration techniques that attempt to *reconstruct* or *recover* an image that has been degraded, and

▶ Enhancement techniques that are mostly heuristic procedures designed to manipulate an image to improve it.

For restoration, knowledge of the degradation phenomenon can be used. Restoration techniques are oriented toward modeling the degradation and applying the *inverse* process in order to recover the original image. This approach usually involves formulating a criteria of goodness. These criteria are different in image processing and synthetic biometric information generation:

▶ In image processing, criteria of goodness yield an optimal estimate of the desired result

$$\overbrace{< \texttt{Image} >}^{\textit{Degradated}} \Rightarrow < \texttt{Inverse filtering} > \Rightarrow \overbrace{< \texttt{Image} >}^{\textit{Original}}.$$

▶ In the synthesis of biometric data, the criteria of goodness means acceptability, or realism of synthetic data

$$\overbrace{< \texttt{Image} >}^{Original} \Rightarrow \underbrace{< \texttt{Degradation} >}_{Controlled} \Rightarrow \overbrace{< \texttt{Image} >}^{Synthetic}.$$

> **Example 2.15** *An inverse filter is a technique for restoring images degraded by processes which can be modeled as linear, shift invariant systems. Let $F(u, v)$ be the Fourier transform of the original image and $M(u, v)$ be the Fourier transform of the degradation:*
>
> $$D(u, v) = F(u, v)M(u, v),$$
> $$F(u, v) = D(u, v)\frac{1}{M(u, v)},$$
>
> *where $D(u, v)$ is the degraded image.*

It follows from this example that $1/M(u, v)$ carries information about the nature of the degradation. In image analysis, restoration techniques are oriented toward *modeling the degradation* and applying the inverse process in order to recover the original image. In the synthesis of topological structures representing biometric data, the degradation function and input image are given. The degradation produces a new image that satisfies the criteria of acceptability of topological structure.

## 2.5.2 Restoration of an image containing additive noise

Consider a degradation process input image $f(x, y)$ with additive noise to produce a degraded image $g(x, y)$:

$$g(x, y) = H[f(x, y)] + \eta(x, y).$$

The objective is to obtain an estimate $\hat{f}(x, y)$ of the original image.

> **Example 2.16** *Let $H$ be a* linear, spatially invariant *process. A model of degradation can be viewed as a model of the generation of synthetic images (Figure 2.11), where $h(x, y)$ is the spatial representation of the degradation function and, $h(x, y) * f(x, y)$ is a convolution.*

Modeling synthetic images in Example 2.16 utilizes unwanted effects of restoration, namely:

▶ Even if $H(u, v)$ is known exactly, it is impossible to recover $F(u, v)$ because of the strong noise component.

▶ Even if term $N(u, v)$ is negligible, dividing it by vanishing values of $H(u, v)$ would dominate restoration estimates.

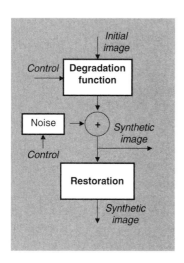

*Model of degradation:*

$$g(x,y) = \overbrace{h(x,y) * f(x,y)}^{Convolution} + \overbrace{\eta(x,y)}^{Noise}$$
$$G(u,v) = \underbrace{H(u,v)F(u,v)}_{Spectrum} + \underbrace{N(x,y)}_{Spectrum\ of\ noise}$$

*The simplest model (inverse filtering):*

$$\hat{F}(u, v) = F(u, v) + \frac{N(u, v))}{H(u, v)}$$

- *The degradation image $g(x,y)$ is represented by the sum of the degradation function and noise function.*
- *Knowledge about the degradation function $H[f(x, y)]$ and noise function $\eta(x, y)$ is essential in modeling.*
- *Reconstruction of the initial image $f(x, y)$ is an estimate $\hat{f}(x, y)$ of the original image.*
- *The estimate $\hat{f}(x, y)$ is close to the original image $f(x, y)$, as the degradation function $H[f(x, y)]$ and noise function $\eta(x, y)$ are defined.*

**FIGURE 2.11**

Using a degradation model for generation of synthetic biometric data (Example 2.16).

## 2.6  Image warping

Warping is a basic technique of biometric topological data synthesis. In this section, the focus is the controlling ability of warping in synthetic biometric data design.

### 2.6.1  Image warping as a topological transformation

Image warping is defined as a topological transformation that changes the spatial configuration of an image. The basis of warping is the curve evolution theory that studies the deformation of curves using geometric measures.

**Example 2.17** *The evolution of the curve along its normal direction can be characterized by the partial differential equation and curve deformations are defined as curvature deformation and constant deformation.*

The simplest warps are rotations, translations, and shears. A warp is separated into two types of data structures: a model of data representation (point-to-point mappings, uniform grid of pixels in the plane, etc.) and user controls data (specification of coefficients, data that have geometric meaning, etc.). One criteria of effectiveness of a user controls object is that it have a minimal number of control parameters. Controlling a warp's evolution over time is crucial in synthesis, for example, synthesis facial expressions. In morphing, the user smoothly varies a deformation. Warping requires a good level of user specification to achieve the desired results.

Two classes of transformations (deformable models) are distinguished:

▶ Parametric and
▶ Nonparametric transformations (deformable models).

Parametric deformable models represent curves and surfaces explicitly in their parametric forms during deformation. This representation allows direct interaction with the model. However, adaptation of the model topology (for example, splitting and merging parts during the deformation) can be difficult. Parametric warping includes Euclidean distance warping (scale, rotation, translation, and their combinations). Euclidean warping is mostly performed backwards by inverse transforms. Nonparametric deformable models can handle topological changes naturally. Their parameterizations are computed after complete deformation.

## 2.6.2 Spatial transforms

Let image $f$ be given by the pixel coordinates $x, y$. Geometric distortion is defined as a transform of image $f$ that produces new image $\widehat{f}$ with coordinates $(\widehat{x}, \widehat{y})$. In formal notation,

$$< Coordinate \ \widehat{x} > = t_x(x, y)$$
$$< Coordinate \ \widehat{y} > = t_y(x, y),$$

where $t_x(x, y)$ and $t_x(x, y)$ are the spatial transformations.

**Example 2.18** *Let* $t_x(x, y) = \frac{x}{2}$ *and* $t_x(x, y) = 2y$, *the distortion is defined as scaling of image* $f(x, y)$ *by factor* $1/2$ *in* $x$ *and by factor* $2$ *in* $y$ *spatial direction. Inverse distortion is defined as recovering of* $f(x, y)$ *from* $\widehat{f} = (\widehat{x}, \widehat{y})$.

In general, inverse spatial transformation does not always exist, i.e., it is not always possible to recover the original image from the deformed image.

There are several approaches to development of direct and inverse spatial transformations, for example, by using control points with known locations in original and distorted images.

> **Example 2.19** *Control points can be defined if distortion is known in algebraic form. Let geometric distortions be defined by a pair of bilinear equations*
>
> $$t_x(x,y) = a_0 + a_1x + a_2y + a_3xy$$
> $$t_y(x,y) = b_0 + b_1x + b_2y + b_3xy.$$
>
> *The coordinates of a distorted image are computed as follows*
>
> $$< Coordinate \; \widehat{x} > = a_0 + a_1x + a_2y + a_3xy$$
> $$< Coordinate \; \widehat{y} > = b_0 + b_1x + b_2y + b_3xy.$$

It follows from the above example that the number of control points is equal to eight and coordinates can be found by solution of these equations. For example, in physical facial expression modeling an exact simulation of all face components is not required, i.e., modeling of neurons, muscles, etc. In practice, a physical face model is based on a few dynamic parameters related to biomechanical description.

> **Example 2.20** *The development of a facial muscle process that is controllable by a limited number of parameters and is nonspecific to a facial topology is based on physical modeling and warping techniques. Muscles can be modeled with several control parameters, for example, direction and magnitude.*

Digital image warping consists of two basic operations:

▶ A spatial transformation to define the rearrangement of pixels, and
▶ Interpolation to compute their values.

> **Example 2.21** *Figure 2.12b,c,d illustrates the radial based (in the polar coordinates $r, \phi$) warp of the original image (a). The MATLAB code is given at the end of Chapter (see Supporting information for problems).*

### 2.6.3    Controlled distortion by affine transform

One of the most commonly used forms of spatial transformations is the *affine transform*: identity, scaling, rotation, shear, (vertical), and shear (horizontal), and translation.

The general representation of an *affine transformation* is

$$[x, y, 1] = [u, v, 1] \begin{bmatrix} a_{11} & a_{12} & 0 \\ a_{21} & a_{22} & 0 \\ a_{31} & a_{32} & 1 \end{bmatrix} \tag{2.1}$$

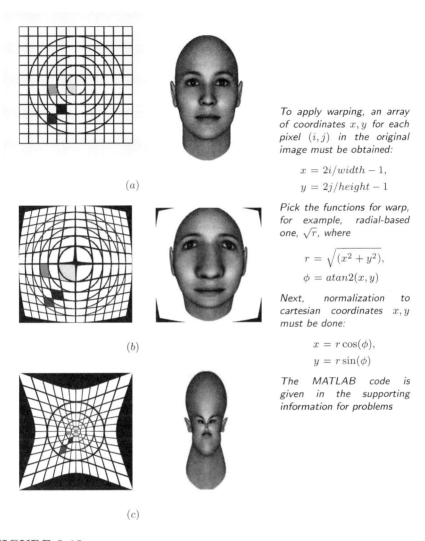

To apply warping, an array of coordinates $x, y$ for each pixel $(i, j)$ in the original image must be obtained:

$$x = 2i/width - 1,$$
$$y = 2j/height - 1$$

(a)

Pick the functions for warp, for example, radial-based one, $\sqrt{r}$, where

$$r = \sqrt{(x^2 + y^2)},$$
$$\phi = atan2(x, y)$$

Next, normalization to cartesian coordinates $x, y$ must be done:

$$x = r\cos(\phi),$$
$$y = r\sin(\phi)$$

(b)

The MATLAB code is given in the supporting information for problems

(c)

**FIGURE 2.12**

Image warping of the original image (a) with the functions $\sqrt{r}$ (b) and $r^p$, $p = 1.5$ (c) respectively.

For affine transformations, the forward mapping functions are

$$x = a_{11}u + a_{21}v + a_{31} \tag{2.2}$$

$$y = a_{12}u + a_{22}v + a_{32} \tag{2.3}$$

This accommodates translations, rotations, scale, and shear. Since the product of affine transformations is also affine, they can be used to perform a general orientation of a set of points relative to an arbitrary coordinate system while still maintaining a unity value for the homogeneous coordinate.

**Translation.** All points are translated to a new position by adding offsets $T_u$ and $T_v$ and $u$ and $v$, respectively (Figure 2.13).

**Rotation.** All points in the $uv$-plane are rotated about the origin through the counterclockwise angle $\theta$ (Figure 2.13).

**Scale.** All points are scaled by applying the scale factor $S_u$ and $S_v$ to the $u$ and $v$ coordinates, respectively (Figure 2.13). Enlargements (reductions) are specified with positive scale factors that are larger (smaller) than unity. Negative scale factors cause the image to be reflected, yielding a mirrored image. If the scale factors are not identical, then the image proportions are altered, resulting in a differentially scaled image.

**Shear.** The coordinate scaling described above involves only the diagonal terms $a_{11}$ and $a_{22}$. Consider the case where $a_{11} = a_{22} = 1$, and $a_{12} = 0$ in Equations 2.1, 2.2, and 2.3. By allowing $a_{21}$ to be nonzero, $x$ is made linearly dependent on both $u$ and $v$, while $y$ remains identical to $v$. A similar operation can be applied along the $v$-axis to compute new values for $y$ while $x$ remains unaffected. This effect, called shear, is therefore produced by using off-diagonal terms (Figure 2.13).

## 2.7    Deformation by interpolation and approximation

In synthesis, controlled deformation of biometric data structures plays a key role. Biometric data can be deformed in spatial dimensions and the spectral domain, where the techniques of interpolation and approximation can be used. The focus of this section is a brief overview of various techniques which are useful in the manipulation of biometric topological data.

### 2.7.1    Parametric and implicit models

Because of the high complexity of initial topological data, two strategies of biometric data design are distinguished: local and global strategies. Requirements for models in these strategies are different.

Original image       Sheared image       Rotated image

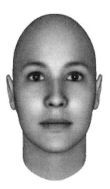  

*The affine transforms:*
*identity, scaling, rotation, shear, (vertical), and*
*shear (horizontal), and translation.*

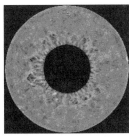

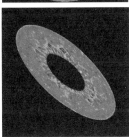

$$Translation: \quad [x, y, 1] = [u, v, 1] \begin{bmatrix} 1 & 0 & 0 \\ 0 & 1 & 0 \\ T_u & T_v & 1 \end{bmatrix}$$

$$Rotation: \quad [x, y, 1] = [u, v, 1] \begin{bmatrix} \cos\theta & \sin\theta & 0 \\ -\sin\theta & \cos\theta & 0 \\ 0 & 0 & 1 \end{bmatrix}$$

$$Scale: \quad [x, y, 1] = [u, v, 1] \begin{bmatrix} S_u & 0 & 0 \\ 0 & S_v & 0 \\ 0 & 0 & 1 \end{bmatrix}$$

$$Shear \ (u\text{-}axis): \quad [x, y, 1] = [u, v, 1] \begin{bmatrix} 1 & 0 & 0 \\ H_v & 1 & 0 \\ 0 & 0 & 1 \end{bmatrix}$$

$$Shear \ (v\text{-}axis): \quad [x, y, 1] = [u, v, 1] \begin{bmatrix} 1 & H_v & 0 \\ 0 & 1 & 0 \\ 0 & 0 & 1 \end{bmatrix}$$

*For example, the transformation matrix of iris*
*image sheared by a factor of 0.5 vertically and*
*horizontally is defined as*

$$T = \begin{bmatrix} 1 & 0.5 & 0 \\ 0.5 & 1 & 0 \\ 0 & 0 & 1 \end{bmatrix}$$

**FIGURE 2.13**
Affine transforms.

> **Example 2.22** *The entire shape $S$ can be modeled via a single system of equations. For complex shapes, a combinatorial explosion usually occurs, which makes this approach impractical. Such an approach is not modular, and unsuitable to tackle problems for which it may be necessary to modify certain parts and leave the other parts unchanged.*

The general approach is to use functions that are of a higher degree than are the linear functions. The functions still generally only approximate the desired shapes, but offer easier interactive manipulation than do linear functions. The higher-degree approximations can be based on one of three models:

▶ *Explicit* models (in the form $y = f(x)$),

▶ *Implicit* models (in the form $f(x, y)$), and

▶ *Parametric* models.

**Implicit model.** The implicit equation may have more solutions than we want. For example, in modeling a half circle using $x^2 + y^2 = 1$, constraints $x \geq 0$ must be added. A shape $S$ of dimension $s$ specified by a function $F : d \rightarrow P$, where $d$ is of dimension $n \geq s$, is defined as the *zero locus* $F^{-1}(0)$ of the function $F$, that is, as the set of zeros of the function $F$: $S = F^{-1}(0) = \{a | a \in d, F(a) = 0\}$. The space $P$ is a vector space, and it has a certain dimension $d \geq n - s$.

> **Example 2.23** *Figure 2.14 illustrates function $F : \mathbf{A}^2 \rightarrow \mathbf{A}$ for a straight line and parabola.*

**Parametric model.** A shape $S$ of dimension $s$ specified by a function $F : P \rightarrow d$, where $d$ is of dimension $n \geq s$, is defined as the *range* $F(P)$ of the function $F$. Thus, the parameter space $P$ also has the dimension $s \leq n$. Every point lying on the shape $S$ is represented as $F(a)$, for some parameter value $a \in P$ (possibly many parameters values).

> **Example 2.24** *The function $F : \mathbf{A} \rightarrow \mathbf{A}^2$ defined such that $F_1(t) = 2t + 1$, $F_2(t) = t - 1$, represents a straight line in the affine plane. Figure 2.15 illustrates the parametric model via function $F : \mathbf{A} \rightarrow \mathbf{A}^3$ for a straight line, parabola and curve in 3D space.*

A *polygon mesh* is a set of connected polygonally bounded planar surfaces. A polygon mesh can be used to represent visual biometric information with curved surfaces, however, the representation is only approximate. The errors in this representation can be made arbitrarily small by using more and more polygons to create a better piecewise linear approximation.

*Parametric polynomial curves* define points on a 3D curve by using

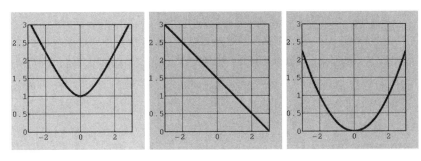

$$F : \mathbf{A}^2 \to \mathbf{A}$$
$$F(x, y) = x^2 + y^2 + 1$$

$$F : \mathbf{A}^2 \to \mathbf{A}$$
$$F(x, y) = 2y - x + 3$$

$$F : \mathbf{A}^2 \to \mathbf{A}$$
$$F(x, y) = 4y - x^2$$

*No real solutions, F defines the empty curve*

*∀ x, y ∈ **A**, defines a straight line (the same line as defined parametrically)*

*∀ x, y ∈ **A**, defines the same parabola as the parametric definition, $y = x^2/4$ for every point on this parabola*

**FIGURE 2.14**
Implicit model (Example 2.23).

▶ Three polynomials in a parameter $t$, one for each of $x, y$, and $z$,

▶ The coefficients of the polynomials selected such that the curve follows the desired path,

▶ Various degree of the polynomials, for example, cubic polynomials that have powers of the parameter up through the third.

> **Example 2.25** *A curve is approximated by a piecewise polynomial curve instead of a piecewise linear curve. Each segment of the overall curve is given by three functions, $x, y$, and $z$, which are cubic polynomials in the parameter $t$, $0 \leq t \leq 1$,*
>
> $$x(t) = a_x t^3 + b_x t^2 + c_x t + d,$$
> $$x(t) = a_y t^3 + b_y t^2 + c_y t + d,$$
> $$x(t) = a_z t^3 + b_z t^2 + c_z t + d,$$
>
> *or in matrix form*
>
> $$\mathbf{Q} = \mathbf{T} \cdot C_{32} = [t^3 \ t^2 \ t^1 \ 1] \begin{bmatrix} a_x & a_y & a_z \\ b_x & b_y & b_x \\ c_x & c_y & c_z \end{bmatrix} = [ \ x(t) \ y(t) \ z(t) \ ]$$

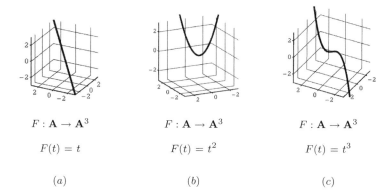

$$F : \mathbf{A} \to \mathbf{A}^3 \qquad\qquad F : \mathbf{A} \to \mathbf{A}^3 \qquad\qquad F : \mathbf{A} \to \mathbf{A}^3$$

$$F(t) = t \qquad\qquad F(t) = t^2 \qquad\qquad F(t) = t^3$$

$$(a) \qquad\qquad\qquad (b) \qquad\qquad\qquad (c)$$

**FIGURE 2.15**
Parametric model: a straight line (a) and parabola (b) in the affine plane, and a curve in the 3D affine space (c) (Example 2.24).

### 2.7.2    Generating curves by approximation

An approximation curve (or surface) is a curve (or surface) which passes near but not necessarily through the given points, while an interpolation is guaranteed to pass through them.

Approximation methods are divided into *control point methods* and *fits*. A fit is a method which tends to minimize the distances between the given points and the generated curves (or surfaces). This method is not very useful in synthesis of biometric topology. Here, control point methods are much more preferable because they allow a user to easily control the shape of the generated curve. The *convex hull* of a set of control points is the convex polygon obtained by connecting the control points. In general, the convex hull of an arbitrary set $S$ is defined as the smallest convex set containing $S$. A convex hull of the region of topological structure (enclosed by the boundary) is a useful computation in synthesis, for example, for robust decomposition of the boundary.

> **Example 2.26** *A   difference   between   approximating   and interpolating methods is illustrated by Figure 2.16.*

The important feature of approximation is that changing control points changes the shape of the curve. Thus, by changing control points, new shapes can be generated.

> **Example 2.27** *Figure 2.17 shows the simplest experiment on synthesis of fingerprint topology by degradation of an original image of a fingerprint.*

One can observe that the designed topology is unacceptable, and that it can be caused by incorrect models of degradation and/or control parameters of degradation.

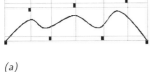

*(a)*

Methods for generating a curve or a surface are numerous and may be classified into two categories

- Interpolations and
- Approximations.

*(b)*

The problem of interpolation:

<u>Given:</u> $N + 1$ data points, $P_i$, $i = 0$ to $N$,
<u>Find:</u> A curve $Q(t)$ passing through these points.

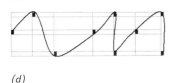

The problem of approximation:

*(c)*

<u>Given:</u> $N + 1$ control points, $P_i$, $i = 0$ to $N$,
<u>Find:</u> A curve $Q(t)$ with a shape specified by the control points but not passing through these points.

*(d)*

**FIGURE 2.16**

A comparison between approximation (a), (b), and interpolation (c), (d) methods (Example 2.26).

## 2.8 Extracting and generating features

This section briefly introduces two aspects of data structure used in image identification and recognition: (a) given data structure, extract features, and (b) given a set of features, design a data structure.

### 2.8.1 Features

*Features* are defined as identifiable components of images. Features are manifestations of biometric data in analysis and synthesis, and they carry the most useful information about the objects. Two classes of feature representations (descriptors) are distinguished: *low-level*, or pixel-level, and *high-level* representations.

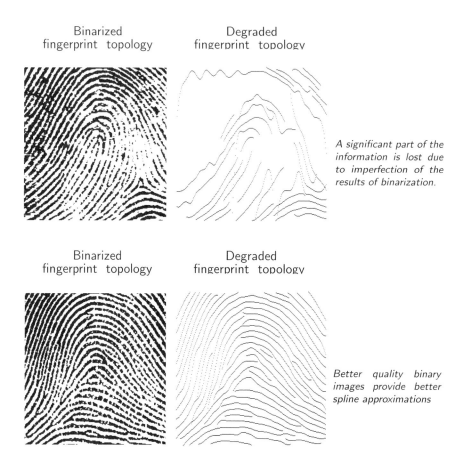

Binarized
fingerprint topology

Degraded
fingerprint topology

*A significant part of the information is lost due to imperfection of the results of binarization.*

Binarized
fingerprint topology

Degraded
fingerprint topology

*Better quality binary images provide better spline approximations*

**FIGURE 2.17**
Design of synthetic fingerprint image by spline approximation of an original (Example 2.27).

Two types of problem are related to features: *feature extraction* and *feature generation*.

**Feature extraction** is the identification of segments which have particular relevance to the task at hand. Geometric features provide structural descriptors of an image. Structural descriptors are used in interpretation of image information and high-level image analysis, in particular, for efficient storage representations of image data. Having located a recognizable feature we can then identify the particular part and may go on to infer its location and orientation.

> **Example 2.28** *The Hough transform, widely used in image analysis and recognition. For example, straight lines are represented in the Hough domain as a colinear set of points. The Hough transform can be used for detection of any given curve in:*
>
> ▶ *Algebraic form $f(\mathbf{x}, \mathbf{p})$, where $\mathbf{x}$ and $\mathbf{p}$ are a point in the image and parameter vector correspondingly, and*
>
> ▶ *Non-algebraic form (the shape must be represented by a lookup table).*

**Feature generation** is a process that is inverse to feature extraction. In feature generation, the focus is descriptors in topology. To achieve acceptable characteristics in generation, features must carry extractable attributes of topology.

> **Example 2.29** *Useful descriptors both in analysis and synthesis are: area, perimeter, diameter, curvature (the rate of change of slope), Euler number, texture (measured by smoothness, regularity, etc.).*

## 2.8.2 Spatial feature extraction and generation

The most important feature extraction and generation techniques are based on shape. Shape offers the most direct correspondence between an image and the objects of interest.

**Template matching** can be applied under controlled conditions to locate features. Template matching is based on correlation. The mask or template must contain an image of the feature being searched for. It must be exactly the same size as the feature and in exactly the same orientation. If the size of the feature or its orientation is unknown, the correlation must be repeated with the template in a variety of orientations and sizes. Frequently this causes a combinatorial explosion. The amount of work required to perform all the necessary correlations increases to such an extent that the method becomes unworkable.

**Parametric shape feature representation.** Instead of trying to match image features with templates on a pixel by pixel basis, it is possible to match them based on parameters derived from the segments themselves. Even the simplest derived parameters, such as perimeter length or area, can be used to discriminate between known objects, if the values are significantly different. These simple measures are clearly dependent on the distance of the object within the field of view. They are called *rotation invariant* parameters.

**Example 2.30** *Below several topological measures which are useful in analysis and synthesis are listed:*

- *The area* of *an object is the total number of pixels it contains.*
- *The perimeter* of *an object can be estimated by counting the total number of pixels on its boundary.*
- *The centre of mass of an object for binary images is the coordinate given by the mean of the x values of all object pixels and the mean of their y values.*
- *A central moments of order p is computed as*

$$\mu_{jk} = \sum_{x=0}^{N-1} \sum_{y=0}^{M-1} (x - \overline{x})_j (y - \overline{y})^k I(x,y)$$

*where $\overline{x}$ is the x coordinate of the object's centre of mass and $\overline{y}$ is its y coordinate.*

**Transform features.** Sometimes it can be easier to recognize features in an image transform than in the image itself. For example, an extended structure that has a characteristic periodicity in the image may show up as a single, well-defined peak in a Fourier transform. This may make it easy to deduce that the object is present in the image. Since the geometry of a correctly assembled part is known, it is possible to predefine specific ares of the image in which to look for the presence of the anticipated features. The frequency of the pick in the transform can also be used.

In addition to these simple, transform domain features, it is possible to define two dimensional shapes in terms of Fourier components. These representations of shape are known as *Fourier descriptors*. Their importance lies in the fact that scaling and rotation tends to affect them in very simple ways, making it easier to compare a standard shape with one from an image than if the comparison is attempted in the spatial domain. As with moments, it is possible to derive Fourier descriptors which are scale- and rotation-invariant.

**Distance transform** is defined as a computation which assigns to each feature pixel of a binary image a value equal to its distance to the nearest non-feature pixel, i.e., measures the distance of points within the object from the boundary. A distance transform

▶ Can be computed recursively and

▶ Can be measured in terms of Euclidean distance, city block, etc.

A *distance skeleton* is defined as a thinned subset derived from the distance transform by extracting the local maxima. A distance skeleton can be considered as an intermediate representation in an *inverse distance transform*:

$<$ Original image $> \Leftrightarrow$    $<$ Distance skeleton $>$
$\Leftrightarrow <$ Distance transform image $>$ .

---

## 2.9  Summary

The *analysis-by-synthesis* paradigm is the crucial point of inverse biometrix. It states that synthesis can verify the perceptual equivalence between original and synthetic objects.   The study of this equivalence can justify the biometrical model chosen during the analysis.  An extension of this model by controlled parameters and user-friendly interfaces is defined as a *generator* of synthetic biometrical data.

1. A *generator* of synthetic biometrical data is defined as a biometrical model. Two main types of models are distinguished: *physical* and *behavioral*. Physical models are used in modeling long-term characteristics, for example, iris and face topologies.   Behavioral models are used in representing short-term characteristics of biometrical data, for example, facial expressions and signatures.

2. Biometrical models can be developed based on various principles. Physical modeling utilizes the physical phenomena of biometric information; for example, in the dermatoglyphics of fingerprints, the process of creating a fingerprint starting from eight weeks of life.   Another approach is to analyze biometrical data structures and formal descriptions. For example, the topological structure of a fingerprint is well-studied and can be described at various levels of abstraction using mathematical equations and the techniques of topological representation and manipulation.

3. Interpolation, approximation, and stochastic modeling are the basic techniques in the design of generators of synthetic biometric information. In synthesis, biometric data are defined by means of topological and morphological configurations.

4. The selection of data structures is a crucial point of biometrical data representation. Topological data structure is related to many biometrics (signature, fingerprints, iris and retina, face and facial expressions). Such data structure can be first decomposed, and then assembled over a library of topological primitives.   This enables the utilization of topological manipulations, signal and image processing methods, and artificial intelligence methods.

5. A biometrical model generates data which can be classified by the selected criterion with a *reasonable level of acceptability* as *acceptable* or *unacceptable* for further processing and usage. Acceptability is defined as a set of characteristics which distinguish original and synthetic data.

Acceptable biometrical data is called *synthetic* biometrical data, based on controllability, topological similarities, etc. Models that approximate original data at a reasonable level of accuracy for the purposes of analysis are not considered generators of synthetic biometric information.

6. There are several strategies for synthetic data generation. Application requirements dictate the choice of appropriate strategies. For example, at the preprocessing phase in identification systems based on the cancellable technique of security, the original data are transformed to their synthetic equivalents which are processed, stored, analyzed, and matched. For the purpose of testing, large samples of synthetic data must be generated using user-defined characteristics.

## 2.10  Problems

In Problems 2.1, 2.2, and 2.3, use affine transforms given in Figure 2.13[‡]. Problems 2.4 and 2.5 address the simplest formal (algebraic) representation of topological primitives used in signature, iris, and fingerprint design. In design, a polar coordinate system is used[§]. Problem 2.7 can be considered as a team project.

**Problem 2.1**   Use the MATLAB function `problem21` (Figure 2.18) to scale the image. Use the following input parameters: vertical scale, horizontal scale, and name of image file. Choose your own image in a format such as *.jpg*. A sample function `problem21(0.5,0.5,'face.jpg')` in initial and resulting images are given below in Figure 2.18.

**Problem 2.2**   Figure 2.19 shows the initial synthetic face image, and the image rotated 60 degrees clockwise ($\theta = 60^0$) and 60 degrees counterclockwise ($\theta = -60^0$). Implement similar affine transforms given the following parameters:
(a) $\theta = 40^0$; (b) $\theta = -30^0$; (c) $\theta = 120^0$.

---

[‡]This section was written by S. Yanushkevich, S. Kiselev, and V. Shmerko

[§]Recall that a polar coordinate system in a plane is defined as a fixed point $O$, called the pole, and a ray emanating from the pole, called the polar axis. Each point $P$ in the plane is associated with a pair of polar coordinates $(r, \theta)$, where $r$ (radial coordinate of $P$) is the distance from $P$ to the pole and $\theta$ (angular coordinate) is an angle from the polar axis to the ray $OP$. The polar coordinates of a point are not unique: if a point $P$ has polar coordinates $(r, \theta)$, then $(r, \theta + n360^\circ)$ and $(r, \theta - n360^\circ)$ are also polar coordinates of $P$ for any nonnegative integer $n$. Details of analytic geometry in calculus can be found in [1].

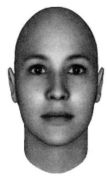

To scale an image by a factor of 0.5 vertically and horizontally, the following transformation matrix is applied:

$$T = \begin{bmatrix} 0.5 & 0 & 0 \\ 0 & 0.5 & 0 \\ 0 & 0 & 1 \end{bmatrix}$$

```
function problem21(hscale,
vscale,nameImageFile)
T1 = [1 0 0; 0 1 0; 0 0 1];
Image = imread(nameImageFile);
resamp = makeresampler('nearest','cubic','fill');
stretch1 = maketform('affine',T1);
OriginalImage = imtransform(Image,stretch1,resamp);
imshow(OriginalImage);
ScaleImage = scale(Image, hscale, vscale);
imshow(ScaleImage);
```

**FIGURE 2.18**
Face image scaling (Problem 2.1).

**Problem 2.3** Figure 2.20 shows the original and sheared image of iris, and also two shared images of face (iris is sheared 2 times horizontally and 3 times vertically), and the face image is shared 2 times horizontally and 3 times vertically in both directions (the corresponding matrices are given in the Figure, as well as a sample MATLAB function `problem23`). Implement shear of a sample with the following parameters:
(a) 4 times horizontally and 3 times vertically
(b) 6 times horizontally and 2 times vertically
(c) 4 times horizontally and 3 times vertically in the opposite direction.

**Problem 2.4** Using the MATLAB function `polar`, create the following curves in polar coordinates:
(a) $r = \cos A\theta$ for $A = 3$ and $\theta = [0, 2\pi]$.
(b) $r = A\sin n\theta$ for $A = 3$, $n = 2$ and $\theta = [0, 2\pi]$.
(c) $r = 2A\cos\theta$ for $A = 2$, $n = 2$, and $\theta = [-\pi, \pi]$.
Table 2.1 shows the sample curves.

**Problem 2.5** Using the MATLAB function `polar`, create the following curves in polar coordinates:
(a) $r = e^{\cos\theta} - A\cos B\theta + \sin B\frac{\theta}{4}$ for $r = A = 2.5$, $B = 9$, and $\theta = [0, 2\pi]$.
(b) $r = A + \cos\frac{\theta}{3}$ for $A = 4$ and $\theta = [0, 2\pi]$. Distinguish parameters that produce curves with symmetry and partially symmetry properties.

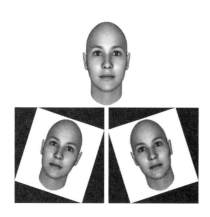

Rotation of the initial image (top) 60 degrees clockwise (left down) is implemented using the following transformation matrix:

$$T = \begin{bmatrix} \cos 60 & \sin 60 & 0 \\ \sin 60 & -\cos 60 & 0 \\ 0 & 0 & 1 \end{bmatrix}$$

The corresponding fragment of the MATLAB code is given below:

```
function newim=rotate(I,theta)
tform=maketform('affine',[cos(theta)
sin(theta) 0; sin(theta)
-cos(theta) 0; 0 0 1]);
newim = imtransform(I,tform);
```

Rotation 60 degrees counterclockwise is shown on the right down

**FIGURE 2.19**
Rotation (Problem 2.2).

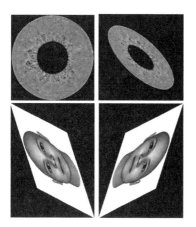

Image sheared by a factor of 2 horizontally and 3 vertically. The transformation matrices for shear with respect to axes $x, Y$ and the opposite direections $(-X and -Y)$ are given below:

$$T = \begin{bmatrix} 1 & 2 & 0 \\ 3 & 1 & 0 \\ 0 & 0 & 1 \end{bmatrix}, \quad T = \begin{bmatrix} 1 & -2 & 0 \\ -3 & 1 & 0 \\ 0 & 0 & 1 \end{bmatrix}$$

```
function
problem23(strImageFile, hshear,
vshear)
Image = imread(strImageFile);
tform = makeform'affine',[1
hshear 0; vshear 1 0; 0 0 1]);
newImage = imtransform(Image,
tform);
Image1 =
imtransform(Image,tform);
imshow(Image), figure,
imshow(Image1);
```

**FIGURE 2.20**
Shear (Problem 2.3).

(c) $r = A + B \sin \theta$ for $A = 1, B = 2$, and $\theta = [-\pi, \pi]$.
Details of analytic geometry in calculus can be found in [1].

**Problem 2.6** This problem addresses the design of fingerprint primitives based on formal (algebraic) representation. Curves called cardioids can be used for this purpose. Using the MATLAB function `polar`, create the simplest fingerprint topological primitives
(a) $r = A\theta$ for $A = 2$ and $\theta = [0, 4\pi]$.
(b) $r = A\sqrt{\theta}$ for $A = 2$ and $\theta = [0, 4\pi]$.
(c) $r = Ae^{B\theta}$ for $B = 2$ and $\theta = [0, 4\pi]$.
(d) $r = \frac{A}{\sqrt{\theta}}$ for $A = 1$ and $\theta = [0, 4\pi]$.
Details of analytic geometry can be found in [1].

**Problem 2.7** This problem addresses image warping technique. There are several algorithms that can be used in design of biometric topological data. In the paper by Heckbert [19], an algorithm for deformation by bilinear Coons patch is given. In addition, Wolberg's warping technique can be used [49].
(a) Figure 2.21 illustrates some warping results.
(b) Figure 2.22 illustrates some periodic warping.

### Supporting information for problems

Two MATLAB functions were used in Example 2.21 (Figure 2.12) and Problem 2.7 (Figure 2.22):

▶ Function `imageWarping3()`,
▶ Function `trans_func()`

The input parameter is an image, for example: `face.jpg`. The MATLAB function for warping `imageWarping3(strImageFile)` is given below.

```
I1 = imread('original_radial_grid1.jpg');I = rgb2gray (I1);
imshow(I),figure
[nrows,ncols] = size(I); % Find image size
[xi,yi] = meshgrid(1:ncols,1:nrows);
imid = round(size(I,2)/2);% Find index of the image center
xt = xi(:) - imid; yt = yi(:) - imid;
[theta,r] = cart2pol(xt,yt);
%Examples of parameters  for 'original_radial_grid1.jpg'
p = 1.5;
 a = 0.069;
  s = a * r.^p;
   r =  r.^p,
   where a is a scale coefficient
%p = 2; a = 0.005;s = a * r.^p;
%p = 3; a = 0.000025; s = a * r.^p;
```

**TABLE 2.1**

Geometrical configurations for topological primitives design (Problem 2.4).

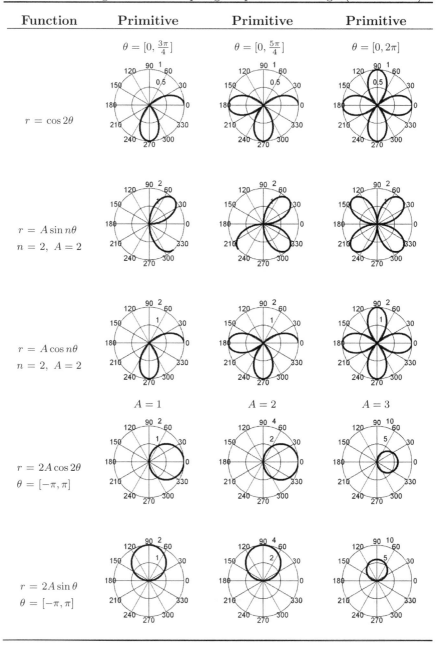

| Function | Primitive | Primitive | Primitive |
|---|---|---|---|
| | $\theta = [0, \frac{3\pi}{4}]$ | $\theta = [0, \frac{5\pi}{4}]$ | $\theta = [0, 2\pi]$ |
| $r = \cos 2\theta$ | | | |
| $r = A \sin n\theta$ $n = 2, \; A = 2$ | | | |
| $r = A \cos n\theta$ $n = 2, \; A = 2$ | | | |
| | $A = 1$ | $A = 2$ | $A = 3$ |
| $r = 2A \cos 2\theta$ $\theta = [-\pi, \pi]$ | | | |
| $r = 2A \sin \theta$ $\theta = [-\pi, \pi]$ | | | |

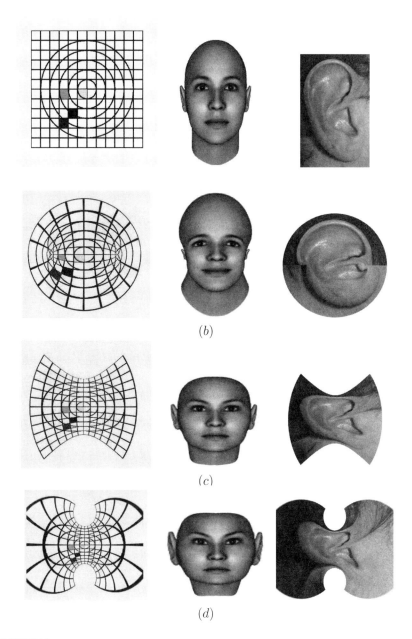

**FIGURE 2.21**
Image warping (Problem 2.7).

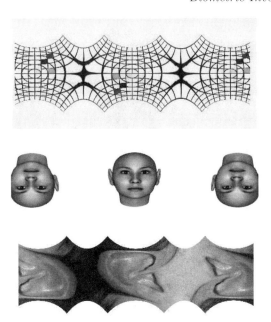

**FIGURE 2.22**
Periodic image warping (Problem 2.7).

The MATLAB function `trans_func()` that converts polar coordinates back to cartesian is given below

```
[ut,vt] = pol2cart(theta,s);
u = reshape(ut,size(xi)) + imid;
v =reshape(vt,size(yi)) + imid;
tmap_B = cat(3,u,v);
resamp = makeresampler('linear','fill');
tformarray(I,[],resamp,[2 1],[1 2],[],tmap_B,[]);
imshow(I_pin,'truesize');
```

## 2.11  Further reading

**Image processing.** The techniques of the inverse problems of biometrics are mostly based on image processing algorithms. There are several excellent textbooks and the handbooks on image processing and pattern recognition. For example, the handbook by Ritter and Wilson [36] introduces the basic concepts of image algebra, enhancement techniques, edge and shape detection algorithms, and morphological transforms. The fundamentals of image processing are introduced, in particular, in books by Gonzalez and Wintz [18], Pavlidids [30], and Serra [39]. The techniques of image synthesis are introduced by Glassner [17] and Hégron [20]. There are many books on computer graphics. For example, the book by Rogers and Adams [40] on mathematical aspects of computer graphics is useful in development of biometric systems. The book by Cormen et al. [4] on state-of-the-art algorithm theory will be an excellent support in understanding the techniques of direct and inverse problems of biometrics.

In this book, the word *morphology* is used in two different sense: (a) to focus on the physical (biological) nature of biometric data and its natural developments; this is "biological" morphology and (b) to focus on extracting features from topological structure (image); this is *mathematical* morphology.

Mathematical morphology is useful in biometric topological structure design [18, 36, 39]. Notice that morphological operations and so-called *parametric differential operators* [5, 23, 53] produce the same topological configurations.

**Image restoration** *Restoration* is an inverse procedure to degradation and plays an essential role in synthetic biometric data design. Restoration attempts to recover an image that has been degraded. It uses a priori knowledge of the degradation process or reasons. In image analysis, restoration techniques are oriented toward modeling the degradation and applying the inverse transform to recover the original image. This approach usually involves formulating a criterion of feasibility that yields an optimal estimate of the desired result. In analysis, enhancement involves heuristic procedures designed to manipulate an image in order to take advantage of the psychophysical aspects of the human visual system. The basics of image restoration technique are discussed in most the books on image processing, for example, in [18, 36, 39].

**Random numbers and stochastic processes generation.** Generation of randomness in synthesized biometric data is relevant to the methods of random number and stochastic process generation. This topic has been introduced, in particular, in books by Rubinstein and Melamed [41] and Peebles [33].

**Image warping.** Digital image warping deals with geometric transformation techniques and is the basis of synthetic biometric data design. Warping in general is not reversible since many-to-one mappings are allowed. Warping techniques for biometric data are widely used to create synthetic biometric data, in particular, in maxilla-craniofacial surgery (appearance of face change when the skull and jaw are altered), forensics, and archaeology (given a skull, design the face). Image warping is performed by:

> *Forward mapping* – each pixel in the source image is mapped into the target image, and

*Backward mapping* – for each target location pixel, pixels in the source image
   which are required to create the target pixel are determined.

There are many algorithms for image warping, in particular, algorithms
based on piecewise affine, bilinear, biquadratic, and bicubic mapping. Some
of them are introduced by Wolberg [49]. A family of parametric surfaces has
been developed by Coons [10] known as *Coons patches*. The bilinear Coons
patch is the class of Coons patches that is defined as a mapping from 2D to
an arbitrary number of dimensions.

Correct warping is based on correctly chosen spatial transformations. For
example, in facial expression synthesis, a spatial transform can be defined
from a physical facial model. This model combines both a muscle process and
a physically-based simulation of skin. Waters has used warping technique in
facial expression modeling [91]. He calculated the deformation functions by
physically modeling facial muscle actions in 3D. In [52], computer aided design
of synthetic faces based on various warping techniques has been introduced.

A review on image-warping methods can be found in a paper by Glasbey
and Mardia [16]. Fractals are useful in image synthesis. Mandelbrot [26]
developed *fractal* geometry. Objects constructed in this geometry are self
similar: each small portion, when magnified, reproduces exactly a larger
portion. A technique for generating random fractals is *fractional Brownian
motion*, which is an extension of the concept of Brownian motion used in
physics. This technique generates a Gaussian stochastic process useful for
approximating a wide range of biometric phenomena.

Milliron et al. [52] presented a framework for warps and deformations. They
also discussed the problem of invertability of warps: only some warps can be
unwarped resulting into the same initial image. Feng and Wah [14] developed
dynamic time warping technique originally used in speech recognition and
applied in the signature modeling. In addition, see results by Jain et al.
[21, 22] and Terzopolous et al. [45].

**Information notation of biometric data.** In this book, the term
"information" is used to characterize the quality of data. Qualitative
estimation of information is often used in artificial intelligence tools and
evolutionary algorithms for analysis, identification, recognition, and synthesis
of biometric data.

The entropy principle of conquering uncertainty has a long history. In
1850, Rudolf Clausius introduced entropy as the amount of disorder in the
system. In 1948, Shannon suggested a measure to represent the information
by a numerical value, nowadays known as *Shannon entropy* (the term
"uncertainty" is interchangable with the term "entropy.") [42]. Shannon
entropy is a reasonable measure in various optimization algorithms, including
in the area of analysis and synthesis of biometric data. The basic idea is that
the bit strings of information are understood as messages to be communicated
from a messenger to a receiver. Each message $i$ is an event that has a

certain probability of occurrence $p_i$ with respect to its inputs. The measure of information produced when one event is chosen from the set of $N$ events is the entropy of message $i$: $-\sum_{i \in N} p_i \log p_i$.

The paper by Pavlidis et al. [31] is an excellent example of how to apply this notation of information to optimization of topological configurations. They formulated the problem of encoding information on some medium using printed technology as a set of the following conflicting requirements (high density of information in code, reliable reading of the code, minimization of the cost of the printing process, etc.)

The information theoretical standpoint on measures of images is based on the following notations:

▶ *Source of information*, a stochastic process where an event occurs at time point $i$ with probability $p_i$. Often the problem is formulated in terms of sender and receiver of information and used by analogy with communication problems.

▶ *Quantity of information*, a value of a function that occurs with the probability $p$ carries a quantity of information equal to $(-\log_2 p)$.

▶ *Entropy*, the measure of the information content. The greater the uncertainty in the source output, the higher is its information content. A source with zero uncertainty would have zero information content and, therefore, its entropy would be likewise equal to zero.

For example, in topological analysis it is reasonable to assume that all combinations of values of pixels occur with equal probability. A value of an image that occurs with the probability $p$ carries a quantity of information equal to $< Quantity\ of\ information > = -\log_2 p$ *bit*, where $p$ is the probability of that value occurring.

Mathematical details of information theory can be found in many books, for example, [8, 27].

The results of entropy-based analysis of natural language processing have been reported by Berger [2]. In machine learning, information theory has been recognized as a useful criterion [32, 55]. Zhong and Ohsuga [55] showed that it is useful to calculate the information in a logical description of data and used these information-theoretical measures in manipulation of data. Image redundancy based on information measures has been studied in many papers. For example, Studholme [44] proposed normalized mutual information as an invariant measure for image registration problems. The authors of [47] evaluated Markov random fields in terms of entropy and used information for redundancy of images.

**Distance transform** The basic idea of distance transform is to approximate the global Euclidean distance by propagation of local distances (between neighbouring pixels). A distance transform (direct and inverse) is a useful computation both in analysis and synthesis. Various aspects of the distance transforms are studied in [3, 6, 7, 35, 37, 39, 43]. The notion of fuzzy distance is formulated by defining the length of a path on a fuzzy subset [29, 34, 38, 46]. For a binary object, a distance transform is a process that assigns a value at each location within the object that is simply the shortest distance between that location and the complement of the object. The notion of a distance transform for fuzzy objects, called *fuzzy distance transform*, is useful in computing data inaccuracies, graded object compositions, or limited image resolution. Notice that the fuzzy distance transform may be useful in fault detection in integrated circuit chips or in computer motherboard circuits, analysis of the dynamics of hurricanes, etc.

**Markov random fields in image synthesis.** A *random field* is defined as a probabilistic measure on the set of all possible configurations[¶]. *Markov random field* is a random field that satisfies particular local properties (positivity and markovianity). It is a spatial generalization of the Markov chain (the time points in the Markov chain are interpreted as sites in the Markov random field). Several features of the Markov random field are applicable to biometric data synthesis:

▶ Representation of images of a stochastic nature,

▶ Relationship to entropy and information measures,

▶ Simple processing,

▶ Flexible control for random field generation,

▶ Relationship to Voronoi diagram technique (neighborhood systems is a crucial point in the Markov random field), and

▶ Relationship to analysis-by-synthesis approach (the parameters for modeling can be easily replaced by the parameters of generation).

In addition, a representation based on a Markov random field can be minimized and used as a quantitative measure of the global quality of an image.

Markov random fields are useful in the generation of visual information, in particular, in image motion, texture modeling, image restoration and surface reconstruction, template deformation, and data fusion. Details on Markov models can be found in [4, 5, 12, 13, 15, 24, 47, 50, 51]. Fundamentals of Markov random fields in image analysis and synthesis were introduced by Li [25].

---

[¶]A finite set of labels in an object, for example, in a graph, is called a *configuration*.

## 2.12 References

[1] Anton H. *Calculus. A New Horisons.* 6th Edition. John Wiley & Sons, New York, 1999.

[2] Berger A, Pietra SD, and Pietra VD. A maximum entropy approach to natural language processing. *Computational Linguistics*, 22(1):39–71, 1996.

[3] Bezdek JC and Pal SK. *Fuzzy Models for Pattern Recognition.* IEEE Press, New York, 1992.

[4] Bobick AF and Wilson AD State-based recognition of gesture. In Shah M and Jain R, Eds., *Motion Based Recognition*, Kluwer, Dordrecht, pp. 201–226, 1997.

[5] Bochmann D and Posthoff Ch. *Binäre Dynamishe Systeme.* Akademieverlag, Berlin, 1981.

[6] Borgefors G. Distance transformations in digital images. *Computer Vision, Graphics and Image Processing*, 34(3):344–371, 1986.

[7] Borgefors G. On digital distance transformation in three dimensions. *Computter Vision and Image Understanding*, 64(3):368–376, 1996.

[8] Brillouin L. *Science and Information Theory.* Academic Press, New York, 1962.

[9] Cormen TH, Leiserson CE, Riverst RL, and Stein C. *Introduction to Algorithms.* MIT Press, Cambridge, MA, 2001.

[10] Coons SA. Surfaces for computer-aided design of space form. *Technical Report MAC-TR-41*, MIT, 1967.

[11] Cross GC and Jain AK. Markov random field texture models. *IEEE Transactions on Pattern Analysis and Machine Intelligence*, 5(1):25–39, 1983.

[12] Derin H and Kelly PA. Discrete-index Markov-type random fields. *Proceedings of the IEEE*, 77(10):1485–1510, 1989.

[13] Dubois E and Konrad J. Estimation of 2-D motion fields from image sequences with application to motion-compensated processing. In Sezan MI and Lagendijk RL, Eds., *Motion Analysis and Image Sequence Processing*, pp. 53–87, Kluwer, Dordrecht, 1993.

[14] Feng H and Wah CC. Online signature verification using a new extreme points warping technique. *Pattern Recognition Letters*, 24:2943-2951, 2003.

[15] Geman S and Geman D. Stochastic relaxation, Gibbs distribution and the Bayesian restoration of images. *IEEE Transactions on Pattern Analysis and Machine Intelligence*, 6(6):721–741, 1984.

[16] Glasbey CA and Mardia KV. A review of image-warping methods. *Journal of Applied Statistics*, 25(2):155–172, 1998.

[17] Glassner AS. *Principles of Digital Image Synthesis*. Morgan Kaufman Publishers, San Francisco, CA, 1995.

[18] Gonzalez R and Wintz P. *Digital Image Processing*. Addison-Weley, Reading, MA, 2nd edition, 1987.

[19] Heckbert P. Bilinear Coons patch image warping. In *Graphics Gems IV*, Heckbert P., Ed. Academic Press, Boston, pp. 438–446, 1994.

[20] Hégron G. *Image Synthesis Elementary Algorithms*. The MIT Press, Cambridge, Massachusetts, London, England, 1985.

[21] Jain AK, Zhong Y, and Dubuisson-Jolly MP. Deformable template models: a review. *Signal Processing*, 71:109–129, 1998.

[22] Jain AK, Zhong Y, and Lakshmanan S. Object matching using deformable templates. *IEEE Transactions on Pattern Analysis and Machine Intelligence*, 18(3):267–278, 1996.

[23] Kucharev GA, Shmerko VP, and Yanushkevich SN. *Parallel Processing in VLSI*. "High School" Publishers, Belarus, 1991.

[24] Künsch H, Geman S, and Kehagias A. Hidden Markov random fields. *Ann. Appl. Probab.*, 5:577–602. 1995.

[25] Li SZ *Markov Random Field Modeling in Image Analysis*. Springer, Heidelberg, 2001.

[26] Mandelbrot B. *The Fractal: Form, Chance and Dimension*. WH Freeman, San Francisko, CA, 1977.

[27] Martin NFG and England JW, *Mathematical Theory of Entropy*. Addison-Wesley, Reading, MA, 1981.

[28] Milliron T, Jensen RJ, Barzel R, and Finkelstein A. A framework for geometric warps and deformations. *ACM Transactions on Graphics*, 21(1): 20–51, 2002.

[29] Mordeson J and Nair P. *Fuzzy Graphs and Fuzzy Hypergraphs*. Physica-Verlag, Berlin, 2000.

[30] Pavlidis T. *Algorithms for Graphics and Image Processing*. Computer Science Press, Rockville, MD, 1982.

[31] Pavlidis T, Szwartz J, and Wang Y. Information encoding with two-dimensional bar codes. *Computer*, pp. 18–28, June 1992.

[32] Principe JC, Fisher III JW, and Xu D. Information theoretic learning. In Haykin S, Ed. *Unsupervised Adaptive Filtering*, Wiley, New York, NY, 2000.

[33] Peebles PZ. *Probability, Random Variables, and Random Signal Principles*, 3-rd edition, McGraw-Hill, New York, 1993.

[34] Punam K., Saha F, Wehrli W, and Gomberg BR. Fuzzy distance transform: theory, algorithms, and applications. *Computer Vision and Image Understanding*, 86:171–190, 2002.

[35] Ragnemalm I. The Euclidean distance transform in arbitrary dimensions. *Pattern Recognition Letters*, 14:883–888, 1993.

[36] Ritter GX and Wilson JN. *Handbook of Computer Vision Algorithms in Image Algebra*. CRC Press, Boca Raton, FL, 2000.

[37] Rosenfeld A and Pfaltz J. Distance functions in digital pictures. *Pattern Recognition*, 1:33–61, 1968.

[38] Rosenfeld A. The fuzzy geometry of image subsets. *Pattern Recognition Letters*, 2:311–317, 1991.

[39] Serra T. Image Analysis and Mathematical Morphology. Academic Press, San Diego, 1982.

[40] Rogers DF and Adams JA. *Mathematical Elements for Computer Graphics*. McGraw-Hill, New York, 1990.

[41] Rubinstein RY and Melamed B. *Modern Simulation and Modeling*, Wiley, New York, NY, 1998.

[42] Shannon C. A mathematical theory of communication. *Bell Systems Technical Journal*, 27:379–423, 623–656, 1948.

[43] Sintorn I-M and Gunilla Borgefors G. Weighted distance transforms for volume images digitized in elongated voxel grids. *Pattern Recognition Letters*, 25:571–580, 2004.

[44] Studholme C, Hill DLG, and Hawkes DJ. An overlap invariant entropy measure of 3D medical image alignment. *Pattern Recognition*, 32:71–86, 1999.

[45] Terzopolous D, Platt J, Barr A, and Fleischer K. Elastically deformable models. *Computer Graphics*, 21(4):205–214, 1987.

[46] Udupa JK and Samarasekera S. Fuzzy connectedness and object definition: theory, algorithms, and applications in image segmentation. *Graphical Models Image Process*, 58:246–261, 1996.

[47] Volden E, Giraudon G, and Berthod M. Information in Markov random fields and image redundancy. In *Proceedings of the 4th*

*Canadian Workshop on Information Theory and Application II*, Springer, Heidelberg, pp. 250–268, 1995.

[48] Waters K. Modeling three dimensional facial expressions. In *Bruce V and Burton M, Eds., Processing Images of Faces*, pp. 202–227, Ablex Publishing Corporation, Norwood, NJ, 1992.

[49] Wolberg G. *Digital Image Warping*. IEEE Computer Society Press, Los Alamitos, CA, 1990.

[50] Yamamoto E, Nakamura S, and Shikano K. Lip movement synthesis from speech based on hidden Markov models. *Speech Communication* 26(1-2):105–115, 1998.

[51] Yang L, Widjaja BK, and Prasad R. Application of hidden Markov models for signature verification. *Pattern Recognition*, 28(2):161–170, 1995.

[52] Yanushkevich SN, Supervisor. Computer aided design of synthetic biometric data. *Technical Report, Biometric Technologies: Modeling and Simulation Laboratory*, University of Calgary, Canada, 2004.

[53] Yanushkevich SN. *Logic Differential Calculus in Multi-Valued Logic Design.* Technical University of Szczecin Academic Publishers, Szczecin, Poland, 1998.

[54] Yanushkevich SN and Shmerko VP. Biometrics for electrical engineers: design and testing. *Lecture notes*. Private communication. 2005.

[55] Zhong N and Ohsuga S. On information of logical expression and knowledge refinement. *Transactions of Information Processing Society of Japan*, 38(4):687–697, 1997.

# 3

# *Synthetic Signatures*

This chapter focuses on synthetic signatures as one of the prominent examples of inverse biometrics. Historically, manually synthesized signatures and forged handwriting have always been a point of concern for handwriting experts and for forensics in general. There are quite a few reports on detecting free-hand forgeries, where a forger makes an attempt to simulate or trace a genuine signature. In the latter case, the forged handwritten object has the proper shape, but differs from a genuine object in the quality and dynamics of the strokes. Free-hand forgeries, on the other hand, differ from a genuine signature with respect to the values of various size and shape features. With recent advances in signature capturing, e.g. low-cost pressure sensitive tablets, handwriting analysis includes another set of features, i.e., kinematic characteristics such as pressure and speed of writing.

Techniques and methods associated with signature processing can be viewed as a combination of *direct* and *inverse* transformations:

▶ Direct transformations are used as a preprocessing step to create libraries of primitives and features by analyzing the topological structure of signatures, and inverse transformations are used to generate a collection of synthetic signatures (Figure 3.1a).

▶ Inverse transformations are used to generate a collection of synthetic signatures, and direct transformations are used as a postprocessing step to check the validity of newly synthesized signatures by performing an analysis of their topological structure (Figure 3.1b).

The chapter is structured as follows. A brief overview of some current trends in signature analysis and synthesis is given in Section 3.1. Section 3.2 explains the basics of signature synthesis. Concepts of automated signature synthesis are introduced in Section 3.3. Section 3.4 introduces the models for signature synthesis, including statistical, $n$-gram, vector recombination, and kinematical models. An algorithm which supports these models is outlined in Section 3.5. After the summary (Section 3.6), a set of problems and recommendations for "Further Reading" are given in Sections 3.7 and 3.8 respectively.

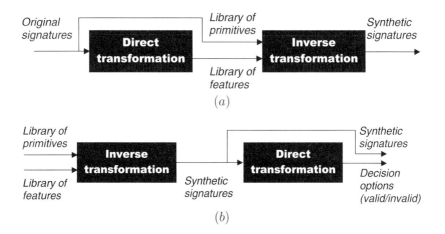

**FIGURE 3.1**
Signature processing techniques and methods: (a) two stage direct and inverse transformations to generate synthetic signatures, (b) two stage inverse and direct transformations to synthesize signatures and check their validity.

## 3.1    Introduction

In this section, we introduce automated signature processing focusing on (i) two types of processing (off-line and on-line), (ii) the information carried by signatures, and (iii) signature acquisition techniques.

### 3.1.1    Off-line and on-line automated signature processing

Some biometric characteristics, such as irises, retinas, and fingerprints, do not change significantly over time, and thus they have low in-class variation. In contrast, the in-class variations for signatures can be large meaning that signatures from a single individual tend to be different. Even forensic experts with long-term experience cannot always make a decision whether a signature is authentic or forged. Signatures are easier to forge than, for example, irises or fingerprints. However, the signature has a long standing tradition in many commonly encountered verification tasks, and signature verification is accepted by the general public as a non-intrusive procedure (hand prints and fingerprint verification, on the other hand, are associated with criminal investigations). It is assumed in automated signature processing that an individual's signature is unique. The state-of-the-art tools for automated signature analysis and synthesis rely partially or totally on methods of artificial intelligence. Some of these methods are covered in this chapter.

There are two types of automated signature processing: *off-line* and *on-line.*

*Off-line* signature processing is used when a signature is available on a document. It is scanned to obtain its digital representation.

*On-line* signature processing relies on special devices such as digitizing tablets with pens and/or gyropens to record pen movements during writing. In on-line processing, it is possible to capture the dynamics of writing, and, therefore, it is difficult to forge.

**Example 3.1** *The 3D representation of a signature is given in Figure 3.2*

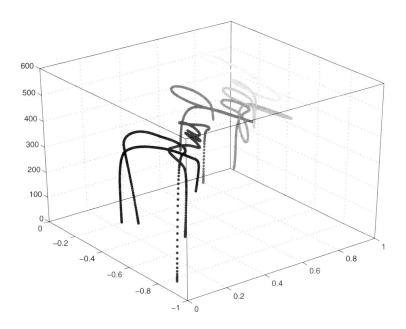

**FIGURE 3.2**

3D view of an on-line signature: the plain curve is given by the two-tuple $(X, Y)$, the pressure is associated with the $Z$ axis, the speed of writing is depicted by the shade of the curve, where darker is slower speed (Example 3.1).

Like signature imitation, the imitation of human handwriting is a typical inverse problem of graphology*. Automated tools for the imitation of

---

*Graphology is the study of handwriting, especially when employed as a means of analyzing handwriting characters.

handwriting have been developed. It should be noted that more statistical data, such as context information, are available in handwriting than in signatures. Some sophisticated systems use context entropy in handwriting analysis.

The simplest method of generating synthetic signatures is based on geometrical models. Spline methods and Bezier curves are used for curve approximation, given some control points. Manipulations of control points give variations of a single curve in these methods.

The following applications benefit from the availability of automated tools for signature synthesis:

▶ Engineering applications for advanced biometric systems where the quantity of data is crucial for decision making.

▶ Systems for testing existing biometric solutions against fraudulent biometric images.

▶ Systems for testing the accuracy, reliability, and speed of multimedia databases including biometric databases.

### 3.1.2    Information carried by signatures

The following evaluation properties are distinguished for synthetic signatures:

*Statistical properties.* Artificially generated signatures should be statistically related to pre-existing collections of signatures.

*Kinematical properties.* Artificially generated signatures should include kinematic characteristics like pressure, speed of writing, etc.

*Topological and spectral properties.* Topological and spectral transformations can be used to extract features from signatures.

*Uncertain properties.* Generated images can be intensively "infected" by noise.

*Filtering properties.* Inlier and outlier data are used for controlling model deformation, and avoiding possible distractions to the model. For example, to extract a character from a handwritten word, the outliers may appear in the form of stroke anomalies, closely cluttered characters or the printed background.

### 3.1.3    Signature acquisition

Signatures are *behavioral biometrics*, and can be obtained by various image acquisition devices. Dynamic signature acquisition records kinematic characteristics of human handwriting such as pressure, speed, acceleration, duration of writing, etc.

**Example 3.2** *Dynamic acquisition devices include numerous tablets, writing pads, etc. There are various pen-enabled devices ranging from pressure sensitive tablets to gyropens. Recently developed devices include handheld computers (PDAs and smart phones), and Tablet PCs. In the study presented here, we use a Wacom® compatible pressure sensitive tablet. It is set to provide 100 samples per second containing values for pen coordinates, pressure, and time stamps.*

**Example 3.3** *On the contrary, scanned signatures lose all kinematic characteristics retaining only geometric ones like shape and color intensity. Scanners, cameras, etc. are examples of* static *acquisition devices.*

Usually images are acquired by a digital pressure-sensitive tablet which records the ordered sequence of points with pressure and time interval characteristics. Since the ordered sequence of points is a parametric description of the shape, it is possible to evaluate curves, angles, and curvature at each point of the shape. In this chapter, signatures and handwriting are considered not only from the geometric viewpoint related to shapes and contours, but also from the viewpoint of pen dynamics and pressure. The adequacy and variability of generated images are assured by using existing databases of signatures and handwriting.

## 3.2    Basics of signature synthesis

To generate signatures with any automated technique, it is necessary to consider (a) the formal description of curve segments and their kinematical characteristics, (b) the set of requirements which should be met by any signature generation system, and (c) possible scenarios for signature generation.

### 3.2.1    Signature geometry

The simplest form of the handwriting is represented by a plane curve. There are three principally used descriptions of plane curves:

*Implicit form:* $f(x, y) = 0$;

*Explicit form:* $y = f(x)$, considered as a function graph;

*Parametric form:* $\mathbf{r}(t) = (x(t), y(t))$, using Cartesian parametric equations $x = x(t)$ and $y = y(t)$.

The latter form has been selected for further plane curve description of handwriting. Such a selection is justified by the nature of data which is acquired in a parameterized format. We use the following characteristics of the plane curve representing handwriting with an arbitrary parameter $t$ (Figure 3.3):

▶ Position,
▶ Velocity,
▶ Speed, and
▶ Acceleration.

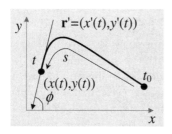

Position: $\mathbf{r}(t) = (x(t), y(t))$;
Velocity: $\mathbf{r}'(t) = (x'(t), y'(t))$;
Speed: $|\mathbf{r}'(t)| = |x'(t), y'(t)|$;
Acceleration: $\mathbf{r}''(t) = (x''(t), y''(t))$.

**FIGURE 3.3**
Characteristics of the plane curve representing handwriting with an arbitrary parameter $t$.

If $t \in [t_0, t] \mapsto \mathbf{r}(t)$ with velocity $\mathbf{v}(t) = \mathbf{r}'(t)$ and speed $|\mathbf{v}(t)|$, then

$$s(t) = \int_{t_0}^{t} |\mathbf{v}(t)| \mathrm{d}t \tag{3.1}$$

is called the arc length of the curve. For a plane curve:

$$s(t) = \int_{t_0}^{t} \sqrt{x'(t)^2 + y'(t)^2} \tag{3.2}$$

The arc length is independent of the parameterization of the curve. The value $\frac{\mathrm{d}s(t)}{\mathrm{d}t} = s'(t) = \sqrt{x'(t)^2 + y'(t)^2}$ indicates the rate of change of arc length with respect to the curve parameter $t$. In other words, $s'(t)$ gives the speed of the curve parameterization.

For convenience we also parameterize the curve with respect to its arc length $s$: $\mathbf{r}(s) = (x(s), y(s))$. The parameterization defines a direction along the curve on which $s$ grows. Let $\mathbf{t}(s)$ be the *unit tangent vector* associated with the direction. Denote by $\phi(s)$ the angle between the tangent at a point $(x(s), y(s))$ and the positive direction of the axis $x$. The *curvature* measures the rate of bending as the point moves along the curve with unit speed and can be defined as

$$k(s) = \frac{\mathrm{d}\phi(s)}{\mathrm{d}s}. \tag{3.3}$$

The curvature is the length of the acceleration vector if $\mathbf{r}(t)$ traces the curve with constant speed 1. A large curvature at a point means that the curve is strongly bent. Unlike the acceleration or the velocity, the curvature does not depend on the parameterization of the curve (Figure 3.4).

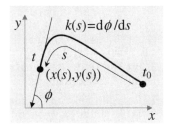

Unit tangent vector: $\mathbf{t}(t) = \frac{(r)'(t)}{|\mathbf{r}'(t)|}$

Curvature:

$$k(t) = \frac{|\mathbf{t}'(t)|}{|\mathbf{r}'(t)|} = \frac{|\mathbf{r}'(t) \times \mathbf{r}''(t)|}{|\mathbf{r}'(t)|^3}$$

**FIGURE 3.4**

The curvature measures.

Curvature possesses the useful beneficial features for signature processing:

▶ Invariant to shifting (transition) and rotation;
▶ Gives a unique representation of the curve up to rigid motion; and
▶ Inversely proportional to scaling: the curve $c\alpha(t)$ has the curvature $\frac{k(t)}{c}$ for $c > 0$.

But the most important feature is that the computation of curvature is a reversible operation: by performing the reverse sequence of steps (numerical integration), it is possible to restore the original image. This fact has been used for the signature generation method presented here.

## 3.2.2 Requirements for signature synthesis

The following requirements for signature synthesis are used to guide the automated signature generation:

*Requirement 1.* Signatures should have a signature-like visual appearance.

*Requirement 2.* Images should incorporate distortions/fluctuations within a class of signatures.

*Requirement 3.* Signatures should have meaningful visual parts (segments).

*Requirement 4.* Images should be invariant to scaling and orientation.

Since the requirements 1, 2 and 3 are of a cognitive nature, their satisfaction should be justified by observations and cognitive experiments. The requirement 4 is of a mathematical nature and its satisfaction should be shown by mathematical arguments. In this chapter, we focus on a method that satisfies all the above requirements.

### 3.2.3    Scenarios for signature generation

Below, the database-driven signature where the database is a collection of signatures and their attributes including labels such as individual names or identifiers. Signature with identical labels form a *class*. There are two applicable scenarios for signature generation that are essential for signature biometrics (see Figure 3.5):

▶ *Intra-class* signature synthesis and

▶ *In-class* signature synthesis.

The first, intra-class scenario, creates a synthetic signature which is independent of the set of sample signatures from the database. The later, in-class scenario, takes into consideration the class label, using signatures from this class to generate new signatures within this class. By following these scenarios, we can populate the database with biometrically significant information. Data sampling is the only distinctive feature to compare these two scenarios (statistical data generation from the entire collection vs. statistical data generation from the set of signatures belonging to the same class).

> **Example 3.4** *A random forgery is a pattern that by design is not related to the original signature; such a forgery is to be expected when a forger does not have ready access to the original signature.*
>
> *Even with an image of the signature available, dynamic information pertinent to a signature is not as readily available to a potential forger as is the shape of signature.*
>
> *Professional forgers can reproduce signatures to fool the unskilled eye. Human experts can distinguish genuine signatures from the forged ones. Modeling the invariance in the signatures and automating the signature recognition process pose significant challenges.*

There are two approaches to signature verification: static and dynamic. In static signature verification, only geometric (shape) features of the signature are used for authentication of identity. Typically, the signature impressions are normalized to a known size and decomposed into simple components (strokes). The shapes and stroke relationships are used as features. In dynamic signature verification, not only the shape features are used for authentication of the signature but the dynamic features are also employed. The signature impressions are processed as in a static signature verification system. Invariants of the dynamic features augment the static features, making forgery difficult since the forger must know not only the impression of the signature but also the way the impression was made.

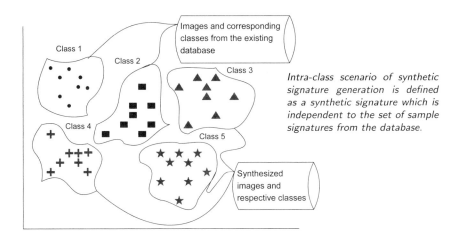

*Intra-class scenario of synthetic signature generation is defined as a synthetic signature which is independent to the set of sample signatures from the database.*

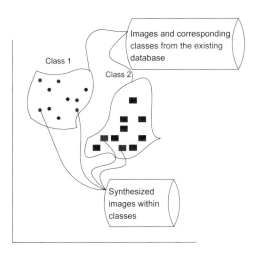

*In-class scenario of synthetic signature generation is defined as the class label, using signatures for this class to generate new signatures within this class.*

**FIGURE 3.5**
Intra-class vs. in-class scenarios in signatures synthesis.

## 3.3 Signature synthesis techniques

This section focuses on the basics of signature synthesis techniques. Known approaches to analysis and synthesis of signatures can be divided into two distinct classes:

▶ *Statistical* approaches use the fact that a signature shape can be described by a limited set of features. These features are statistically characterized.

▶ *Rule-based* approaches assume that a signature is the composition of a limited number of basic topological primitives which can be formally described. This composition can be implemented by a set of rules and/or grammars.

Further, we employ only statistical approaches to signature synthesis. One can consider the following ways of generating synthetic signatures:

*Deformation technique.* Applying reversible transformations
    `< Original signature >` ⟺ `< Synthetic signature >`

*Assembling technique.* Assembling topological primitives
    `< Topological primitives >` ⟹ `< Synthetic signature >`

*Cancellable technique.* Applying irreversible transformations
    `< Original signature >` ⟹ `< Synthetic signature >`

The choice of an appropriate technique for signature generation is dictated by a particular application.

### 3.3.1   Deformation technique

Figure 3.6 illustrates the design of a signature synthesis system based on the deformation technique. The main phases of the design process are the following:

*Phase 1.* Preprocess the original signature.

*Phase 2.* Choose appropriate deformation rules and control parameters for the deformation.

*Phase 3.* Transform the original signature into its deformable equivalent.

*Phase 4.* Use the synthesized signature for matching.

> **Example 3.5** *Application of some deformation operators to a signature is illustrated by Figure 3.7, where the original signature is given in parametric form $(x(t), y(t))$.*

### 3.3.2   Assembling technique

Figure 3.8a depicts the design of a signature synthesis system based on the assembling technique. The main phases of the design process are the following:

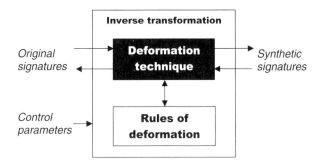

**FIGURE 3.6**
Design of synthetic signatures based on the deformation technique.

*Phase 1.* An arbitrary set of two topological primitives is taken from the database of primitives.

*Phase 2.* Based on appropriate assembling rules, a macroprimitive is created by merging the two initial primitives.

*Phase 3.* The next topological primitive is taken, and then added to the macroprimitive by applying appropriate assembling rules.

*Phase 4.* Control parameters direct the process of "growing" the macroprimitive. This process is stopped by a predefined criterion from the set of control parameters.

### 3.3.3 Cancellable technique

The cancellable technique can be used for signature distortion in a repeatable and usually non-reversible manner. The main phases of the cancellable technique are described by the following set of phases (Figure 3.8b):

*Phase 1.* Preprocess the original signature.

*Phase 2.* Choose an appropriate transform and control parameters for cancellability.

*Phase 3.* Transform the original signature into its cancellable equivalent.

The obtained cancellable signature can then be used for matching. The cancellable technique enables a high level of security for (i) biometric databases because cancellable data (synthetic signatures) are stored instead of the original signatures, and (ii) data processing because cancellable data (synthetic signatures) are matched instead of original signatures.

**The center of mass:**

$$(x_c, y_c) = \left( \frac{1}{n} \sum_{t=1}^{n} x(t), \frac{1}{n} \sum_{t=1}^{n} y(t) \right)$$

**Translation:** $(x(t), y(t))$ *are converted to relative coordinates with respect to the center of mass:*

$$(x(t), y(t)) = (x(t) - x_c, y(t) - y_c)$$

**Scaling:** *vertical and horizontal deforming are*
$y_n(t) = y(t) \cdot g_V$
$x_n(t) = x(t) \cdot g_H$ *correspondingly*

**Rotation:**
$x_n(t) = -x(t) \cdot \sin \alpha + y(t) \cdot \cos \alpha$
$y_n(t) = x(t) \cdot \cos \alpha + y(t) \cdot \sin \alpha,$
*where* $\alpha$ *is a rotation angle*

**FIGURE 3.7**
Translation, scaling, and rotation operators (Example 3.5).

## 3.4   Statistically meaningful synthesis

In this section, signature synthesis methods for various modeling techniques are described. These methods are based on statistical data collected from pre-existing databases of signatures.

The first step in a statistically-driven synthesis is *learning*. Learning from the pre-existing collection of signatures (or signature database) allows the accumulation of a priori knowledge about image structures, detailed curve behaviors, and other important characteristics, including those of kinematic

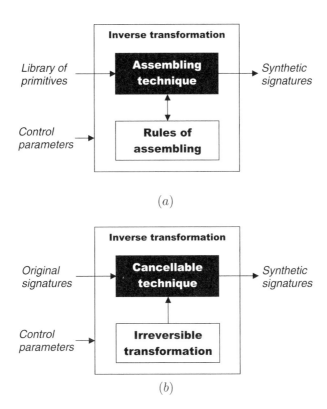

**FIGURE 3.8**
Design of synthetic signatures based on (a) the assembling technique, and (b) the cancellable technique.

nature if it is an on-line signature. A pre-existing online database of signatures can be used to generate this statistical data.

## 3.4.1 Preprocessing: learning

The learning stage results in multiple distributions with distribution function $\Phi(z)$ giving the probability that a standard normal variate assumes a value $z$. The following statistically meaningful characteristics can be acquired through the learning process from the pre-existing database of original signatures (Figure 3.9):

▶ Distribution of the number of segments,
▶ Distribution of the segments' length,
▶ Distribution of the length and orientation of inter-segment connections, and

▶ Distribution of the angle $\Phi(\alpha)$ and curvature $\Phi(k)$ characteristics along the contingent segments.

The distribution of curvature can be replaced with the distribution of length's intervals $\Phi(\Delta s)$. To improve the visual quality of the curve, it is important to consider sequential distributions in the form of Markov chains or $n$-grams.

Other distributions of various kinematic characteristics include pressure $\Phi(P)$, time intervals $\Phi(\Delta t)$, etc. These distributions are used to complement different geometric characteristics.

> **Example 3.6** *Figure 3.10 shows the distribution of the segments' lengths for the entire database of signatures.*

## 3.4.2   Random number generation

The image synthesis requires a generator of random numbers. There are several generators used in the system:

- ▶ Basic random number generators (statistically independent), and
- ▶ Statistically dependent random number generators or *surrogate generators.*

Since the first generator is based on existing built-in tools[†], let us describe the latter in more detail. Surrogate data are artificially generated data which mimic the statistical properties of real data. Surrogates have the same distribution function as real data. The surrogate generator accepts a distribution of values $y = \Phi(x)$ and a step interval $\Delta x$ as its input parameters, and returns random numbers based on this distribution (Figure 3.11).

> **Example 3.7** *Figure 3.12 displays the original distribution of the number of segments (left panel) which was used for the random number generator, and the distribution of generated random numbers (right panel). It can be seen that the distribution of randomly generated numbers based on the surrogate generator reflects the original distribution. This fact allows the integration of the surrogate generator as one of the tools in the statistically meaningful synthesis.*

## 3.4.3   Geometrically meaningful synthesis

The main question in a geometry-driven synthesis is how to preserve the geometric structure of the curve. One possible solution is based on learning

---

[†]Computer-generated random numbers are sometimes called pseudorandom numbers, while the term "random" is reserved for the output of unpredictable physical processes. Computer-generated pseudorandom numbers are assumed to be uniformly distributed.

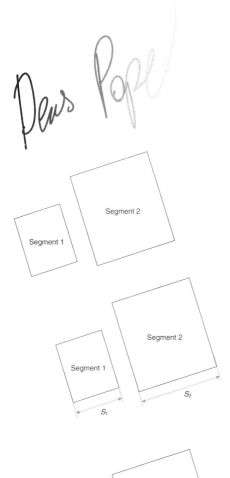

Original signature from the database of signatures. Distribution of the number of segments $\Phi(N_{segments})$. It is a simple and common observation that visual parts (segments) determine the overall visual appearance of the signature. For example, the original signature can be segmented to $N_{segments} = 2$ segments

Distribution of the segment's length $\Phi(S_{segment})$. In addition to the distribution of the segment's length across all segments $\Phi(S_{segment})$, it is necessary to collect information about the dependence of the segments' length on the segment's number $C_{segment}$: $\Phi(S_{segment}, C_{segment})$. In this case, the segmented signature has $S_{segment1} = S_1$ and $S_{segment2} = S_2$

Distribution of the length $\Phi(L_{inter-segment})$ and orientation $\Phi(\phi_{inter-segment})$ of inter-segment connections. In addition to the distribution of characteristics of all inter-segment connections for the entire image, it is necessary to collect information about the dependence of these characteristics on the inter-segments' number $C_{inter-segment}$: $\Phi(L_{inter-segment}, C_{inter-segment})$ and $\Phi(\phi_{inter-segment}, C_{inter-segment})$. In this case, the segmented signature has $L_{inter-segment} = L_1$ and $\phi_{inter-segment} = \phi_1$.

**FIGURE 3.9**
Statistically meaningful characteristics are acquired through the learning process from the pre-existing database of original signatures.

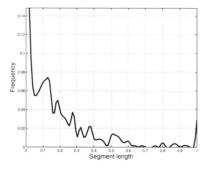

*A series of experiments with the above distribution of the segments' lengths revealed that a segment's length significantly depends on the segment's number. The proper substitution algorithm in terms of automated synthesis needs to consider this statistical dependance.*

**FIGURE 3.10**

Distributions of the segments' length for the preexisting signature database (Example 3.6)

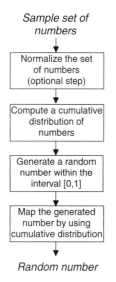

*Step 1* (Optional) *Perform normalization of values if necessary to guarantee the range of cumulative values* $[0, 1]$.

*Step 2 Build a cumulative distribution* $z = \Phi_c(x)$: $z_n = \int_{i=0}^{i=n} \Phi(x_i)$, *where* $n = 0 \ldots N$.

*Step 3 Generate a uniformly distributed random number* $r_1$ *within the interval* $[0; 1]$. *Map this random number by using the cumulative distribution* $x = \Phi_c^{-1}(z)$, *and take the ceiling value:* $x_1 = \lceil \Phi_c^{-1}(r_1) \rceil$.

*Step 4* (Optional) *Generate a random number* $r_2$ *within the interval* $[0; \Delta x]$, *and return the final value* $x_2 = x_1 - \Delta x + r_2$. *This optional step generates additional fluctuation of values, and can be selected if necessary.*

**FIGURE 3.11**

Algorithm for random number generation.

principles and $n$-grams which model information about the curve structure. Curvature is defined by

$$k = \frac{\mathrm{d}\phi(s)}{\mathrm{d}s}.$$

Original distribution of the number of segments for the random number generator

The result of random number generation based on the distribution of the number of segments

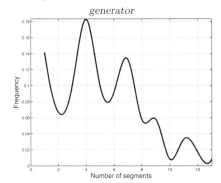 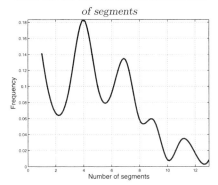

**FIGURE 3.12**

Surrogate generator: original distribution (left panel) and randomly generated (right panel).

For a plane curve,

$$k = \frac{(x(t)'^2 + y(t)'^2)^{3/2}}{x(t)'y(t)'' - x(t)''y(t)'}$$

The curvature in the equations above can be either positive or negative.

The representation of the curve is based on its intrinsic properties such as its arclength $s$, and its curvature $k$. For every curve there is a dependance of the form $F(s, k) = 0$. Such an equation is called the *natural equation* of the curve. This representation does not depend on any coordinate system and relates two geometric properties of the curve.

Two curves with the same natural equation can differ only by their location in the plane so that the shape of the curve is determined uniquely by its natural equation. To justify this statement, let us assume that two curves, I and II, have the same natural equation

$$k = f(s). \tag{3.4}$$

To show that the two curves are congruent, we move one to the other so that the points from which arclengths are measured are identical. Then we rotate one curve so that both have the same tangent and the same orientation at this point. Details are shown in Figure 3.13.

According to Equation 3.4, $k_1 = k_2$ for all $s$, where

$$\frac{d\phi_1}{ds} = \frac{d\phi_2}{ds}.$$

In addition, for $s = 0$,

$$x_1 = x_2; \ y_1 = y_2; \ and \ \phi_1 = \phi_2.$$

From the above equations it follows that the angles $\phi_1$ and $\phi_2$ can differ by an additive constant, but for $s = 0$ they are equal. It can be shown that

$$\frac{dx_1}{ds} = \cos \phi_1 = \cos \phi_2 = \frac{dx_2}{ds}$$

$$\frac{dy_1}{ds} = \sin \phi_1 = \sin \phi_2 = \frac{dy_2}{ds}$$

Hence, the two curves are congruent (Figure 3.13).

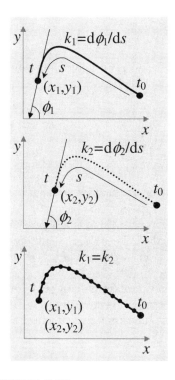

The subscripts 1 and 2 refer to the two curves I and II respectively:

- Coordinates of points on the curves $(x_1, y_1)$ and $(x_2, y_2)$.
- Slope angles of the curves with the x-axis $\phi_1$ and $\phi_2$.
- Curvatures $k_1$ and $k_2$.

The first curve defined by angles and curvature.
The second curve defined by angles and curvature.
The two congruent curves with the same curvature

$$\cos \phi_1 = \cos \phi_2 = \frac{dx_1}{ds} = \frac{dx_2}{ds}$$

$$\sin \phi_1 = \sin \phi_2 = \frac{dy_1}{ds} = \frac{dy_2}{ds}$$

**FIGURE 3.13**
Two congruent curves.

Next, a parametric representation from the natural Equation 3.4 is derived. Equation 3.4 implies that $d\phi/ds = f(s)$. So,

$$\phi = \int_0^s f(u) \ du + \phi_0,$$

where $\phi_0$ is a constant. Integrating with

$$\mathrm{d}x = \cos\phi\,\mathrm{d}s, \; \mathrm{d}y = \sin\phi\,\mathrm{d}s,$$

we obtain

$$x = \int_0^s \cos\phi\,\mathrm{d}s + x_0, y = \int_0^s \sin\phi\,\mathrm{d}s + y_0,$$

where $x_0$ and $y_0$ are constants. It is easy to show that rotations of the curve change only $\phi_0$, and translations change only $x_0$ and $y_0$.

The preprocessing part of the synthesis method uses sequence processing techniques to gather essential information about the sequential nature of the curve. In the natural equation, $F(s, k) = 0$, it is important to recognize some patterns of how the curve behaves and changes. In digitized cases, a curve is described by a selected parameterization technique, for example, the unit tangent vector can be a function of $t$: $\phi(t)$. In such cases, we can compute the curvature using the chain rule:

$$\frac{\mathrm{d}\phi}{\mathrm{d}t} = \frac{\mathrm{d}\phi}{\mathrm{d}s}\frac{\mathrm{d}s}{\mathrm{d}t}.$$

**Basic method of curve generation.** The *basic method* of generating curves based on various distributions (given below) is controlled by the length of the curve. Figure 3.14 shows the distribution of interval lengths and angles for the database of signatures used in our study.

*Step 1.* Generate lengths of intervals along the trajectory so that the total length is the given length of the curve.

*Step 2.* For each interval with its length $s$, generate the curvature $k$ or angle $\phi$ based on the statistical distribution of curvatures, their derivatives and the spatial distributions of $k(s)$. Repeat Steps 1-2 until the desired total arc length is obtained.

> **Example 3.8** *Figure 3.14 gives an example of statistical distributions generated from the preexisting database of original signatures.*

**Method based on the *n*-gram model.** The basic method outlined above has been extended to improve the visual appearance of the generated signatures using *n*-gram structures and modeling. The sequential nature of curve characteristics, length intervals, angles and curvatures allows us to design a model which encompasses the probability of possible combinations of values and their transition from one to another. Thus, in our study we used digram (a pattern of two adjacent values) and trigram (a pattern of three adjacent values) models to simulate the sequential behavior of curves. Models based on digrams proved to be more efficient than those based on

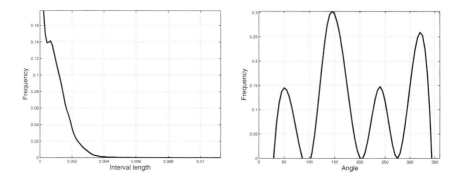

**FIGURE 3.14**
Distributions of interval lengths (left panel) and angles (right panel) for the preexisting signature database.

hidden Markov models in terms of storage and run-time. In addition to these general properties assessed by the final outcome, the digram models can be built with less computational resources than hidden Markov models.

The generation method produces an angle value $\phi_i$ based on digram model statistics and a previously generated value $\phi_{i-1}$. The same methodology is applied to other sequential characteristics, such as curvature and length intervals. Formally, the method can be expressed as:

*Step 1.* Randomly select the initial values for a length interval and angle (or curvature).

*Step 2.* Build the distribution of values for length intervals and angles (or curvatures) from the corresponding digram models taking previous values as reference parameters.

*Step 3.* Randomly generate current values for length intervals and angles (or curvatures) based on distributions from Step 2, and repeat Steps 2-3 until the total arc length is reached.

> **Example 3.9** *Some samples of generated signatures are given in Figure 3.15. The digram model $\Theta(\phi, A)$ for angle sequences is defined on the alphabet $A$ of angle values (in our experiments $A = 0 \ldots 359$) and contains the frequencies of all pairs $< \phi_{i-1} \; \phi_i >$ acquired at the learning stage. Figure 3.16a shows the angle digram model for the entire signature database. Figure 3.16b shows the segment length model for the signature database. Evidently, only a discrete set of angles were permitted.*

**FIGURE 3.15**
Three signatures generated by the digram model.

**Method based on vector recombination.** In order to generate images with a high percentage of natural looking parts, we introduce a method based on repositioning image sub-parts randomly selected from the pool of training images. Since the curve structure is described in sequential forms, the repositioning implies vector recombination. Thus the resulting signature with the number of segments $N_{segments}$ and the their lengths $S_{segment}$ can be composed of $N_{segments}$ pieces taken from different signatures with the total length $S_{segment}$. This method is outlined in the following series of steps:

*Step 1.* The sequential representation of interval lengths and angles (or curvatures) is treated as a pool of sequences.

*Step 2.* Randomly select the start point within the pool of sequences and select as many points as the length of the current segment. This guarantees the mutual correspondence between interval lengths and angles (or between interval lengths and curvatures).

*Step 3.* Repeat Step 2 for the given number of segments.

> **Example 3.10** *Some samples of generated signatures are given in Figure 3.17. Experimental results demonstrate that the method based on vector recombination is fast, and visually accurate, but allows inconsistencies of the inter-segment parts. Figure 3.18 shows the distribution of inter-segment lengths and angles used in this vector recombination model.*

One can observe that the segment-to-segment connection is predominantly straight without significant changes in inter-segment angles.

### 3.4.4 Kinematically meaningful synthesis

By following the different synthesis approaches described above (basic, *n*-gram, and recombination), it is possible to produce not only the shape of the image, but also to generate kinematic characteristics in the form of sequences. Here the kinematically meaningful synthesis of pen pressure, time intervals, and speed of writing are considered.

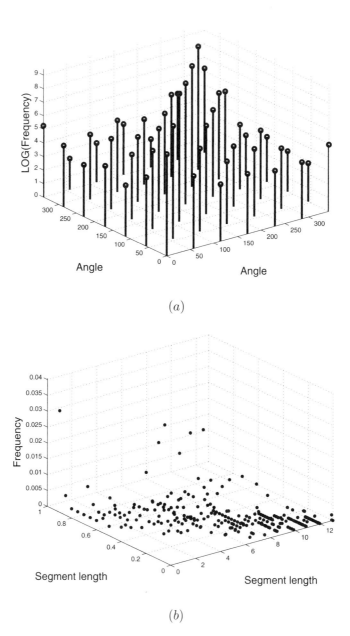

(a)

(b)

**FIGURE 3.16**

An angle digram model for the preexisting signature database (a) and a segment length digram signature database (b).

**FIGURE 3.17**
Three synthetic signatures generated based on the recombination model.

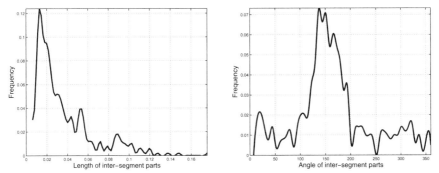

**FIGURE 3.18**
Distributions of inter-segment lengths $L_{inter-segment}$ (left) and angles $\phi_{inter-segment}$ (right) for the preexisting signature database.

> **Example 3.11** *Some samples of generated signatures are given in Figure 3.19. Figure 3.20 gives the distributions of pressure levels and time intervals based on statistical data from the signature database. Two peaks for the level 0 and for the level 511 (on a $[0 \ldots 511]$ scale) were caused by inconsistencies of the image acquisition equipment for high pressure and low pressure measurements (an off-the-shelf pressure sensitive tablet was used).*

Shape, speed of writing, and pressure patterns are a powerful combination of measurements for detecting forged signatures. Applying pressure information in signature recognition works quite well for a number of signatures; however for others, it increases the false rejection rate. From observations, the pressure patterns are not always as stable as the position and velocity patterns for the data sets, and because not all signature data contain pressure information, they are sometimes ignored for signature verification, but are used during preprocessing and segmentation.

**FIGURE 3.19**

Two synthetic signatures generated based on the kinematic model: pressure levels (left panel) and speed (right panel) are visualized in grayscale.

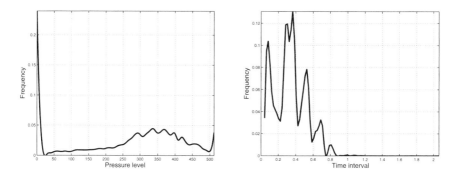

**FIGURE 3.20**

Distributions of pressure levels (left panel) and time intervals (right panel) for the preexisting signature database.

**Example 3.12** *The pressure information given by pressure sensitive tablets or screens proves to be valuable in preprocessing and signature segmentation. In the preprocessing stage, most unintentional pen-down sample data can be identified by a very small pressure reading and can be removed. The pressure information also gives valuable hints on where the handwriting motion breaks. By combining the occurrences in handwriting pressure reaching a local minimum with vertical or horizontal pen-velocity reaching a zero-crossing and the curvature value approaching a local peak, the accuracy of signature segmentation is improved.*

### 3.4.5 Postprocessing: validation

During the validation stage, the generated signature is compared to the existing database of signatures in order to evaluate false identification and false error rates. If the validation stage returns inappropriate results in terms of error rates, the generated signature is discarded and not included in the final solution. The visual appearance can be checked manually, and this is especially valuable for the in-class generation scenario.

## 3.5 Implementation

This section describes the final algorithm for signature synthesis. Experimental results and evaluation metrics are given to demonstrate the accuracy and efficiency of signature generation models.

**Algorithm for signature generation.** The final algorithm consists of the following steps:

*Step 1.* Select a sample (training) set of signatures from the database.

*Step 2.* Generate statistical data from the sample set.

*Step 3.* Applying the random number generator and considering various distributions from the sample set, determine the number of segments, the length and orientation of inter-segment intervals, etc.

*Step 4.* Applying the random number generator and considering one of the outlined scenarios, generate the interval's lengths and curvatures (or the interval's lengths and angles) along the trajectory.

*Step 5.* Applying the random number generator and considering various distributions from the sample set, determine kinematic characteristics along the trajectory.

*Step 6.* Perform postprocessing validation of the obtained image.

Input data are signatures selected from either the entire database for the intra-class scenario or class-specific signatures for the in-class scenario. The basic algorithm implements the statistically meaningful synthesis by analyzing a sample set of signatures (Step 1) in order to build necessary statistical distributions (Step 2). The set of statistical distributions includes distributions of (i) the number of segments, (ii) the segment's length, (iii) the length and orientation of inter-segment connections, (iv) the angle and curvature characteristics along the contingent segments; and (v) the pressure and time intervals. The surrogate generator is used to randomize various signature characteristics (Steps 3-5). Adding a noise (an optional step in the surrogate generator) is aimed to increase the variability of signature characteristics. The result is a generated signature described geometrically and kinematically. Such a signature possesses similar statistical characteristics as the sample set. The postprocessing validation is necessary to evaluate deviations between the generated signature and the existing database of signatures.

Exploiting the basic principles of signature generation, the following extensions are suggested:

*Generator 1:* The generator is based on adequate distributions of visual parts and kinematic characteristics.

*Generator 2:* The generator relies on $n$-gram models for sequence generation.

*Generator 3:* The generator is based on sequence processing and generating visual attributes based on preexisting sequences and their recombination.

**Experimental results.** The preexisting database of handwritten signatures has been used to generate necessary statistical distributions.[‡] An example below provides a study of intra-class and in-class scenarios in signature synthesis.

### Example 3.13

> Intra-class scenario. *Figure 3.21 shows a sample collection of synthesized signatures generated solely using the n-gram model for the intra-class scenario. Figure 3.22 depicts a collection of synthesized signatures based on the vector recombination model.*
>
> In-class scenario. *Figure 3.23 shows the output of in-class signature generation. A collection of signatures is formed by using the deformation model (the original signature is placed in the upper left corner).*

**Observation: intra-class scenario, $n$-gram model.** Though visual analysis can reveal numerous unexpected direction changes for curves, all signatures from this sample passed postprocessing validation. In particular, various effects of the $n$-gram model can be observed:

*Synthetic signatures A1-A4, B1-B4, and C1-C4.* In this set, there are many direction changes. The location and the length of segments are visually normal.

---

[‡]The database contained 120 signatures collected from 25 different individuals. The signatures were acquired through pen based interfaces (i.e., pressure sensitive tablets) that provide 100 samples per second containing values for pen coordinates, pressure, and time stamps.

*Synthetic signatures D1-D4.* In this set, the generated lines are predominantly straight.

**Observation: intra-class scenario, vector recombination model.** The collection given in Figure 3.22 has more meaningful visual interpretation for signature images, however some visual parts of generated signatures can be misclassified because of recombination. The recombination model presents a trade-off between the signature-like visual appearance and the uniqueness of synthesized signatures making postprocessing validation a necessity. In particular, various effects of *n*-gram model can be observed:

*Synthetic signatures A1, A3, C2, D3-D4.* In this set, the generated lines are predominantly straight. The transition from one line to another is smoother than in the *n*-gram model.

*Synthetic signatures A2, A4, B1-B4, C1, C3-C4, D1-D2.* This set is characterized by many direction changes. The transition from one segment to another is smoother than in the *n*-gram model.

**Observation: in-class scenario, deformation model.** Figure 3.23 shows the output of the deformation model, in-class scenario. It is observed that by increasing the degree of deformation applied to the original signature A1, the signature transforms gradually to D4.

**Metrics.** The following system of performance metrics were used in the experiments:

*Accuracy of visual appearance.* The expert visual analysis is performed for signatures generated by signature synthesis methods. Visual appearance is checked by cognitive experiments for all three generation strategies. The robustness of the system is evaluated with respect to noise and shape variations.

*Accuracy of postsynthesis comparison.* There are two widely accepted characteristics for measuring accuracy: the false acceptance rate (percentage of inappropriate images accepted) and the false reject rate (percentage of appropriate users rejected).

*Speed of generation and postsynthesis comparison.* The speed of signature synthesis and validation is evaluated. This determines the scalability of the algorithm with respect to existing and newly generated databases of signatures.

Experiments reveal the following characteristics:

*Accuracy of visual appearance.* All three generators embedded into the signature generation algorithm produced acceptable results in terms

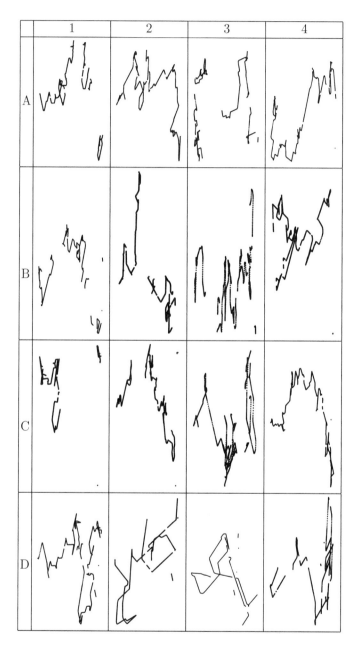

**FIGURE 3.21**

Intra-class scenario (*n*-gram model) in synthetic signatures generation.

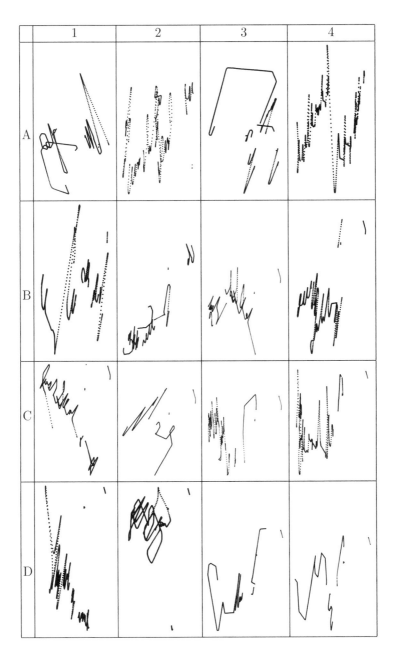

**FIGURE 3.22**

Intra-class scenario (vector recombination) in synthetic signatures generation.

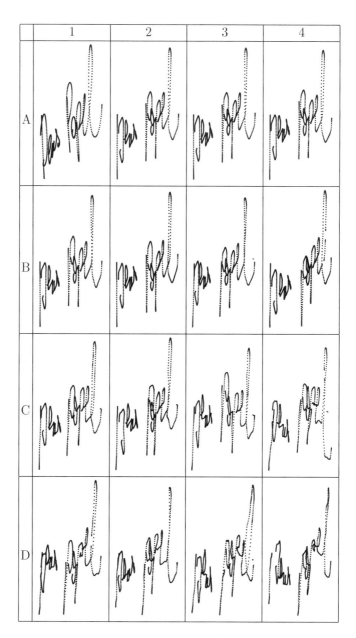

**FIGURE 3.23**

In-class scenario (the original signature is in the upper left corner, A1) in synthetic signatures generation.

of visual appearance. Strokes have not only straight but also cursive shapes. Their relationships are also natural. Since the generation has been based on statistical data only, curves and shapes do not contain any information about language.

*Accuracy of postsynthesis comparison* Both intra-class and in-class scenarios have been checked through the validation stage with the following outcome: 6% of signatures were rejected for the intra-class scenario, and 4% of were rejected for the in-class scenario. A similarity threshold of 95% was set up to reject signatures for the intra-class scenario, and accept signatures for the in-class scenario.

**Further parameter adjustments** Adding extra parameters such as speed, acceleration, curvature, etc. results in an improvement in accuracy characteristics of the system. Biometric fusion of several methods (fingerprint, voice, handwriting, etc.) can improve the results of verification and identification. Even combining methods within the same biometric method, like the fusion of on-line and off-line signature processing, can make the system more robust and efficient.

## 3.6 Summary

Handwritten signatures are commonly used to individualize the contents of a document. Signature verification is based on visual inspection of two signatures and its acceptance or rejection. The hypothesis that each individual has handwriting that is distinct from the handwriting of another individual is widely used in practice. Several factors like age, gender, mood, writing instrument, changes over time, etc. may influence handwriting style.

1. The crucial point of inverse biometrics, the *analysis-by-synthesis* paradigm, is used in the generation of synthetic signatures as well as in the verification of signatures. The analysis-by-synthesis paradigm in signature processing states that the signature synthesis can verify the perceptual equivalence between original signatures and synthetic signatures.
2. Signatures are described by geometric and kinematic characteristics with the requirements to generate both. The design of curves plays an essential role in the construction of synthetic biometric objects.
3. *Generator* of synthetic biometric data is defined as a model of biometric data. Two main types of models are distinguished: *static* and *dynamic* models. Static models are used in modeling long-term behavioral characteristics, for example, iris and face topologies. Dynamic models

are used for representation of short-term characteristics of biometric data, for example, facial expressions and signatures.

4. Two scenarios for signature generation are reasonable: intra-class and in-class. The first, intra-class scenario, allows us to obtain a synthetic signature which is independent to the set of sample signatures from the database. The later, in-class scenario, takes into consideration the class label and already available signatures for this class to generate new signatures within this class.

5. Data generated by various models are classified as *acceptable* or *unacceptable* for further processing and use. The classification criteria must provide a *reasonable level of acceptability*. The acceptability is defined as a set of characteristics which distinguish original and synthetic data. Acceptable biometric data are called *synthetic* based on the controllability criteria of topological similarities, etc. The model that approximates original data at reasonable levels of accuracy for the purpose of analysis is not considered as a generator of synthetic biometric information.

6. Methods and algorithms of signature generation can be applied in the following areas:

▶ Engineering applications for advanced biometric systems where the quantity of data is crucial for decision making.

▶ Environments for testing existing biometric solutions against fraudulent biometric images.

▶ Environments for testing the accuracy, reliability, and speed of multimedia databases including biometric databases.

▶ Other applications including document automation, information encryption, and Internet search engines.

## 3.7 Problems

Problems from 3.1 through 3.10 can be solved using MATLAB. Problems 3.11, 3.12, and 3.13 can be considered as hands-on projects.

**Problem 3.1** A curve is given in its implicit form: $2 \cdot x^2 + 3 \cdot y^2 = 0$. Write the curve's description in explicit, $y = f(x)$, and parametric forms, $(x(t), y(t))$. *Hint:* many parametric equations can be provided for the given curve, select any simple equation.

**Problem 3.2** A curve is given in its parametric form: $x(t) = 2 \cdot t^2 + 5 \cdot t - 2$ and $y(t) = t^3 + 4 \cdot t$. Write the parametric equations for the following characteristics of signature: (a) Position $\mathbf{r}(t)$; (b) Velocity $\mathbf{r}'(t)$; (c) Speed $|\mathbf{r}'(t)|$; (d) Acceleration $\mathbf{r}''(t)$.

*Hint:* the parametric equation for the position $\mathbf{r}(t)$ is required for outlining equations for velocity, speed and acceleration (see Subsection 3.4.3).

**Problem 3.3** (Continuation of Problem 3.2) Provide the parametric description for the arc length $s(t)$ and curvature $k(t)$. Plot $s(t)$ and $k(t)$ using MATLAB.

*Hint:* compute the unit tangent vector $\mathbf{t}'(t)$ prior to curvature computation.

**Problem 3.4** (Continuation of Problem 3.3) Parameterize the curve with respect to its arc length $s$. Provide the parametric description for the curvature $k(s)$. Plot $k(s)$ and $\mathbf{r}(s)$ using MATLAB.

*Hint:* first find the parametric description of the curve in terms of $x(s)$ and $y(s)$.

**Problem 3.5** Using MATLAB, design two functions for:

($a$) The basic random number generator;

($b$) The surrogate random number generator.

*Hint:* Use MATLAB function **rand** for pseudorandom number generation.

**Problem 3.6** Using a sample of the vector representation provided below, design a MATLAB function that reads such a format and visualizes the curve in a 2D plot:  162 165 198 677; 174 156 305 679; 193 150 379 681; 216 149 422 683; 236 150 446 685; 252 156 454 687; 261 166 468 689; 263 183 483 691; 255 205 499 693; 240 231 505 696; 222 256 505 698; 201 278 511 700; 179 292 511 702; 155 297 511 704; 132 291 511 706; 116 279 511 708; 110 268 499 710; 115 263 409 712; 130 262 17 714; 199 308 157 735; 199 312 280 736.

Values are $x$ and $y$ coordinates, pressure and time stamps accordingly.

*Hint:* $x$ and $y$ coordinates, pressure $p$ and time are vectors; consider the parametric description $x(t)$, $y(t)$ and $p(t)$.

**Problem 3.7** (Continuation of Problem 3.6) Using MATLAB, visualize the given vector representation as a 3D plot considering pressure as the third coordinate and speed as the curve color.

*Hint:* use **plot3c or scatter3** MATLAB functions.

**Problem 3.8** (Continuation of Problem 3.7) Using the given vector representation, write the parametric equations for the following characteristics of signature: (a) Position $\mathbf{r}(t)$; (b) Velocity $\mathbf{r}'(t)$; (c) Speed $|\mathbf{r}'(t)|$; (d) Acceleration $\mathbf{r}''(t)$.

Plot these equations using MATLAB.

*Hint:* the parametric equation for the position $\mathbf{r}(t)$ is required for outlining equations for velocity, speed and acceleration (see Subsection 3.4.3).

**Problem 3.9** (Continuation of Problem 3.8) Using the given vector representation, provide the parametric description for the arc length $s(t)$ and curvature $k(t)$. Plot $s(t)$ and $k(t)$ using MATLAB.

*Hint:* compute the unit tangent vector $\mathbf{t}'(t)$ prior to curvature computation.

**Problem 3.10** (Continuation of Problem 3.9) Using the given vector representation, parameterize the curve with respect to its arc length $s$. Provide the parametric description for the curvature $k(s)$. Plot $k(s)$ and $\mathbf{r}(s)$ using MATLAB.

*Hint:* first find the parametric description of the curve in terms of $x(s)$ and $y(s)$.

**Problem 3.11** Design an application for a tablet using the Wintab library (see Manual for solutions) to capture the trajectory of signatures along with kinematic characteristics such as pressure and pen velocity. The application should be capable of recording signatures and human handwriting as a vector file preserving all the attributes necessary for image restoration. The vector format from Problem 3.6 can be used as a model.

**Problem 3.12** (Continuation of Problem 3.11) Using the generated vector file, implement the signature synthesis algorithm for basic image recombination. Results can be visualized in MATLAB by converting the vector file to any raster formats such as BMP, JPEG, GIF, etc.

**Problem 3.13** (Continuation of Problem 3.11) Using the set of generated vector files, collect statistical data in the form of the following distributions: the number of segments, angle, pressure, inter-segment distances, etc. Visualize all the distributions using MATLAB.

## Supporting information for problems

MATLAB includes many functions to help with parameterized image analysis and synthesis. Here is an overview of some useful MATLAB functions.

**MATLAB function** `load`. Loads formatted data from a file. In parameterized descriptions of curves, vector files are formatted to store coordinates, pressure, and other curve characteristics. The function returns the contents of a file in a variable (matrix). The syntax is

```
Variable = load('curve.file');
```

**MATLAB function** `plot`. Plots vector versus vector on the same graph. This is a basic MATLAB function, for plotting. Other commands for 2D plotting are `scatter`, `polar`, `bar`, `stairs`, etc. The syntax of `plot` is

```
Handle = plot(Vector1, Vector2);
```

**MATLAB function** `plot3`. Plots 3D line plot. The function displays a three-dimensional plot of a set of data points. Other commonly used functions for 3D graphics include `scatter3`, `surf`, `mesh`, etc. The syntax of `plot3` is

```
Handle = plot3(Vector1, Vector2, Vector3);
```

**Example: vector file visualization.** `RR = load('signature.file');`

```
X = RR(:,1); Y = RR(:,2);
P = RR(:,3); T = RR(:,4);
h = plot(X, -Y,'.');
set(h, 'markers',3);
set(h, 'MarkerSize',10);
```

**Example: 3D visualization.** `RR = load('signature.file');`

```
X = RR(:,1); Y = RR(:,2);
P = RR(:,3); T = RR(:,4);
h = scatter3(X,-Y,P,7,T);
set(h, 'markers',3);
set(h, 'MarkerSize',10);
colormap(gray);
view(3);
```

**Example: format transformation.**

```
set(gcf,'PaperUnits','centimeters','paperposition',[0 0 20 15])
print -depsc picture.eps
print -djpeg75 picture.jpg
print -dtiff picture.tiff
```

The book by Seul et al. [31] includes algorithms, codes, and many practical examples on image processing which can be useful in solving the above problems.

## 3.8   Further reading

The fundamentals of statistical shape modeling are introduced in [10, 37]. Methods of image processing are directly related to most problems of signature analysis and synthesis [17, 31].

**State-of-the-art analysis of signatures** is introduced by Fairhurst [11], Jain et al. [20], Plamondon and Srihary [28], Srihari et al. [38]. An overview by Ammar [2], Plamondon and Lorette [24], Leclerc and Plamondon [18], Tappert et al. [39], and [27] are useful for understanding the trends in signature analysis and synthesis. In [33], signature analysis and synthesis are discussed in relation to other biometrics. Several papers in [34] have focused on applications of signatures in banking and security.

One of the first papers on modelling of dynamic signature forgery and complexity measure of topological structure of handwriting is paper by Brault and Plamondon [4].

Development of the models for generating synthetic signatures started from reconstruction of various characteristics from the model that aimed for the validation of the model (analysis-by-synthesis paradigm). For example, Doermann and Rosenfeld [8] have discussed this problem with respect to reconstruction of interfering strokes.

**Signature generation based on deformation.** Another algorithm for signature generation based on deformation has been introduced by Oliveira et al. [22]. This algorithm includes the following steps: *step 1:* acquisition of original signatures; *step 2:* preprocessing: filtering, thresholding, thinning, shrinking (removing any undesirable pixels from the image); *step 3:* determination of characteristic points (at the ends of each trace segment or the ones at intersections of two or more lines); *step 4:* determination of trace segments (segmentation), signature is separated into segments; *step 5:* calculation of the center of mass for each segment that enables individual processing of each trace in relative coordinates with origin in the center of mass.

The convolution principle has been used as a model for all deformations. The deformation is defined as $h(x) = \int_{-\infty}^{+\infty} f(u)g(x-u)du$, where $h(x)$ is the deformation, $f(x)$ is a sample function, $g(x)$ called *instrument* (deforming polynomial). A proper choice of $g(x)$ can modify the characteristics of $f(x)$ as desired.

**Generation synthetic signatures based on human motor system modeling.** Hollerbach [13] has introduced theoretical bases of handwriting generation based on oscillatory motion model. In Hollerbach's model, handwriting is controlled by two independent oscillatory motions superimposed on a constant linear drift along the line of writing. If these parameters are fixed, the model generates a cycloidal motion. By modulation of the cycloidal motion parameters, an arbitrary handwriting can be generated. There are many papers on extension and improving Hollerbach's model, for example [25, 36]. In particular, Singer and Tishby [36] have improved control of Hollerbach's model by so-called motor control symbols which can be interpreted as a high level coding of the motor system.

**Other strategies for synthetic handwriting generation.** Idea by Wang et al. [40] is to combine a shape and physical models in synthetic handwriting generation. They proposed various interesting solutions of typical problems that is worthy to discuss in more detail. One of the most difficult phases in development of the model is to construct the baseline of generated synthetic letters which are aligned and juxtaposed in a sequence. The problem is to concatenate letters with their neighbors to form a cursive handwriting, but directly connecting adjacent letters will produce some discontinuities in the handwriting trajectory. Possible solution, i.e., smooth connection, is to deform the head part and tail part of the trajectory of each character. The drawback of this simple solution is that the deformed letters must be considered as new letters which are not similar to those in the training set. To solve this problem, a so-called *delta-log normal model* [26] was applied. This model can produce smooth connections between characters, but also can ensure that the deformed characters are consistent with models.

Choi et al. [6] proposed to generate character shapes by Bayesian networks. By collecting handwriting examples from a writer, a system learns the writers' writing style. This generator is implemented by Bayesian network based on-line handwriting recognizer. Also, possible applications of character generations are discussed. In particular, authors of [6] considered personalized font in chatting service: create a font for person and every message will be converted to their own writing style.

An approach to synthetic handwriting is proposed by Guyon [12] based on generation handwriting from collected handwritten glyphs. Each glyph is a handwriting sample of two or three letters. When synthesizing a long word, this approach simply juxtaposes several glyphs in sequence and does not connect them to generate fluent handwriting.

There are many modifications aimed to improve a model for various scenarios of applications, for example, [3, 5, 14, 19, 23, 29, 30, 42]. Details of the technique introduced in Section 3.4 one can find in [21]. Practical aspects of online database design for image data have been discussed in [1].

**Hidden Markov model in synthetic handwriting generation.** Hidden Markov model is defined as a doubly stochastic process governed by (a) underlying Markov chain with a finite number of states and (b) a set of random functions; each function is associated with one state. The model is hidden in the sense that only sequence can be observed; the underlying state which generated each symbol is hidden. Details of application of hidden Markov model in handwriting can be found, for example, in [9, 41].

**Stability modeling** is the crucial factor in analysis. Synthesis algorithms are not so critical for variation. More over, the variation must be incorporated into a model for generation of synthetic signatures. Stability modeling is the focus, in particular, [7, 15]. If stability parameters are taken into account, a model is represented by (a) topological description and (b) stability information. For example, Huang and Yan [15] have proposed a stability modeling technique of handwritings in the context of on-line signature verification. In this technique, reliability measures of the extracted signature features are incorporated into the signature segmentation, model building, and verification algorithms. The stable handwriting features are emphasized in the signature matching process, while style variations for less stable features are intentionally tolerated.

**Devices for signature identification and generation.** Shmerko [32, 35] introduced various practical aspects of using signatures in banking, including requirements for devices for signature and handwriting identification. Munich and Perona [20] have introduced class so-called visual input for pen-based computers. Nikitin and Popel [21] have developed signature recognizing tools.

**Motor control modeling** is related to physical modeling approach that states that the pen movements produced during cursive handwriting are the result of so-called "motor programs" controlling the writing apparatus, i.e., handwriting is generated by human motor system. This system can be described by so-called *spring* muscle model near equilibrium. A *spring* muscle model assumes that muscle operates in the linear small deviation regions and can be described by differential equation. Solution of this equation is represented by various cycloidal trajectories controlled by *spring* constants and cycloid parameters. Handwriting can be represented by a slowly varying dynamical system with control parameters corresponding to cycloidal parameters. Generation of a handwriting is defined as inverse dynamic problem.

**Shape simulation techniques** based on geometrical modeling of handwriting. For example, each letter can be modeled by a B-spline curve controlled by several control points. Linear combinations of control points represent the stroke. In this way, the problem of learning the writing style is converted to learning the particular distributions of control points. Synthetic signature generation consists of several phases: generation master-letters, adding a noise to increase the variability of geometrical configuration, and assembling letters in stroke or signature. In this approach, learning and training are essential phase of model development.

## 3.9    References

[1] Alexander S, Strohm J, and Popel D. Online image database with content searching capabilities. In *Proceedings of the Midwest Instruction*

*and Computing Symposium*, Morris, MN, April 2004.

[2] Ammar M. Progress in verification of skillfully simulated handwritten signatures. *Pattern Recognition and Artificial Intelligence,* 5:337–351, 1993.

[3] Bastos LC, Bortolozzi F, Sabourin R, and Kaestner C. Mathematical modelation of handwritten signatures by conics. *Revista da Sociedade Paranaese de Matematica*, Curitiba, 18(1-2):135–145, 1998.

[4] Brault JJ and Plamondon RA. Complexity measure of handwritten curves: modelling of dynamic signature forgery. *IEEE Transactions on Systems, Man and Cybernetics*, 23:400–413, 1993.

[5] Bruckstein AM, Holt R, Netravali A, and Richardson T. Invariant signatures for planar shape recognition under partial occlusion, *CVGIP: Image Understanding*, 58(1):49–65, 1993.

[6] Choi H, Cho SJ, and Jin Kim JH. Generation of handwritten characters with Bayesian network based on-line handwriting recognizers, In *Proceedings of the 17th International Conference on Document Analysis and Recognition*, Edinburgh, Scotland, pp. 995-999, Aug. 2003.

[7] Dimauro G, Impedovo S, Modugno R, and Pirlo G. Analysis of stability in hand-written dynamic signatures, In *Proceedings of the 8th International Workshop on Frontiers in Handwriting Recognition*, Ontario, Canada, pp. 259–263, Aug. 2002.

[8] Doermann DS and Rosenfeld A. The interpretation and reconstruction of interfering strokes. In *Frontiers in Handwriting Recognition III*, CEDAR, SUNY Buffalo, New York, pp. 41–50, May 1993.

[9] Dolfing JGA, Aarts EHL, and Van Oosterhout JJGM. On-line signature verification with hidden Markov models. In *Proceedings of the 14th International Conference on Pattern Recognition*, Brisbane, pp. 1309–1312, 1998.

[10] Dryden IL and Mardia KV. *Statistical Shape Analysis*. Wiley, New York, NY, 1998.

[11] Fairhurst MC. Signature verification revisited: promoting practical exploitation of biometric technology. *Electronics and Communication Engineering Journal*, 9(6):245253, 1997.

[12] Guyon I. Handwriting synthesis from handwritten glyphs. In *Proceedings of the 5th International Workshop on Frontiers of Handwriting Recognition*, Colchester, UK, pp. 309-312, 1996.

[13] Hollerbach JM. An oscilation theory of handwriting. *Biological Cybernetics*, 39:139–156, 1981.

[14] Huang K and Yan H. On-line signature verification based on dynamic segmentation and global and local matching. *Optical Engineering*, 34(12):3480–3487, 1995.

[15] Huang K and Yan H. Stability and style-variation modeling for on-line signature verification. *Pattern Recognition*, 36:2253–2270, 2003.

[16] Jain AK, Griess FD, and Connell SD. On-line signature verification, *Pattern Recognition* 35(12):2963–2972, 2002.

[17] B. Jähne. *Digital Image Processing.* Springer, Heidelberg, 2002.

[18] Leclerc F and Plamondon R. Automatic signature verification. *International Journal on Pattern Recognition and Artificial Intelligence*, 8(3):643–660, 1994.

[19] Munich ME and Perona P. Visual identification by signature tracking. *IEEE Transactions on Pattern Analysis and Machine Intelligence* 25(2):200–217, 2003.

[20] Munich ME and Perona P. Visual input for pen-based computers. *IEEE Transactions on Pattern Analysis and Machine Intelligence* 24(3):313–328, 2001.

[21] Nikitin A and Popel D. Signmine algorithm for conditioning and analysis of human handwriting. In *Proceedings of the International Workshop on Biometric Technologies*, Calgary, Canada, pp. 171–182, 2004.

[22] De Oliveira C, Kaestner C, Bortolozzi F, and Sabourin R. Generation of signatures by deformation, In *Proceedings of the BSDIA97*, Curitiba, Brazil, pp. 283–298, Nov. 1997.

[23] Parizeau M and Plamondon R. A comparative analysis of regional correlation, dynamic time warping, and skeletal tree matching for signature verification. *IEEE Transactions on Pattern Analysis and Machine Intelligence*, 12(7):710–718, 1990.

[24] Plamondon R and Lorette G. Automatic signature verification and writer identification, the state of the art. *Pattern Recognition*, 22(2):107–131, 1989.

[25] Plamondon R and Maarse FJ. An evaluation of motor models of handwriting. *IEEE Transactions PAMI*, 19:1060–1072, 1989.

[26] Plamondon R and Guerfali W. The generation of handwriting with delta-lognormal synergies. *Bilogical Cybernetics*, 78:119–132, 1998.

[27] Plamondon R, Lopresti D, Schomaker LRB, and Srihari R. On-line handwriting recognition. In Webster JG, Ed., *Encyclopedia of Electrical and Elecronics Engineering*, 15:123–146, Wiley, New York, NY, 1999.

[28] Plamondon R and Srihari SN. Online and off-line handwriting recognition: a comprehensive survey. *IEEE Transactions on Pattern Analysis and Machine Intelligence*, 22(1):63–84, 2000.

[29] Rhee TH, Cho SJ, and Kim JH. On-line signature verification using model-guided segmentation and discriminative feature selection for skilled forgeries. In *Proceedings of the 6th International Conference on Document Analysis and Recognition*, Seattle, Washington, pp. 645–649, Sept. 2001.

[30] Sato Y, and Kogure K. On-line signature verification based on shape, motion and writing pressure. In *Proceedings of the 6th International Conference on Pattern Recognition*, Munich, Germany, pp. 823–826, 1982.

[31] Seul M, O'Gorman L, and Sammon MJ. *Practical Algorithms for Image Analysis: Description, Examples, and Code*. Cambridge University Press, Cambridge, 1999.

[32] Shmerko V, Kochergov E., Michaylenko S, Yanushkevich S. Systems for personal identification using methods of signature and handwriting verification. In *Proceedings of the 3rd IEE International Conference on Pattern Recognition and Information Analysis*, Minsk, Belarus, vol. 3, pp. 25–35, 1995.

[33] Shmerko V, Perkowski M, Rogers W, Dueck G, and Yanushkevich S. Bio-technologies in computing: the promises and the reality. In *Proceedings of the International Conference on Computational Intelligence Multimedia Applications*, Australia, pp. 396–409, 1998.

[34] Shmerko V and Yanushkevich S, Eds., *Banking Information Systems*. Academic Publishers of Technical University of Szczecin, Poland, 1997.

[35] Shmerko V. Experience of designing and using signature and handwriting person identification systems. In *The British Computer Society Computer Security Specialist Group Annual Conference*, London, March 1995.

[36] Singer Y, and Tishby N. Dynamic encoding of cursive handwriting. *Biological Cybernetics*, Springer, 71(3):227–237, 1994.

[37] Small CG. *The Statistical Theory of Shape*. Springer, Heidelberg, 1996.

[38] Srihari SN, Cha SH, Arora H, and Lee S. Individuality of Handwriting. *Journal of Forensic Science*, 47(4):1–17, 2002.

[39] Tappert CC, Suen CY, and Wakahara T. The state of the art in on-line handwriting recognition. *IEEE Transactions on Pattern Analysis and Machine Intelligence*, 12:787–808, 1990.

[40] Wang J, Wu C, Xu YQ, Shum HY, and Li L. Learning based cursive handwriting synthesis. *Proceedings of the 8th International Workshop on Frontiers in Handwriting Recognition*, Ontario, Canada, pp. 157–162, August 2002.

[41] Yang L, Widjaja BK, and Prasad R. Application of hidden Markov models for signature verification. *Pattern Recognition*, 28(2):161–170, 1995.

[42] Yanushkevich SN, Supervisor. Computer aided design of synthetic signatures. *Technical Report, Biometric Technologies: Modeling and Simulation Laboratory*, University of Calgary, Canada, 2004.

[43] Yanushkevich SN and Shmerko VP. Biometrics for electrical engineers: design and testing. Chapter 4: CAD of signature analysis and synthesis. *Lecture notes*, University of Calgary, Canada, 2005.

# 4

# *Synthetic Fingerprints*

In this chapter, techniques for generating synthetic fingerprints are introduced. There are several motivations for developing techniques for synthetic fingerprint design, in particular, improving testing of fingerprint systems, enhancing the security of a database of fingerprints, building training systems, and watermarking algorithms.

This chapter provides the basics of synthetic fingerprint design. These basics include state-of-the-art automated fingerprint analysis techniques and approaches to solving the inverse problem, that of generating synthetic fingerprints. As an introductory example, Figure 4.1 illustrates two fingerprint-like topologies generated by computer-aided design techniques based on topological and physical paradigms.*

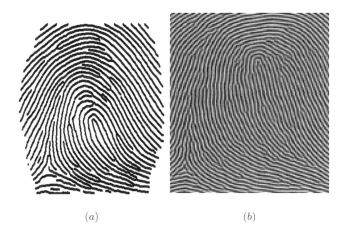

(a)                              (b)

**FIGURE 4.1**
Synthetic fingerprint topologies generated by topological (a) and physical (b) techniques.

---

*These two techniques are introduced in the "Further Reading" Section as Cappelli's and Kuecken's synthetic fingerprint generators respectively.

The chapter is organized as follows. In Section 4.1, two approaches to synthetic fingerprint design are discussed: non-automated and automated. The focus of Section 4.2 is fingerprint topology: a global and local representation. This is the base for topological decomposition, which produces topological primitives. Feature extraction is considered in Section 4.3. The primitives constitute a library of primitives (Section 4.4). An example of a topological primitive generator is discussed in Section 4.5. One of the possible approaches for generating synthetic fingerprints that utilizes the properties of the orientation map is discussed in Section 4.6. Another approach to generate topology similar to the fingerprint, is introduced in Section 4.7 and Section 4.8. The synthesis of realistic fingerprint images is a multiphase design process. In Section 4.9, these phases of the design process are introduced. Manipulation of topological components using a library of primitives results in a so-called master-fingerprint, which is a binary image of a synthetic fingerprint. Realistic prints are achieved by smoothing, adding noise, and other stochastic information to the master-fingerprint. After a brief summary (Section 4.10), problems are given in Section 4.11. Useful details can be found in the "Further Reading" Section.

## 4.1   Introduction

The methods of fingerprint processing presented here can be viewed as a combination of *direct* and *inverse* problems (Figure 4.2):

▶ The group of *direct* problems is aimed at analyzing the topological structure of the fingerprint. The result is the set of features that are used in various systems for decision making (Figure 4.2a).

▶ The group of *inverse* problems addresses the synthesis of fingerprints; the methods adopted from inverse problems give the rules for manipulating synthetic fingerprint primitives to produce a new, synthetic fingerprint (Figure 4.2b).

▶ Generating synthetic fingerprints employs both of these processes (Figure 4.2c):

(*a*) First based on inverse methods, samples of fingerprint-like topologies are generated, including various effects that make the fingerprints realistic.

(*b*) Then direct methods (analysis) are used to decide which generated fingerprints are acceptable, i.e., the acceptable synthetic fingerprints are chosen to be included in the final set.

Approaches to designing synthetic fingerprints can be *non-automated* or *automated*.

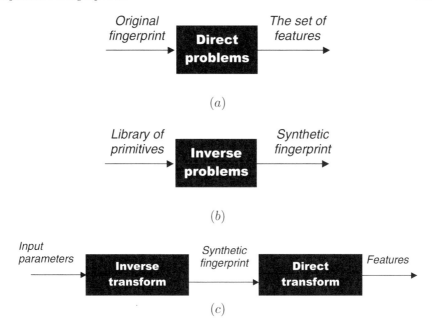

**FIGURE 4.2**
Direct problem (a), inverse problem (b), and schematic representation of synthetic fingerprint generation (c).

**Non-automated design technique.** Much before computerized fingerprint acquisition and processing, criminals (and forensic experts) used quite natural techniques to tamper with fingerprints[†].

> **Example 4.1** *The simplest examples of synthetic fingerprints created by non-automated techniques are as follows:*
>
> (a) *Latent prints on a surface are enhanced using a fine brush and toner (printer cartridge). Then the prints are lifted with adhesive tape. The resulting images are transferred to a transparency material on a photocopier.*
>
> (b) *A mold is created from plaster, then it is filled with silicone, and a pounder is used to make the synthetic finger wafer-thin.*

**Automated design technique.** There are various strategies for automated synthesis of fingerprints. One strategy is based on assembling primitives

[†]D. Willis and M. Lee. Six biometric devices point the finger at security. Network Computing, June 1998

$$< \texttt{Primitives} > \; \Rightarrow \; < \texttt{Synthetic fingerprint} > .$$

In this approach, both original (extracted from original fingerprints) or synthetic primitives can be used. Another approach is based on the transformation of an original fingerprint into its synthetic equivalent:

$$< \texttt{Original fingerprint} > \; \Rightarrow \; < \texttt{Synthetic fingerprint} > .$$

Figure 4.3 shows the assembly technique design process. The main steps of the process are as follows:

*Step 1.* Two topological primitives are chosen arbitrarily from the database of primitives.

*Step 2.* Based on the appropriate rules, a macroprimitive is built by combining the two initial primitives.

*Step 3.* The next primitive is taken, and added to the macroprimitive using the appropriate rules.

*Step 4.* The process of "growing" the macroprimitive is performed by repeating step 3 and is regulated by control parameters. This process is stopped by applying appropriate criteria.

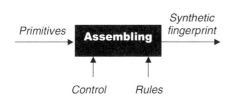

Design of synthetic fingerprints based on the assembling process:

▶ Assembly procedure
▶ Rules of assembling
▶ Control parameters
▶ Library of topological primitives

**FIGURE 4.3**

Synthetic fingerprint design based on the assembling paradigm.

In the second approach, the transformation of an original fingerprint into a synthetic equivalent, the so-called *cancellable* technique is introduced. The main phases of the cancellable technique are as follows (Figure 4.4):

*Phase 1.* Given an original fingerprint, choose an appropriate transform, and control the parameters of cancellability. The existence of an inverse transform is not a necessary requirement.

*Phase 2.* Transform the original fingerprint into its cancellable analog.

*Phase 3.* Use the cancellable fingerprint for matching against the one from database.

The cancellable technique is useful for the following reasons:

(*a*) Security of the database; because cancellable not original data is stored.

(*b*) Security of processing; because cancellable not original data is matched.

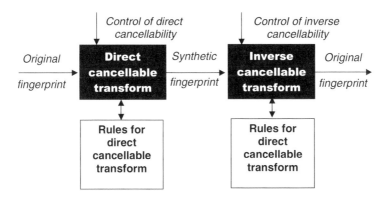

**FIGURE 4.4**
Synthetic fingerprint design based on the cancellable transform.

Two *data* representations are useful in fingerprint analysis and synthesis (Figure 4.5):

▶ *Topology*, that is, data representation in topological space, and

▶ *Frequency* or *spectrum*, that is, representation in the Fourier domain.

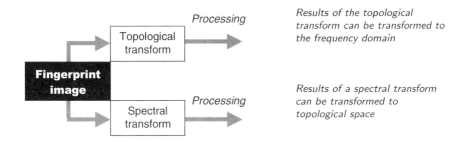

**FIGURE 4.5**
Topological and spectral data structure are distinguished in fingerprint analysis and synthesis.

Topological and frequency representations have a relationship, i.e., the topological structure of a fingerprint can be transformed into the frequency domain and vice versa:

$$< \texttt{Topological space} > \ \Leftrightarrow \ < \texttt{Frequency domain} > .$$

## 4.2 Modeling a fingerprint

The aim of automated fingerprint synthesis is to create a realistic image of a fingerprint without copying a real fingerprint. This new image is an average model of a human fingerprint. While analysis is aimed at the identification and recognition of fingerprints, synthesis is the reverse task. Most analysis methods do not use inverse transforms. The synthesis procedure produces many images, most of which do not resemble an acceptable fingerprint topology. Unacceptable synthetic images can be corrected by manipulating their topology.

### 4.2.1  Topology

Fingerprint topology can be designed based on

▶ *Morphological* models, and
▶ *Geometrical* models.

Morphological models describe how fingerprint patterns are formed over time in the embryo. This is modeling at the physical level. Geometrical models are aimed at constructing images based on the rules obtained from the analysis of fingerprints. The general topology is characterized at the global level of a fingerprint description.

> **Example 4.2** *Examples of the global behavior of a fingerprint topology are:*
>
> (a) *Ridges, e.g., a ridge in a landscape raises itself above the rest of the area.*
> (b) *Valleys (or furrow), e.g., a valley in a landscape lies lower than the rest of the area.*
> (c) *The core, the "center" of the print; in a whorl pattern, the core point is found in the middle of the spiral/circles; in a loop pattern, the core point is found in the top region of the innermost loop.*

In an image, ridges show up as dark parts and valleys show up as light parts. The core and delta points shown in Figure 4.6 are critical starting points in forming a global description and synthetic fingerprint design. At the local topological level, fingerprints are described by a set of specific topological features called *minutiae* points. Minutiae points include ridge terminations, bifurcations, independent ridges, dots or islands, lakes, spurs, and crossovers.

> **Example 4.3** *The two most prominent minutiae used in fingerprint analysis and synthesis of fingerprints are ridge termination (ending) and ridge bifurcation. They form the primitives such as delta, core, whorl and arch (Figure 4.6).*

(a)

(b)

(*a*) A delta occurs when ridge bundle primitives meet at angle of approximately 120°.

(*b*) A core pattern: several ridges are intervening from the delta to the core.

(*c*) A whorl is formed by two ridge bundles. The whorl has two deltas, while the loops have one delta. A loop can also be formed from two ridge bundles.

(*d*) An arch pattern: several deltas are merged, and four ridges end up at the perpendicular ridge

(c)

(d)

**FIGURE 4.6**
The local level topology: the four most often occurring primitives.

In synthesis, local topological representation is a result of the topological decomposition of a fingerprint, i.e., representation by a set of topological primitives in topological space.

## 4.2.2   Information carried by the fingerprint

Evaluation of how realistic a synthetic fingerprint is made using certain measures to compare it with real fingerprints.

**Categories of information.** In the analysis and synthesis of fingerprints, the following terms describe the type of information that can be used as measures:

▶ *Global* and *local*, *spatial* and *spectral* information. For example, depending on the quality of an image, efficient processing can be achieved at:

(*a*) The local level in the spectral domain instant of a spatial one or

(*b*) The global level in the spatial domain instant of a spectral one.

▶ *Recoverable* and *unrecoverable* information. When performing a direct and then reverse transform, in some cases the original information can be recovered, and in some cases it is unrecoverable.

▶ *Uncertain* information is defined as information that cannot be used to make a decision. For example, a fingerprint image, intensively "infected" by noise, carries uncertain information. Zero data is a boundary case of uncertainty.

▶ *Redundant* and *insufficient* information. Redundant information is removed from a fingerprint to make fingerprint matching efficient. If the optimization performed is not satisfactory, there may be insufficient information to make a decision.

**Information in an image.** The analysis of a fingerprint includes the removal of redundant information, leaving features useful for recognition and decision making. Hence, the uncertainty, or entropy in an image decreases through analysis. The goal of synthesis is to construct a synthetic fingerprint that appears "as real as possible". In the assembling approach, synthesis starts with near to zero-entropy primitives and uncertain topology. In the synthesis-by-deformation approach, synthesis starts with some templates ("master images") corresponding to minimal entropy. In general, the design of synthetic fingerprints is a stochastic process because of the random composition of topological primitives. Moreover, additive and multiplicative noise is incorporated into models in some phases of the design. Hence, entropy increases through out fingerprint synthesis.

**Uncertainty.** A *latent* fingerprint is a record on a surface of the friction ridges found on fingers. In order for a recognition system to be able to identify the owner of a fingerprint, the fingerprint must be enhanced first. This phase is called information *extraction*. Uncertain information in a fingerprint might come from a combination of different chemicals that originate from natural secretions, blood, or other contaminants. Some contaminants found in fingerprints are caused by contact of the finger with different materials in the environment. Latent fingerprints can be found on all types of surfaces. The condition of the surface also contributes to the quality of a print. Such surface characteristics include dryness, wetness, dirtiness, and tackiness or stickiness. When creating a synthetic fingerprint, *uncertain* information is recognized as flexibility, i.e., the possibility to manipulate certain parameters to generate new fingerprints.

**Redundant information.** During analysis, a captured fingerprint image contains a lot of information *redundant* from the feature point of view, such as features resulting from scars, too dry or moist fingers, or incorrect pressure. The procedures that remove or decrease redundant information, include, in particular:

*Normalization*: the colors of the image are spread evenly throughout the gray scale making it much easier to compare with other images, and to determine the image quality.

*Binarization*: the gray scale image is transformed into a binary image. Either a global or local threshold value is used for each pixel.

*Image enhancement*: the image is smoothed using filtering to match the pixels nearby so that no point in the image differs from its surroundings to a great extent. For example, by low pass filtering an image, errors and incorrect data are removed, and it simplifies the acquisition process of patterns or minutiae. Filters may be oriented (adapted) according to the local ridge directions and are matched to their spatial frequency in the image. Wiener and anisotropic filters, hierarchical filter-banks, learning strategies, etc. are examples of knowledge-based techniques for fingerprint image enhancement.

*Quality markup*: redundant information needs to be removed from the image before further analysis can be performed, and specific features of the fingerprint can be extracted. Therefore segmentation, i.e. separating the fingerprint image from the background, is needed. Furthermore, any unwanted minutiae (which can appear if the print is of bad quality) need to be removed.

> **Example 4.4** *Figure 4.7 illustrates the effect of removing redundant information from the fingerprint.*

## 4.3 Extraction of features

Redundancy removal combined with other processing results in the extraction of features. This is a vital part of matching for identification and pattern recognition. Studying the features plays a key role in the design of synthesis procedures. The following features are critical for analysis as well as for synthesis:

*Minutiae points* have been used by fingerprint examiners for ages.

*Core points* are used as reference points for the coordinate system, and the distance and angle from the core point is calculated and used to locate each minutiae point. For identification, a certain number of minutiae points should match.

Original    Equalization   Weiner filtering   Conversion      Thinning

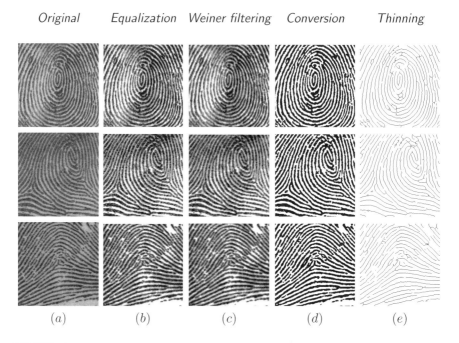

(a)            (b)            (c)            (d)            (e)

**FIGURE 4.7**
Original gray-scale fingerprint (a), fingerprint image after applying histogram equalization (b), fingerprint image by using Weiner filter (c), binary image after the block by block conversion (d), and thinned binary image (e).

The *correlation* between corresponding pixels is computed for different alignments between two images (e.g. various displacements and rotations).

*Ridge features* include the gradient information describing line direction, i.e., topological primitives and the spatial of fingerprint ridges, local orientation and frequency, ridge shape, and texture information. In very low-quality fingerprint images, it can be difficult to extract the minutiae points, and using the ridge pattern for matching is preferred.

## 4.4   Library of topological primitives

In this section, a library of topological primitives from which synthetic fingerprints can be constructed is introduced. Topological primitives play a key role in the analysis and synthesis of fingerprints. Primitives are small portions of information. They can be easily identified and recognized in topological (Cartesian, polar) space and frequency domains.

### 4.4.1 Terminology

The following terminology is used in this chapter:

*Primitive:* an elementary topological component of a fingerprint image,

*Macroprimitive:* a topological structure consisting of at least two primitives,

*Decomposition:* a partition of a fingerprint image into components,

*Composition:* assembling of a fingerprint image from primitives and macroprimitives into larger macroprimitives,

*Constraints:* limitations of topological composition,

*Design rules:* the set of rules for assembling a fingerprint image using primitives or microprimitives.

### 4.4.2 Topological primitive and macroprimitive

A topological primitive is the simplest topological element of a fingerprint from which a topological *composition* of a fingerprint is derived. Table 4.1 shows certain primitives extracted from a fingerprint by topological *decomposition.*

Fingerprint synthesis is the *inverse* decomposition, or a *composition*, of topological primitives. This composition is based on various *topological constraints* and *design rules*, as arbitrarily combining primitives results in unrealistic images.

A composition of two primitives using topological constraints and design rules produces a *macroprimitive.* An extension of a macroprimitive is defined as a topological composition of this macroprimitive and a primitive from the library.

> **Example 4.5** *In Figure 4.8, two topological primitives are composed in various ways. They are examples of acceptable and unacceptable synthesized signatures. In other words, not all compositions produce useful macroprimitives, even if the topological constraints and design rules are applied. By pruning the set of images as they are grown, more realistic effect can be achieved, then by pruning at the end.*

## 4.5 Local generators: fingerprint primitives

In some phases of design, it is useful to manipulate a formal model of a topological primitive. To accomplish this, a topological data structure may

**TABLE 4.1**
Example library of topological primitives for fingerprint design.

| Type | Rule of expansion |
|------|-------------------|
| | Ridge meets a bifurcation with fork appearing from left |
| | Ridge meets a bifurcation with a fork appearing from right |
| | Ridge crossover |
| | Ridge returns to starting-point without any event occurring |

be replaced with an algebraic form. A formal model of a topological primitive is represented as:

▶ An algebraic equation describing the form, and
▶ Some control parameters describing the instance of the form.

A model or data structure representing fingerprints should have:

▶ Minimal encoding/decoding errors,
▶ Reasonable control parameters (local and global topological regulations),
▶ It will have, based on the nature of its construction both local (small size) and global (large size) primitives.

This section describes two strategies for formalizing the description of a fingerprint: *approximation* and *interpolation*.

Synthetic images generated using approximation and interpolation of a sample fingerprint can be combined to generate derivative images by warping ridges, warping minutiae, etc.

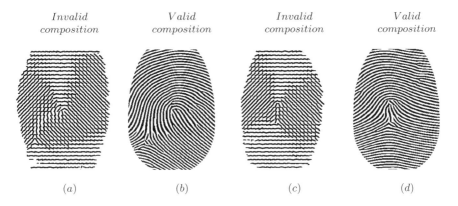

| Invalid | Valid | Invalid | Valid |
| composition | composition | composition | composition |
| (a) | (b) | (c) | (d) |

**FIGURE 4.8**
Invalid (a,c) and valid (b,d) topological compositions of fingerprint primitives (Example 4.5).

### 4.5.1 The simplest generator of primitives based on the curve model

Any model for approximating a line and then warping it, or distorting it, by manipulating control points, is useful as a primitive, for instance, the Bezier curve or splines. A Bezier curve models a line given four points, two of which are the endpoints of the line, and two of which control the initial slope of the line at each end. Another interpolation method is a Hermite curve. The interpolation task is accomplished with both Bezier and Hermite curves by deriving appropriate control points.

### 4.5.2 The simplest generator of primitives based on the spline model

A cubic spline is constructed of piecewise third-order polynomials which pass through a set of $m$ control points. The second derivative of each polynomial is commonly set to zero at the endpoints providing boundary conditions that complete the system of equations. This produces a so-called "natural" cubic spline. For example, the $i$-th piece of an $n$ piece cubic spline can be described by equation $Y_i(t) = a_i + b_i t + c_i t^2 + d_i t^3$, where $t$ is a parameter from the interval [0,1] and $i = 0, 1, ..., n - 1$.

> **Example 4.6** *Figure 4.9 shows the results of an approximating original fingerprints, using the MATLAB* `spline` *function. The initial gray-scale image is binarized. Then ridges are extracted, and dividing to pieces. The spline approximation is applied to each piece.*

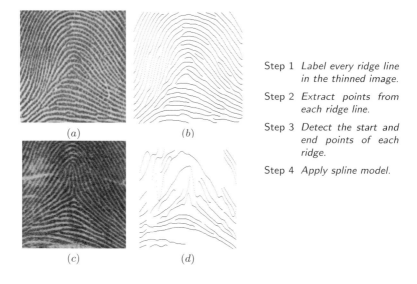

Step 1  *Label every ridge line in the thinned image.*

Step 2  *Extract points from each ridge line.*

Step 3  *Detect the start and end points of each ridge.*

Step 4  *Apply spline model.*

**FIGURE 4.9**
Derivation of the spline model (b),(d) from the original image (a),(c).

This experiment demonstrates the effect of information loss as a result of approximation. Figure 4.9d shows that the algorithms does not work well for a print with cuts and line intarraptions.

## 4.6    Global generators: orientation map

In this section, the orientation map is discussed. The orientation map is the basic model for analysis and synthesis of fingerprints. A binary image of a fingerprint can be transformed into an orientation map. This transformation is called a *direct orientation transform*. The *orientation map* (also called *orientation field*) represents an intrinsic topology of fingerprint images, since it defines invariant coordinates for ridges and valleys in a local neighborhood (Figure 4.10). By viewing a fingerprint image as an oriented texture, a number of methods have been proposed to estimate the orientation field of fingerprint images. After conversion of the image into the spectral domain using the direct polar transform, certain modifications or distortion of the image can be performed, and then the inverse transformation can be implemented, resulting in a new direction map.

The orientation map represents an intrinsic topology of fingerprint images and defines invariant coordinates for ridges and valleys in a local neighborhood

**FIGURE 4.10**
The orientation map.

## 4.6.1 Parameters of orientation map

Techniques of deriving an orientation map include:

▶ Construct *the gradient vector* in topological space,
▶ Compute the *local orientation*, and *local ridge orientation* in topological space, and
▶ Compute the *local correction* in the frequency domain (low-pass filtering).

**The gradient vector** characterizes the gradients, or difference between two neighboring points. Let $(i, j)$ be a pixel of a binary image in the topological space of a fingerprint, and $f(i, j)$ be this pixel's value in terms of the level of gray (the level value is an integer in the range from 0 to 255). The first and the second gradient vector given a pixel $(i, j)$ are defined by the equations:

$$f_x(i,j) = f_{i-1,j-1} - f_{i-1,j+1} + 2(f_{i,j-1} - f_{i,j+1})$$
$$+ f_{i+1,j-1} - f_{i+1,j+1},$$
$$f_y(i,j) = f_{i-1,j-1} - f_{i+1,j-1} + 2(f_{i-1,j} - f_{i+1,j})$$
$$+ f_{i-1,j+1} - f_{i+1,j+1}.$$

It follows from this definition that the gradient vector carries information about ridge line orientation and neighboring pixels' values.

> **Example 4.7** *Figure 4.11 contains the table of gradients. For example, '121' is the gradient with respect to x at pixel* $(50, 82)$:
>
> $$f_x(50, 82) = f_{49,81} + 2f_{50,81} + f_{51,81} - f_{49,83} - 2f_{50,83} - f_{51,83}$$
> $$= 121.$$

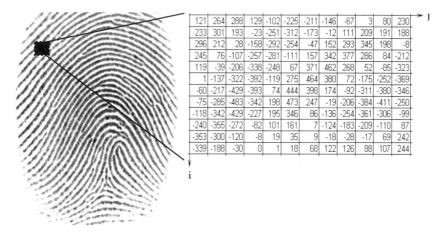

**FIGURE 4.11**
Example of gradient values (Example 4.7).

**The local orientation** of each block centered at pixel $(i,j)$ is defined. The local orientation represents a direction orthogonal to the dominant direction of the Fourier spectrum for a $w \times w$ (for example, $w=8$) window around the pixel.

The ridge orientation $\theta(i,j)$ at the block with the center at pixel $(i,j)$ is calculated by

$$\theta(i,j) = -\frac{1}{2}\, tan^{-1}\frac{V_x(i,j)}{V_y(i,j)}.$$

Here, the components are orientations with respect to axes $x$ and $y$ correspondingly:

$$V_x(i,j) = \sum_{u=1}^{w}\sum_{v=1}^{w}\partial_x(u,v)\partial_y(u,v)$$

$$V_y(i,j) = \sum_{u=1}^{w}\sum_{v=1}^{w}(\partial_y^2(u,v) - \partial_x^2(u,v))$$

The local ridge orientation $O$ is derived by

$$O(i,j) = -\frac{1}{2}\, \tan^{-1}\frac{\Phi_x(i,j)}{\Phi_y(i,j)}$$

where

$$\Phi_x(i,j) = \cos 2\theta(i,j), \quad \Phi_y(i,j) = \sin 2\theta(i,j)$$

are components of the vector field.

> **Example 4.8** *Figure 4.12 illustrates the orientation and local ridge orientation calculation on the gray-scale image.*

**Input data:**
The pixel (48,160) from the fingerprint fragment (marked with star).

**Step 1.** $i = 48$, $j = 160$, and $w = 8$ (block surrounded by a frame in the image). Assume that the values of gradients have been calculated from the pixel values and given in the tables. Therefore,

$$V_x(48, 160) = \sum_{u=48-4}^{48+4} \sum_{v=160-4}^{160+4} f_x(u, v) f_y(u, v)$$

$$= 2219094$$

$$V_y(48, 160) = \sum_{u=40-4}^{40+4} \sum_{v=160-4}^{160+4} (f_y^2(u, v) - f_x^2(u, v))$$

$$= 7902554$$

**Step 2.** Compute orientation $\theta$.

$$\theta(48, 160) = -\frac{1}{2} \tan^{-1} \left( \frac{V_x(40, 160)}{V_y(40, 160)} \right) = -0.13688$$

**Step 3.** Calculate local ridge orientation O.

$$O(48, 160) = -\frac{1}{2} \tan^{-1} \left( \frac{\cos(2\theta(48, 160))}{\sin(2\theta(48, 160))} \right) = 0.6485$$

| i\j | 156 | 157 | 158 | 159 | 160 | 161 | 162 | 163 | 164 |
|---|---|---|---|---|---|---|---|---|---|
| 44 | -80 | -120 | -102 | -76 | -63 | -48 | -41 | -2 | 99 |
| 45 | -4 | -38 | -16 | 16 | 34 | 40 | 47 | 45 | 46 |
| 46 | 15 | -53 | -2 | 110 | 174 | 174 | 153 | 103 | 35 |
| 47 | -43 | -139 | -43 | 200 | 299 | 231 | 172 | 126 | 32 |
| 48 | -81 | -166 | -54 | 200 | 217 | 66 | 49 | 102 | 26 |
| 49 | -33 | -105 | -25 | 128 | 21 | -140 | -50 | 92 | 37 |
| 50 | 33 | -52 | -51 | -9 | -127 | -223 | -104 | 42 | 31 |
| 51 | 52 | -15 | -70 | -96 | -151 | -183 | -119 | -11 | 29 |
| 52 | 27 | 7 | -27 | -53 | -62 | -65 | -49 | -5 | 26 |

| i\j | 156 | 157 | 158 | 159 | 160 | 161 | 162 | 163 | 164 |
|---|---|---|---|---|---|---|---|---|---|
| 44 | -264 | -264 | -242 | -214 | -191 | -184 | -186 | -206 | -298 |
| 45 | 169 | 98 | 40 | 32 | 74 | 130 | 165 | 149 | 55 |
| 46 | 504 | 414 | 301 | 271 | 348 | 442 | 497 | 515 | 486 |
| 47 | 380 | 413 | 436 | 479 | 518 | 497 | 457 | 487 | 570 |
| 48 | -14 | 101 | 221 | 298 | 261 | 132 | 44 | 72 | 167 |
| 49 | -265 | -141 | -50 | -49 | -130 | -206 | -229 | -222 | -187 |
| 50 | -423 | -373 | -334 | -369 | -418 | -408 | -399 | -431 | -431 |
| 51 | -343 | -408 | -438 | -465 | -467 | -425 | -396 | -428 | -490 |
| 52 | -89 | -145 | -186 | -200 | -191 | -166 | -143 | -158 | -223 |

The fragments of MATLAB code:
**Orientation:**

```
for u = 1:w
  for v = 1:w
    Vx = Vx + px(u,v)*py(u,v);
    Vy = Vy + (py(u,v)^ 2-px(u,v)^ 2);
  end end
orient = -0.5*atan2(Vx,Vy)
```

**Local orientation O:**

```
Phix(i,j) = cos(2*orient(i,j));
Phiy(i,j) = sin(2*orient(i,j));
W = fspecial('gaussian', size, sigma);
for u = -size/2:size/2
  for v = -size/2:size/2
    Phixprime(i,j) = Phixprime(i,j)
      + W(u,v)*Phix(i-ux,j-vw);
    Phiyprime(i,j) = Phiyprime(i,j)
      + W(u,v)*Phiy(i-ux,j-vw);
  end end
O(i,j) = -0.5*atan2(Phixprime(i,j),Phiyprime(i,j))
```

**FIGURE 4.12**

Example of the calculation of orientation and local ridge orientation (Example 4.8).

The other characteristics derived from the orientation parameters are the *sin* component $\epsilon(i,j)$ of the local orientation, $O$, the difference of sums of these components, $D$, and the total difference, $C$:

$$\epsilon(i,j) = \sin(O(i,j)),$$
$$D(i,j) = \sum_{R_I} \varepsilon(i,j) - \sum_{R_{II}} \varepsilon(i,j),$$
$$C_i(i,j) = D(i,j-1) + D(i,j) + D(i,j+1).$$

**Example 4.9** *Figure 4.13 illustrates the sin components, the difference D and the total difference C.*

**Local corrections.** Due to the presence of noise, corrupted ridge and valley structures, minutiae, etc. in the input image, the estimated local ridge orientation, $\theta(i,j)$, may not always be correct. Since local ridge orientation varies slowly in a local neighborhood where no singular points appear, a low-pass filter (specifically, a two-dimensional Gaussian low pass filter) can be used to modify the incorrect local ridge orientation. Formal details are given in the example below.

**Example 4.10** *Figure 4.14 illustrates the low-pass filtering performed on the gray-scale image.*

### 4.6.2   Direct orientation transform

The *direct orientation* transform produces a representation of a fingerprint called an *orientation map* or *orientation field*. This transformation is aimed at simplifying feature extraction. This transform produces unrecoverable topological data, meaning some information is lost from the original image and cannot be recovered.

**Example 4.11** *Figure 4.15 illustrates deriving an orientation field for a given gray-scale fingerprint image.*

## 4.7   Polar transform of orientation map

Fingerprint topology is characterized by concentric ridges around central points. The distances between ridges and the central points can be easily measured and manipulated if a polar transform is first applied. In this section, the polar transform of an orientation map is discussed with the purpose of creating a representation where new maps can be generated easily. The first step is to locate a reference point and to determine the original orientation of the fingerprint image.

**Input data:** *pixel (86,124) from the marked fragment, and the calculated values of the sum gradients* $V_x(86, 124) = 6279531$ *and* $V_y(86, 124) = -135792$.

**Step 1.** *Calculate orientation* $\theta$:

$$\theta(86, 124) = -\frac{1}{2}\tan^{-1}\frac{V_x(86, 124)}{V_y(86, 124)} = -0.79621$$

**Step 2.** *Compute local ridge orientation* $O$

$$O(86, 124) = -\frac{1}{2}\tan^{-1}\frac{\cos(2\theta(86, 124))}{\sin(2\theta(86, 124))} = -0.011$$

**Step 3.** *Calculate* $\varepsilon$:
$\varepsilon(86, 124) = \sin(O(86, 124)) = -0.011$
**Step 4.** *Given the square region shown on the picture, calculate* $D(85, 123)$, $D(85, 124)$ *and* $D(85, 125)$. *For example:*
$D(85, 125) = \sum_{R_I}\varepsilon(i, j) - \sum_{R_{II}}\varepsilon(i,j) = 55.896$.
*Similarly,* $D(85, 123) = 48.425$, $D(85, 124) = 53.375$
**Step 5.** *Calculate* $C_i$:
$C_i(86, 124) = 48.425 + 53.375 + 55.896 = 155.175$

*The MATLAB code for calculation of* $\varepsilon, D$ *and* $C$ *is given below:*
**Epsilon:**
```
epsilon(i,j) = sin(O(i,j));
```

**Parameter D:**
```
b = 16;
for x = i-b:i
    for y = j:j+b
        if y < j + tan(orient(i,j))(x-i)
            R2(i,j) = R2(i,j) + ε(x,y);
        elseif y > j + tan(orient(i,j))(x-i)
            R1(i,j) = R1(i,j) + ε(x,y);
        end
    end
end
D(i,j) = R1(i,j) - R2(i,j)
```

**Parameter C:**
```
C(i,j) = D(i-1,j-1)+D(i-1,j)+D(i-1,j+1);
```

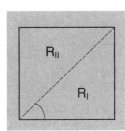

**FIGURE 4.13**
Example of the calculation of $\epsilon, D$ and $C$ (Example 4.9).

## 4.7.1 Extraction of reference and center point

The most commonly used reference point is the core point. A core point is defined as the point at which a maximum direction change is detected in the orientation field of a fingerprint image or the point at which the directional field becomes discontinuous.

An orientation field can be used to determine the reference point. For this, a

(a)

(b)

The low-pass filtering procedure is the following:

the orientation image is converted into a continuous vector field, which is defined as follows:

$$\Phi_x(i,j) = \cos 2\theta(i,j)$$
$$\Phi_y(i,j) = \sin 2\theta(i,j),$$

where $\Phi_x$ and $\Phi_y$ are the $x$ and $y$ components of the vector field, respectively. With the resulting vector field, the low-pass filtering can then be performed as follows:

$$\Phi_x'(i,j) = \sum_{u=-W_\Phi/2}^{W_\Phi/2} \sum_{v=-W_\Phi/2}^{W_\Phi/2} W(u,v)\Phi_x(i-uw,j-vw)$$

$$\Phi_y'(i,j) = \sum_{u=-W_\Phi/2}^{W_\Phi/2} \sum_{v=-W_\Phi/2}^{W_\Phi/2} W(u,v)\Phi_y(i-uw,j-vw),$$

where $W$ are components of a two-dimensional low-pass (Gaussian) filter. Note that the smoothing operation is performed at the block level.

The MATLAB code of the two-dimensional Gaussian low pass filter is given below:

```
filter = fspecial('gaussian', size, sigma);
filteredimage = imfilter(original, filter, 'conv');
```

**FIGURE 4.14**

Illustration of low pass filtering of the image (a); the resulting image is (b) (Example 4.10).

filter to detect the maximum direction change of the ridge flow can be applied to the orientation field.

> **Example 4.12** *Figure 4.16 illustrates how to determine the reference and center points from the orientation map.*

The *sin* component possesses an attractive characteristic in that it reflects the local ridge direction. A perfectly horizontal ridge has a *sin* component equal to 0. On the other hand, the ridge's *sin* component is equal to 1 if it is perfectly vertical. Due to the discontinuity property, the *sin* component value always changes abruptly in areas near a reference point.

To find the absolute center, several *barycenters* must be detected, each of them depends on the choice of the end points. In the fingerprint processing, the barycenter is defined as the center of each singular region, i.e., regions

**Input data:** *An enhanced gray-scale fingerprint image*

**Step 1.** *Divide the image into blocks of size $w \times w$; in this example, $w \times w = 8 \times 8$*

**Step 2.** *Compute the gradients $f_x(i, j)$ and $f_y(i, j)$ at each pixel $f(i, j)$*

**Step 3.** *Estimate the orientation $\theta(i, j)$ of each block centered at pixel $f(i, j)$*

**Step 4.** *Apply a low-pass filtering at the block level*

**Step 5.** *Compute the local ridge orientation $O(i, j)$ at each $(i, j)$*

**Output data:** *Orientation field*

**FIGURE 4.15**
Algorithm for calculating the orientation field (Example 4.11).

of rapid direction change. Note that in the algorithm described in Example 4.12, ridge end points are used, though the other minutiae points can be used as well. To do this, an average center$(x_{centre}, y_{centre})$ is constructed (Figure 4.17).

## 4.7.2 Polar transform

Cartesian coordinates are converted into polar coordinates by the equations:

$$r_i = \sqrt{(x_i - x_r)^2 + (y_i - y_r)^2}$$
$$\theta_i = \tan^{-1}(y_i - y_r, x_i - x_r) - orient_r$$

where $(x_i, y_i)$ are the cartesian coordinates of each end point in the orientation map; $(r_i, \theta_i)$ are the polar coordinates of each end point in the orientation map; $(x_r, y_r)$ are the Cartesian coordinates of the reference point; $orient_r$ is the reference point orientation. After extracting the reference point and applying

**Input data** *Orientation field*
**Step 1** *Compute the* sin *component* $\varepsilon(i,j)$ *of the orientation field:*

$$\varepsilon(i,j) = \sin(O(i,j))$$

**Step 2** *Initialize a 2D array* $Ci$ *and set all its entries to 0.*
**Step 3** *Scan the* sin *component map in an row-by-row, column-by-column manner. For each* sin *component* $\varepsilon(i,j)$, *if* $O(i,j) < O_{threshold}$ *and* $O(i-1,j) > \pi/2$ *and* $O(i+1,j) < \pi/2$ *then*

$(i)$ *Compute the difference* $D$.

$(ii)$ *Compute the* $C_i(i,j)$.

**Step 5** *Find the maximum value and assign its coordinate to the core, i.e., the reference point:* $ls_i$ *is the coordinates of the core (denoted by "+")* $ds_i$ *are the coordinates of the delta (denoted by "o")*

**Step 6** *Note: the orientation* $\theta$ *at each pixel*

$$\theta = \frac{1}{2}\left[\sum_{i=1}^{n_d} g_{ds_i}(arg(z-ds_i)) - \sum_{i=1}^{n_c} g_{ls_i}(arg(z-ls_i))\right]$$

*where*

$$g_k(\alpha) = \bar{g}_k(\alpha_i) + \frac{\alpha - \alpha_i}{2\pi/L}(\bar{g}_k(\alpha_{i+1}) - \bar{g}_k(\alpha_i))$$

*for* $\alpha_i < \alpha < \alpha_{i+1}$, $\alpha_i = -\pi + \frac{2\pi i}{L}$, $L = 8$

**FIGURE 4.16**
Algorithm for calculating the reference points (Example 4.12).

the polar transform, we can get the polar parameters $r_s$ and $\theta_s$ of each end point in the orientation map as shown in Figure 4.18.

### 4.7.3   Creation of a new orientation map

In a polar coordinate representation, the end points are rotationally and transitionally invariant with respect to their reference point.

> **Example 4.13** *An example of the simplest deformation is given in Figure 4.19.*

In Example 4.13, the deformation of the angle for $\theta_i$, the increment randomly selected to be positive or negative in different parts. The objects in

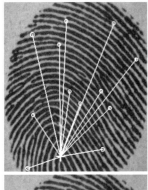

Calculation of the average center $(x_{centre}, y_{centre})$:

**Step 1** *Randomly select M points, each point is the center point of a block in the orientation map (here M = 12)*

**Step 2** *Calculate the distance between M points and the reference point so that we can get one barycenter*

**Step 3** *Repeat step 1 and 2 N times and get N barycenters. Then we can calculate the center point by:*

$$x_{centre} = \sum_{i=0}^{N} x_i / N,$$

$$y_{centre} = \sum_{i=0}^{N} y_i / N$$

*where $(x_i, y_i)$ is the Cartesian coordinates of barycenters*

**FIGURE 4.17**
Deriving of the distances between $M$ points and the reference points and the barycenter.

the orientation can rotate clockwise or counterclockwise. Point $A$ can rotate to $A'$ if the increment is positive or to $A''$ if the increment is negative. Here, $\theta_1'' < \theta_1 < \theta_1'$. This rule ensures that we can get a new orientation map randomly.

> **Example 4.14** *The distance $r_i$ of each end point in polar coordinates as well as the difference between them, where the x-axis presents the number of points, can be changed slightly, and the new orientation map can be derived. Both polar coordinate parameters are distances and angles of each end point, where x-axis is angle from $0°$ to $360°$ and y-axis is the distance (Figure 4.20).*

The above example confirms that it is possible to manipulate the polar parameters in a reasonable range and obtain a new orientation field. For example, for the *orientation* $< 0$, we shorten the distance $r_1$ to $r_1'$ and lengthen $r_2$ to $r_2'$ as well as change the angles from $\theta_1$ to $\theta_1'$, $\theta_2$ to $\theta_2'$. The new end points form the lines in the new orientation field.

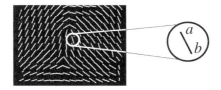

**FIGURE 4.18**

End points in the orientation map.

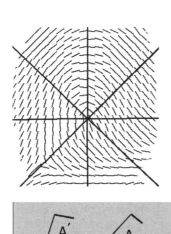

Given: *The original orientation map and the reference point.*

Step 1 *Divide the map into 8 parts in polar coordinates.*

Step 2 *Apply deformations to the end points by changing polar parameters $(r_i, \theta_i)$; note: these parameters vary depending on the chosen part. After rotation, the distance $r_i$ changes so that the direction of the original orientation field is preserved. For each point in the original orientation map:*

*if orientation $< 0$*

▶ *Decrease the distance $r_1$ to $r'_1$.*
▶ *Increase the distance $r_2$ to $r'_2$.*

*else if orientation $> 0$*

▶ *Decrease the distance $r_1$ to $r'_1$.*
▶ *Increase the distance $r_2$ to $r'_2$.*

*This rule ensures that we can get an acceptable orientation map similar to a real one. Here, $|\theta'_i - \theta_i| = \frac{1}{5}$ radian and $|r'_i - r_i| = 8$.*

Step 3 *Coordinates of all end points are changed. New lines are drawn through every two new end points.*

**FIGURE 4.19**

Creation of a new orientation map (Example 4.13).

## 4.8    Generating synthetic fingerprint images from an orientation map

In this section, we overview an approach to generation of synthetic fingerprint images based on enhancing an initial image by filtering adjusted with respect

Conversion of Cartesian coordinates into polar coordinates is implemented using the following parameters:

$$r_i = \sqrt{(x_i - x_r)^2 + (y_i - y_r)^2}$$
$$\theta_i = tan^{-1}(y_i - y_r, x_i - x_r) - orient_r$$

where

Original orientation map

$(x_i, y_i)$ are the Cartesian coordinates of each end point,

$(r_i, \theta_i)$ are the polar coordinates of each end point,

$(x_r, y_r)$ are the Cartesian coordinates of the reference point,

$orient_r$ is the reference point orientation.

The algorithm for designing a new orientation map is as follows:

Polar coordinates

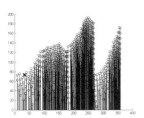

Input data: the initial orientation map

Step 1. Divide the orientation map into 8 parts in the polar coordinates. The origin point of the polar coordinate is the reference point

Step 2. Apply the deformation to end points by changing polar parameters $(r_i, \theta_i)$. For $\theta_i$, randomly select increment to be positive or negative in different parts. The distance $r_i$ is deformed.

New polar coordinates

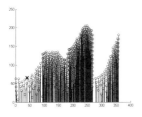

Output data: a new orientation map

The MATLAB code to generate new orientation map based on polar transform is as follows:
Given $(0 < \theta < \pi/4)$ (the first quadrance):

```
if degree(i, 1) >= 0 && degree(i, 1) < pi/4
    incre = betarnd(1, 1, 1, 1);
    Newdist(i, 1) = dist + incre * 20;
    Newdegree(i, 1) = degree(i, 1) + incre;
end
```

Synthetic orientation map

Given $(3\pi/4 < \theta < \pi)$ (the fourth quadrance):

```
if degree(i, 1) >= 3 * pi/4 && degree(i, 1) < pi
    incre = betarnd(1, 1, 1, 1);
    Newdist(i, 1) = dist - incre * 20;
    if Newdist(i, 1) < 0
        Newdist(i, 1) = 0;
    end
    Newdegree(i, 1) = degree(i, 1) - incre;
end
```

**FIGURE 4.20**

Generation of a synthetic orientation map (Example 4.14).

to additional information such as, in particular, orientation map. The problem is formulated as follows: given an orientation map, generate binary ridge pattern, i.e., recreate a master fingerprint.

**Application of Gabor filter for image enhancement.** It has been shown in studies (see "Further Reading" Section) that a realistic fingerprint ridge pattern can be derived by expanding an initial image given by the "seeds" or points, and additional data such as the density map and the orientation map. This is accomplished through a 2D Gabor filter which parameters are "controlled" by local ridge orientation and frequency. A complex 2D Gabor filter in the image domain $(x, y)$ consists of real and imaginary parts,

$$G(x, y) = \exp\left\{-0.5\left(\frac{x'}{\sigma_x}\right)^2 - 0.5\left(\frac{y'}{\sigma_y}\right)^2\right\} \exp\left\{-2\pi i(u_0 x' + v_0 y')\right\},$$

where $(x', y') = (x\cos\theta + y\sin\theta, -x\sin\theta + y\cos\theta)$ specify the rotated coordinates, $(u_0, v_0)$ specify modulation that has spatial frequency $\sqrt{u_0^2 + v_0^2}$ and orientation $\arctan(v_0/u_0)$, and $\sigma_x$ and $\sigma_y$ are standard deviations (variances) along the axes $X$ and $Y$ respectively.

The real component of the filter's real part has been adopted as the most suitable for the considered task of fingerprint image processing and is the following:

$$G(x, y, \theta, f) = \exp\left\{-0.5\left(\frac{x'}{\sigma_x}\right)^2 - 0.5\left(\frac{y'}{\sigma_y}\right)^2\right\} \cos\left\{2\pi f x'\right\},$$

where $\theta$ is the orientation at pixel $(x, y)$, and $f$ is a frequency. In the simplest case considered below to illustrate the Gabor filter application, a fixed ridge thickness is used instead of frequency, $f$, for the sake of simplification.

**Algorithm for master fingerprint generation.** The orientation map of a fingerprint is the initial data that specify the parameters required to adjust the 2D Gabor filter. The main steps are as follows:

▶ Create a fragment of the image (at the center of the image) of the black pixels with white pixel at the center
▶ Apply Gabor filter adjusted to the orientation of the given orientation map
▶ Apply a threshold operator to binarize the image
▶ Repeat filtering and thresholding for a newly appointed pixel.

Illustration to the algorithm is given in Figure 4.21. It should be noted that:

▶ In the presented algorithm, only one central point has been chosen to start. The advanced approached uses many randomly generated points, or seeds,

Phase 1

Phase 2

Phase 3

Phase 4

Phase 5

The algorithm for designing a new orientation map is as follows:

**Input data:** The $256 \times 256$ image of black pixels (except for a white pixel at the center) and the $256 \times 256$ orientation map

**Step 1.** Consider $15 \times 15$ image block at the center of the $256 \times 256$ image of black pixels except for the central white one

**Step 2.** Given $f = 10$ apply $15 \times 15$ Gabor filter

**Step 3.** Apply threshold (for this example, threshold $=200$) and convert the image to a binary one

**Step 4.** Choose another central point of filtering and repeat steps 1 to 3. Note: the chosen sequence of points is the following: the the point on the left from the initial white pixel, the point on the right, upper point and lower point

The shown pictures demonstrate five phases of the generation of a synthetic fingerprint, each takes 50 iterations of the repeated steps 2 to 4

**Output data:** a binary synthetic fingerprint image

**FIGURE 4.21**

Synthetic fingerprint image generation using the orientation map and the Gabor filter.

▶ The frequency $f$ can be varied to produce more realistic results,

▶ The square shape of the resulting image has been chosen to illustrate the techniques of application of the 2D Gabor filter. In the advanced implementations described below and in the "Further Reading" Section, the oval-like shape of the fingertip is used.

The application of the above technique to generation of the macroprimitives of the fingerprints is illustrated in Figure 4.22.

ORIENTATION   SYNTHETIC   ORIENTATION   SYNTHETIC
     MAP             IMAGE           MAP             IMAGE

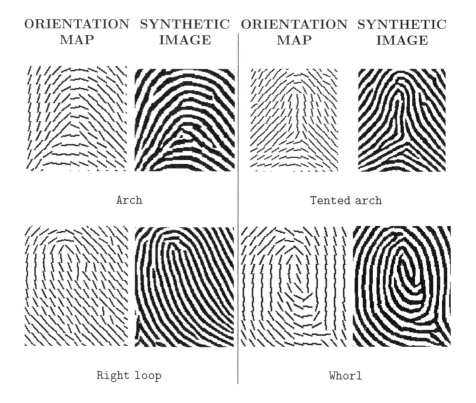

                Arch                                    Tented arch

         Right loop                                      Whorl

**FIGURE 4.22**
Examples of other synthetic fingerprint classes.

## 4.9   Other global topological models

In this section, automated fingerprint topology design methods using previously introduced techniques are presented. It is important that characteristics of the generated topological structures must be similar to the original fingerprints. An example of commercially available software is SFinGe, the Synthetic Fingerprint Generator by Cappelli (see details of a SFinGe in the "Further Reading" Section). In this section, the following topological models are considered:

▶ The shape model,

▶ The orientation map model, and

▶ The density map model.

These models create a binary representation of so-called *master fingerprint*, which has topological characteristics similar to an original fingerprint. The master fingerprint is comprised of:

```
< Master fingerprint >  ⇒  < Synthetic shape model >
                        ⇒  < Orientation map >
                        ⇒  < Frequency model >
                        ⇒  < Ridge pattern model >
```

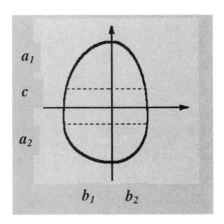

The fingerprint image is divided into 6 parts:

▶ Parameter $a_1$ describes the height of the top two parts of the fingerprint image,
▶ Parameter $c$ describes the middle two parts of the height and $a_2$ is the bottom,
▶ Parameters $b_1$ and $b_2$ describes the width of the left and the right side fingerprint shape respectively.

**FIGURE 4.23**
The fingerprint shape model (Example 4.15).

To create a synthetic orientation field, the *core* and *delta* points must be specified or randomly selected. It must be combined with the shape model.

> **Example 4.15** *Figure 4.23 illustrates the standard shape model that is characterized by its geometrical parameters.*

Another component in the process of designing a synthetic fingerprint is generating a density map model that carries information about the local distances between ridges.

> **Example 4.16** *Figure 4.24 illustrates derivation of a density map. In the task of synthesis, the density map may be randomly created.*

## 4.9.1 Master fingerprint

The previous sections have described how to generate a master fingerprint. The master fingerprint is a binary pattern that describes topology, but does not yet possess important characteristics of real images such as noise, sensor

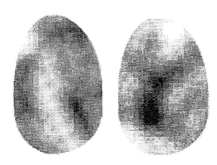

A density map:

Is the inverse of the number of ridges per unit length measured in the direction which is orthogonal to the ridge orientation,

Is calculated in the discrete areas and arranged into an array,

Uses smoothing by a $4 \times 4$ averaging box.

**FIGURE 4.24**
The density map model (Example 4.16).

parameters etc. The master fingerprint image is an acquisition *technology independent* topological structure, it can be designed by any topological method (assembling over the library of primitives, cancellable transforms, "growing" from several patterns, etc.) without the limitation of the physical effects of material, skin, and sensors.

### 4.9.2 Simulation of noise and other conditions

**Pressure and skin condition simulation.** To create a realistic image, additional information must be added to a master fingerprint image: noise, skin conditions, material effects, pressure effects, sensor effects, etc.

$< \text{Master fingerprint} > \Rightarrow \quad < \text{Dry/wet skin simulation} >$

$\Rightarrow \quad < \text{High/low pressure simulation} >$

$\Rightarrow \quad < \text{Placement over the sensing device} >$

$\Rightarrow \quad < \text{Noise simulation} >$

$\Rightarrow \quad < \text{Placement rotation} >$

To implement these conditions, models of these effects and techniques for incorporating them into a master fingerprint must be developed. In particular, the following operators are used in these techniques:

▶ Erosion operator to simulate low pressure on dry skin,

▶ Dilation operator to simulate high pressure on dry skin,

▶ Rendering operator,

▶ Translation operator to model shift of the image due to placement over the sensing device, and

▶ Rotation operator to simulate placement angle.

> **Example 4.17** *Figure 4.25 shows an effect of using erosion and dilation operators to simulate pressure and skin conditions.*

*Low pressure or dry skin simulation*

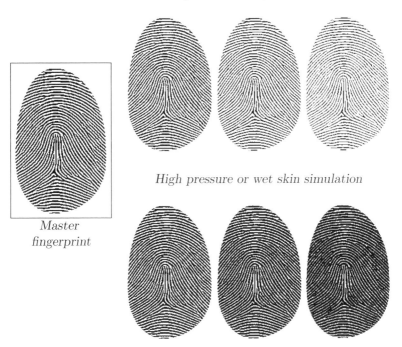

*High pressure or wet skin simulation*

Master
fingerprint

**FIGURE 4.25**
Applying different levels of erosion and dilation to the same master fingerprint
(Example 4.17).

**Placements over the sensing element simulation.** To distinguish
different impressions of the same master fingerprint, one important step is
to apply non-linear distortion to describe skin deformations due to a finger
being rolled over a sensing element.

> **Example 4.18** *Figure 4.26 shows the effect of using
> distortions to simulate possible placements of the finger against
> the sensor.*

**Distortion caused by noise.** Real fingerprint images are gray-scale with
noise due to the sensing element and its condition, the presence of small
pores within the ridges, the presence of scratches, etc. Thus we also need to
introduce noise to the synthetic fingerprint images.

> **Example 4.19** *Figure 4.27 shows the effect of adding salt-
> and-pepper noise to the master fingerprint to simulate a
> realistic, noisy image.*

| Master fingerprint | Deformation of fragment 1 | Deformation of fragment 2 |

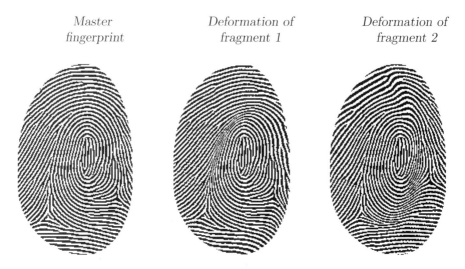

**FIGURE 4.26**

Fingerprint skin deformation by non-linear distortion for different parts of a fingerprint (Example 4.18).

**Position on the sensing surface and background.** Fingerprints are not always perfectly centered in the image due to varying position on the sensing surface.

> **Example 4.20** *Figure 4.28 shows the results of dark and medium background simulation using a Karuhnen-Loeve transform.*

A different approach is to model a fingerprint based on a dermatoglyphics approach. Dermatoglyphics is the study of epidermal ridges on fingers, palms, and soles. In this approach, all the phases of fingerprint formation starting at four months pregnancy are modelled, as well as the properties of skin surface. This is discussed in the "Further Reading" Section.

## 4.10   Summary

A synthetic fingerprint is an image with an acceptable fingerprint-like topology. This image is the result of creating a master-fingerprint with realistic topology and adding an appropriate texture.

1. There are various practical applications for synthetic fingerprints, in particular:

Additive noise          Low pass smoothing

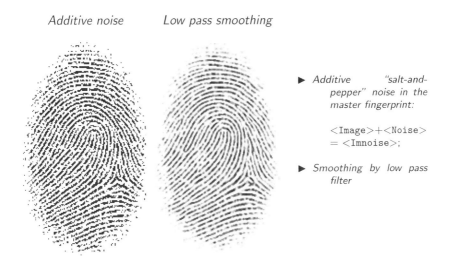

▶ Additive "salt-and-pepper" noise in the master fingerprint:

$$<\text{Image}>+<\text{Noise}> = <\text{Imnoise}>;$$

▶ Smoothing by low pass filter

**FIGURE 4.27**
Adding noise to a fingerprint image (Example 4.19).

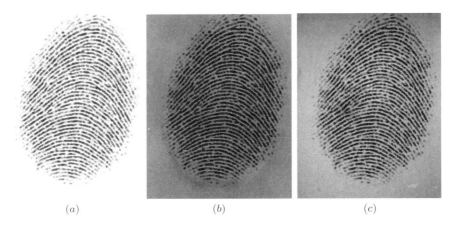

(a)                    (b)                    (c)

**FIGURE 4.28**
Fingerprint background simulation: a fingerprint image with a white background (a), a gray-scale like background called a capacitive background simulated using the Karuhnen-Loeve transform (b), and an optical background (c) (Example 4.20).

> ***Testing fingerprint systems.*** In this application, synthetic fingerprints can extend an original database. This can improve testing a system for fingerprint verification or identification by:

(*a*) Creating a large test databases, and

(*b*) Varying the quality of the print to simulate various effects in recording systems.

**Security of a database.** The idea is to store a synthetic equivalent of the original fingerprint in a database. In this application, original fingerprints are transformed to synthetic fingerprints before storing them in a database. The inverse transform is then applied to recover an original fingerprint from a synthetic fingerprint, i.e., the following direct and inverse transforms are used:

< Original fingerprint > ⇔ < Synthetic fingerprint >.

This is the key idea of so-called cancellable biometrics.

**Training.** Synthetic fingerprints with controlled parameters are used for training personnel that deal with fingerprints.

**Watermarking.** Changing parameters of a synthetic fingerprint allows adjusting the parameters of a protected item. In such watermarking, a synthetic image can be used instead of a real one.

**Medicine.** Fingerprints are useful in express-diagnostics. For example, some genetic defects can be recognized by typical changes in fingerprint patterns (Downs and Turner syndrome, in particular). Also fingerprints carry useful information for ethnic study.

2. Normally unwanted effects of signal transforms, like them being non-cancellable, can be used in synthetic fingerprint design. These controlled effects can be used as:

▶ An extension of flexibility in design, or

▶ Uncertainty that can be replaced by useful (synthetic) information.

3. There are three approaches to computer-aided design of synthetic fingerprints:

▶ *Physical*, based on the modeling of physical process of fingerprint development through the first months of life,

▶ *Topological* modeling based on analysis of topological structure, and

▶ *Topological-physical* modeling that combines both techniques.

4. The *master fingerprint* is the basic intermediate fingerprint-like topology in most techniques. A master fingerprint can be designed using the library of topological primitives

< Primitives > ⇒   < Master fingerprint >
                    ⇒ < Synthetic fingerprint > .

By adding appropriate texture and other user-defined attributes, a master fingerprint can be converted into a realistic (acceptable) synthetic fingerprint image.

## 4.11  Problems

The sample files to be used in the problems can be found in the text.  Small databases of fingerprints can be downloaded from http://bias.csr.unibo.it/fvc2004/download.asp.  Problem 4.6 can be considered as a team project.

**Problem 4.1**  Given the gray-scale fragment of a fingerprint image in Figure 4.29.
(a) Calculate the orientation $\theta$ and the local ridge orientation $O$ at the pixel $f(i,j)$: (2,19); (4,2); (21,8); (18,16).
(b) Given the pixel (8,4) and $w = 6$, calculate: $V_x(50, 155)$; $V_y(50, 155)$; (c)
$\theta(50, 155)$; $O(50, 155)$.
(c) Given the pixel $f(18, 16)$, calculate: $\varepsilon(i, j)$; difference $D$; $C_i(i, j)$.
(d) Given the pixel $f(8, 4)$, implement low-pass filtering at the given pixel for density $f = 8$ and $f = 6$.

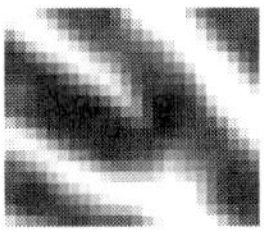

A fragment of a gray-scale fingerprint image

| | 1 | 2 | 3 | 4 | 5 | 6 | 7 | 8 | 9 | 10 | 11 | 12 | 13 | 14 | 15 | 16 | 17 | 18 | 19 | 20 | 21 | 22 | 23 | 24 |
|---|---|---|---|---|---|---|---|---|---|---|---|---|---|---|---|---|---|---|---|---|---|---|---|---|
| 1 | 250 | 251 | 239 | 205 | 152 | 99 | 79 | 85 | 75 | 72 | 123 | 198 | 245 | 254 | 254 | 254 | 242 | 198 | 133 | 88 | 84 | 88 | 85 | 97 |
| 2 | 232 | 247 | 252 | 243 | 206 | 143 | 98 | 88 | 82 | 78 | 101 | 153 | 206 | 241 | 254 | 254 | 238 | 184 | 120 | 90 | 84 | 86 | 87 | |
| 3 | 165 | 213 | 245 | 254 | 245 | 211 | 160 | 117 | 94 | 92 | 108 | 130 | 161 | 198 | 230 | 251 | 254 | 254 | 230 | 172 | 116 | 89 | 79 | 78 |
| 4 | 91 | 140 | 195 | 231 | 244 | 240 | 220 | 182 | 136 | 116 | 119 | 118 | 124 | 144 | 177 | 218 | 249 | 254 | 250 | 221 | 171 | 122 | 88 | 73 |
| 5 | 54 | 73 | 111 | 161 | 205 | 232 | 243 | 230 | 194 | 163 | 138 | 114 | 105 | 108 | 130 | 173 | 221 | 252 | 254 | 254 | 234 | 185 | 121 | 77 |
| 6 | 59 | 58 | 60 | 83 | 129 | 178 | 219 | 238 | 234 | 215 | 170 | 119 | 94 | 87 | 98 | 124 | 164 | 215 | 253 | 254 | 254 | 229 | 167 | 104 |
| 7 | 82 | 82 | 61 | 48 | 60 | 97 | 147 | 193 | 224 | 238 | 217 | 159 | 101 | 81 | 81 | 85 | 105 | 163 | 231 | 254 | 254 | 254 | 219 | 160 |
| 8 | 104 | 109 | 84 | 47 | 37 | 54 | 74 | 106 | 152 | 202 | 226 | 191 | 122 | 77 | 66 | 62 | 71 | 115 | 188 | 247 | 254 | 254 | 245 | 216 |
| 9 | 101 | 102 | 83 | 51 | 37 | 43 | 45 | 46 | 69 | 118 | 177 | 199 | 154 | 78 | 39 | 43 | 55 | 77 | 138 | 216 | 254 | 254 | 254 | 251 |
| 10 | 127 | 99 | 76 | 57 | 49 | 47 | 46 | 41 | 41 | 56 | 98 | 154 | 160 | 92 | 31 | 34 | 56 | 72 | 109 | 171 | 227 | 254 | 254 | 254 |
| 11 | 198 | 161 | 120 | 85 | 61 | 47 | 46 | 51 | 51 | 43 | 47 | 84 | 116 | 89 | 40 | 40 | 64 | 76 | 88 | 118 | 169 | 223 | 253 | 254 |
| 12 | 241 | 226 | 200 | 156 | 103 | 65 | 53 | 60 | 62 | 48 | 37 | 52 | 69 | 64 | 56 | 66 | 85 | 94 | 88 | 85 | 115 | 175 | 226 | 252 |
| 13 | 229 | 239 | 239 | 223 | 181 | 129 | 90 | 74 | 64 | 52 | 48 | 61 | 72 | 72 | 81 | 103 | 110 | 104 | 96 | 86 | 93 | 132 | 176 | 218 |
| 14 | 168 | 194 | 220 | 238 | 233 | 208 | 171 | 134 | 98 | 70 | 68 | 83 | 92 | 98 | 127 | 162 | 152 | 120 | 109 | 105 | 94 | 100 | 123 | 166 |
| 15 | 110 | 134 | 164 | 202 | 233 | 244 | 239 | 221 | 184 | 144 | 128 | 132 | 138 | 145 | 178 | 212 | 192 | 156 | 143 | 124 | 90 | 76 | 90 | 119 |
| 16 | 95 | 95 | 100 | 135 | 189 | 226 | 244 | 248 | 240 | 225 | 211 | 205 | 205 | 208 | 226 | 239 | 222 | 202 | 178 | 130 | 84 | 68 | 75 | 86 |
| 17 | 88 | 72 | 61 | 78 | 124 | 173 | 208 | 232 | 247 | 252 | 250 | 249 | 250 | 251 | 254 | 254 | 254 | 243 | 202 | 137 | 84 | 64 | 74 | 77 |
| 18 | 94 | 65 | 53 | 62 | 83 | 110 | 142 | 176 | 209 | 232 | 240 | 244 | 249 | 254 | 254 | 254 | 254 | 252 | 216 | 150 | 92 | 68 | 76 | 83 |
| 19 | 139 | 92 | 71 | 71 | 74 | 77 | 91 | 118 | 152 | 175 | 187 | 195 | 214 | 244 | 254 | 254 | 254 | 254 | 237 | 191 | 140 | 96 | 80 | 87 |
| 20 | 206 | 167 | 142 | 125 | 104 | 82 | 73 | 80 | 98 | 108 | 107 | 113 | 144 | 195 | 231 | 249 | 254 | 254 | 253 | 238 | 200 | 144 | 110 | 105 |

**FIGURE 4.29**
Values of the pixels for the Problem 5.1.

**Problem 4.2**  Consider a gray-scale fragment of a fingerprint with a ridge meeting a bifurcation with a fork.  Calculate the parameters $r$ and $\theta$ of the

polar transform, given the reference point $R$ and two end points $A$ and $B$:
(a) R(188,186), A(156,163), B(167,151);
(b) R(18,6), A(27,16), B(26,15).

**Problem 4.3** Apply the polar transformation to the point $R$ ((a) and (b) of Problem 4.2). Use the result of Problem 4.2 and choose reasonable $\Delta r$ and $\Delta \theta$.

**Problem 4.4** Consider a gray-scale fragment and the corresponding values of the pixels as shown in Figure 4.29. Derive the orientation map by using MATLAB.

**Problem 4.5** The simplest fingerprint primitives given in Figure 4.30a have been created using the MATLAB function `polar`.
(a) Generate these primitives in MATLAB using the parameter $R$ shown in this Figure
(b) Apply the distance transform,
(c) Generate the gray-scale fingerprint fragments by deriving Voronoi diagram of the given primitives.

**Problem 4.6** Table 4.2 contains the samples of fingerprint primitives, their orientation maps and synthetic orientation maps (generated using the approach described in Section 4.7.3). Generate a reasonable new orientation map from the orientation map obtained in Example 4.4. After you create the new orientation map, generate a new fingerprint using the 2D Gabor filter as described in Section 4.8.

**Problem 4.7** Figure 4.31 contains the results of application of 2D Gabor filter to the the images shown in Figure 4.31. The MATLAB function `problem410(strImageName)`, where `strImageName` is the input image. Use your input images and apply Gabor filter with the following parameters:
(a) $S_x = 2, S_y = 4, f = 7.5, \theta = \pi/5$
(b) $S_x = 1, S_y = 2, f = 4, \theta = \pi/4$
(c) $S_x = 1, S_y = 2, f = 4, \theta = \pi/3$.

The MATLAB codes to apply 2D Gabor filter to an input image with sample parameters is given below:

```
function problem47(strImageName, Sx, Sy, f, theta)

I = imread(strImageName);
[G,gaborOut]=gaborFilter(I,2,4,16,pi/3);
imshow(I),figure,imshow(uint8(gabout));
```

The MATLAB codes for the 2D Gabor filter is shown here:

```
function [G,gaborOut] = gaborFilter(I,Sx,Sy,f,theta); if
isa(I,'double')~=1
    I = double(I);
```

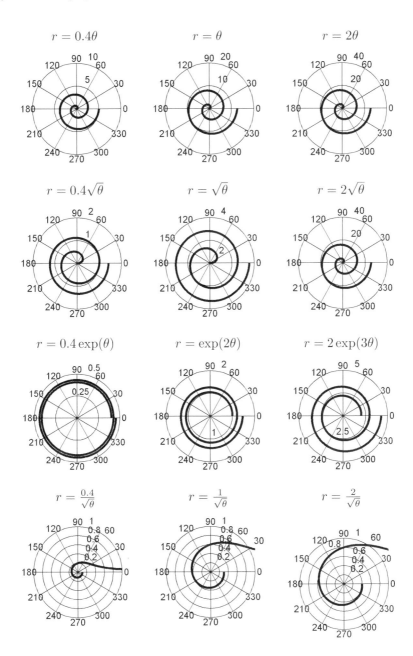

**FIGURE 4.30**

The simplest fingerprint primitives for Problem 4.5.

**TABLE 4.2**

The samples of fingerprint primitives for Problem 4.5.

| Primitive | Orientation map | Synthetic orientation map | Comment |
|---|---|---|---|
| | | | Primitive: ridge runs out of sight without any event occurring |
| | | | Primitive: ridge meets a bifurcation with a fork appearing from right |
| | | | Primitive: ridge ends along the ridges, its orientation map |
| | | | Primitive: ridge ends on the left, its orientation map |
| | | | Primitive: ridge bifurcates |
| | | | Primitive: ridge returns to starting-point without any event occurring |

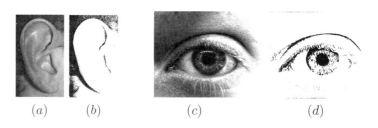

(a)        (b)              (c)                    (d)

**FIGURE 4.31**

Processing images by Gabor filter (Problem 4.7).

```
end
for x = -fix(Sx):fix(Sx)
    for y = -fix(Sy):fix(Sy)
        xPrime = x * cos(theta) + y * sin(theta);
        yPrime = y * cos(theta) - x * sin(theta);
```

```
        G(fix(Sx)+x+1,fix(Sy)+y+1)=exp(.5*((xPrime/Sx)^2+(yPrime/Sy)^2))*...
cos(2*pi*f*xPrime);
    end
end

Imgabout = conv2(I,double(imag(G)),'same'); G3 = real(G); Regabout =
conv2(I,double(real(G)),'same'); gabout = sqrt(Imgabout.*Imgabout +
Regabout.*Regabout);
```

The function `gaborFilter` has the following input parameters: $I$ is an input image, `Sx` and `Sy` are variances along $x$ and $y$-axes, $f$ is the frequency of the sinusoidal function, `theta` is the orientation of Gabor filter, and `gaborOut` is the output (filtered) image.

## 4.12  Further reading

**Basics of fingerprint techniques.** Fundamentals that are useful in the development of fingerprint algorithms can be found, in particular, in books by Rao [36] and Jolliffe [21]. In O'Gorman's papers [33, 34], an overview of the fingerprint verification technique can be found. Advances in fingerprint technology are introduced in the book by Lee and Gaenslen [25]. The handbook by Maltoni et.al. [29] is an excellent reference to state-of-the-art algorithms used for fingerprint recognition.

**Fingerprint image enhancement** processing is an essential step used in the calculating an orientation map from fingerprints. Typically, enhancement is a two phase process that removes unwanted information from a fingerprint image [1, 13, 14]. Enhancement processing must be able to:

▶ Resolve fine scale structures in clear areas,

▶ Reduce the risk of enhancing noise in blurred or fragmented areas,

▶ Produce a reliable and adaptively detailed estimate of the ridge orientation field and ridge width.

A method based upon nonstationary directional frequency domain filtering has been proposed in [39]:

(*a*) In the first phase, the noise is removed by directional filtering (orientation is everywhere matched to the local ridge orientation). The result is a directionally smoothed fingerprint image.

(*b*) In the second phase, an enhanced fingerprint image is produced by thresholding.

Another two phase strategy has been proposed by Almansa and Lindeberg [2]: a shape-adapted smoothing based on second moment descriptors is applied in the first phase (interrupted ridges to be joined without destroying essential singularities) and, in the second phase automatic scale selection based on normalized derivatives is used (estimates local ridge width and adapts the amount of smoothing to the local amount of noise).

Fingerprints are often represented as an orientation map during fingerprint analysis and synthesis [40]. Derakhshani et al. [10] studied the problem of vitality in fingerprint techniques. The problem of classifying fingerprints was the focus of papers [12, 18]. Twins fingerprints are discussed by Jain et al. [20]. Artificial intelligence paradigms are often used in fingerprint technology. For example, Luo and Tian [27] discussed knowledge-based fingerprint image enhancement.

Various techniques are proposed in [8, 11, 16, 17, 19, 32, 37, 38] for fingerprint processing.

**Non-automated fingerprint design technique.** Artificial fingers or fake fingers can be used to attacking fingerprint-based authentication systems. Matsumoto et al. [31] experiments on "gummy" fingers that can be easily made. In experiments, these synthetic fingers were accepted by fingerprint identification systems with extremely high rates. Putte and Keuning [35] studied two strategies in synthetic fingerprint design: duplication of a fingerprint with co-operation (with its owner) and without co-operation (a print of the finger can be obtained, for example, from a glass). The author discussed a distinction between dummy material that is not alive and the epidermis of a finger and scanner characteristics.

**State-of-the-art synthesis of fingerprints.** Cappelli [6] and Kuecken [23] reported two different approaches to synthetic fingerprint generation. The development of some particular techniques for fingerprint synthesis have been reported by Yanushkevich and Shmerko [43, 44]. Justification for using synthetic fingerprints in testing can be found in [30].

**Cappelli's generator of synthetic fingerprints.** The work by Cappelli et al. introduced in [3, 4, 5, 6, 7], can be recommended for the detailed study of fingerprint synthesis and modeling based on image processing methods. Their commercially available synthetic fingerprints generator called SFinGe, has been employed in the experimental work during the preparation of this book. In SFinGe, various models of fingerprints are used, in particular, shape, directional map, density map, and skin deformation models. SFinGe uses initial isolated points and enhance the image through Gabor filter controlled by the local ridge orientation and frequency of ridge appearance. In this way, a master fingerprint is created. To add realism to the image, erosion, dilation, rendering, translation, and rotation operators are used. Figure 4.32 illustrates several steps of Cappelli's generator of synthetic fingerprints (the pictures have been generated using the package SFinGe).

*Step 1*
*Initial image*

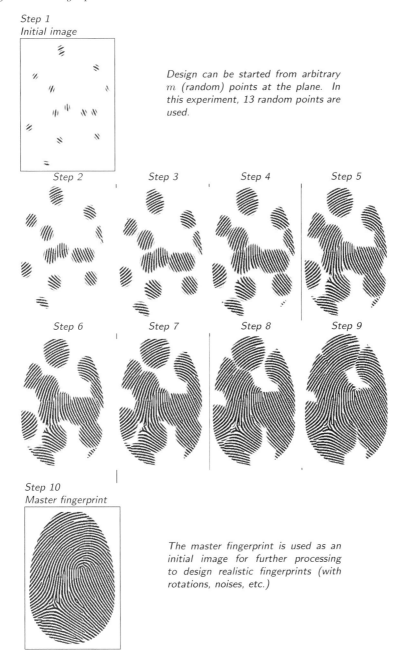

Design can be started from arbitrary
$m$ (random) points at the plane. In
this experiment, 13 random points are
used.

*Step 2*  *Step 3*  *Step 4*  *Step 5*

*Step 6*  *Step 7*  *Step 8*  *Step 9*

*Step 10*
*Master fingerprint*

The master fingerprint is used as an
initial image for further processing
to design realistic fingerprints (with
rotations, noises, etc.)

**FIGURE 4.32**

Cappelli's generator of synthetic fingerprints: "growing" a fingerprint.

**Kuecken's generator of synthetic fingerprints based on the embryological process.** The work by Kuecken [23, 24] can be recommended for a detailed study of natural fingerprint formation and modeling based on state-of-the-art dermatoglyphics, a discipline that studies epidermal ridges on fingerprints, palms, and soles. Many possible mechanisms for fingerprint pattern formation have been considered including cell proliferation phenomena, the mechanical interaction between the extracellular matrix and fibroblasts in the dermis, reaction-diffusion models and others. Kuecken's modeling approach started from an idea originating from Kollmann (1883) that was promoted by Bonnevie in the 1920s and 1930s. According to this hypothesis:

▶ Fingerprint patterns are formed due to differential growth of the basal layer of the epidermis.

▶ Differential growth of the basal layer of the epidermis induces compressive stress in the basal layer and leads to buckling.

This way primary ridges arise that define the future surface pattern. Kuecken analyzed this buckling process using the well-known von Karman equations. These equations describe the mechanical behavior of a thin curved sheet of elastic material:

*Linear* analysis gave the result that curvature effects are not important for the buckling process and the ridges form perpendicular to the greatest stress.

*Nonlinear* analysis of the corresponding amplitude equations shows that hexagon (dot) patterns are another possible solution of the equation for certain parameter regimes.

In fact, dot patterns are found on the palms of some marsupials. The analysis presented above implies that the direction of the ridges is determined by the stress field in the basal layer. Kuecken postulates that this stress field itself arises due to two factors:

(*a*) Resistance at the nail furrow and the major flexion creases produces stress on the basal layer.

(*b*) Stress arises due to the well-observed geometric change of the fingertip pad at the time of fingerprint formation.

Kuecken's computer simulations are able to produce most common fingerprint pattern types: whorl, loop and arch (Figure 4.33).[‡] Whorls are observed for highly rounded fingertip pads, loops for less prominent and slightly asymmetric pads and arches occur for flat pads. These relations have already been seen in numerous empiric studies and found, for the first time,

---

[‡]The help by M. Kuecken in reiteration of experiments is acknowledged.

a consistent explanation in Kuecken's model. The simulations produce an almost periodic pattern that has many features of real fingerprints. However, certain effects are clearly not biological and should be addressed in a more detailed study.

**Gabor transform.** The 2D Gabor filter was introduced by Daugman [8], and become quite popular in image processing and representation, in particular 2D Gabor wavelet theory has been developed in Lee [26].

The modified 2D Gabor filter called even-symmetric real component of the original 2D Gabor filter was adapted by Jain and Farrokhnia [15] for fingerprint image enhancement. Another modification of the method has been offered in Yang [42]. Interestingly, the same real component filter appeared to be useful in the inverse problem, synthesis of the fingerprint topology [4, 6].

**Voronoi diagram technique in fingerprint design** is considered in Chapter 7.

## 4.13  References

[1] Ailisto H and Lindholmy M. A review of fingerprint image enhancement methods. *International Journal of Image and Graphics* 3(3):401–424, 2003.

[2] Almansa A and Lindeberg T. Fingerprint enhancement by shape adaptation of scale-space operators with automatic scale selection. *IEEE Transactions on Image Processing*, 9(12):2027–2042, 2000.

[3] Cappelli R, Lumini A, Maio D, and Maltoni D. Fingerprint classification by directional image partitioning. *IEEE Transactions on Pattern Analysis and Machine Intelligence*, 21(5):402–421, 1999.

[4] Cappelli R, Erol A, Maio D, and Maltoni D. Synthetic fingerprint-image generation. In *Proceedings of the 15th International Conference on Pattern Recognition*, 3:475–478, Barcelona, Spain, 2000.

[5] Cappelli R, Maio D, and Maltoni D. Modelling plastic distortion in fingerprint images. In *Proceedings of the 2nd International Conference on Advances in Pattern Recognition*, pp. 369–376, Rio de Janeiro, Brazil, 2001.

[6] Cappelli R. Synthetic fingerprint generation. In Maltoni D, Maio D, Jain AK, and Prabhakar S, Eds., *Handbook of Fingerprint Recognition*, pp. 203–232, Springer, Heidelberg, 2003.

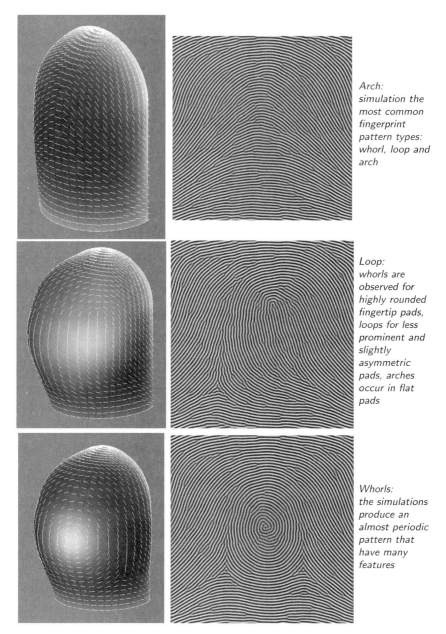

*Arch:
simulation the
most common
fingerprint
pattern types:
whorl, loop and
arch*

*Loop:
whorls are
observed for
highly rounded
fingertip pads,
loops for less
prominent and
slightly
asymmetric
pads, arches
occur in flat
pads*

*Whorls:
the simulations
produce an
almost periodic
pattern that
have many
features*

**FIGURE 4.33**
Kuecken's synthetic 3D (left) and 2D (right) fingerprint design based on
physical modeling (reprinted from [24] with permission from Elsevier).

[7] Cappelli R. SFinGe: synthetic fingerprint generator. In *Proceedings of the International Workshop on Modeling and Simulation in Biometric Technology*, Calgary, Canada, pp. 147–154, June 2004.

[8] Coetzee L, and Botha EC. Fingerprint recognition in low quality images. *Pattern Recognition*, 26(10):1441–1460, 1993.

[9] Daugman JG. Uncertainty relation for resolution in space, spatial frequency, and orientation optimized by two-dimensional visual cortical filters. *Journal of the Optical Society of America*, 2:1160–1169, 1985.

[10] Derakhshani R, Schuckers SAC, Hornak L, and O'Gorman L. Determination of vitality from a non-invasive biomedical measurement for use in fingerprint scanners. *Pattern Recognition*, 36(2):383–396, 2003.

[11] Erol A, Halici U, and Ongun G. Feature selective filtering for ridge extraction. In: Jain LC, Halici U, Hayashi I, Lee SB, and Tsutsui S, Eds., *Intelligent Biometric Techniques in Fingerprint and Face Recognition*, pp. 195–215, CRC Press, Boca Raton, FL, 1999.

[12] Fitz AP and Green RJ. Fingerprint classification using a hexagonal fast Fourier transform. *Pattern Recognition*, 29(10):1587–1597, 1996.

[13] Hong L, Wan Y, and Jain A. Fingerprint image enhancement: algorithms and performance evaluation. *IEEE Transactions on Pattern Analysis and Machine Intelligence*, 20(8):777–789, 1998.

[14] Hung DCD. Enhancement and feature purification of fingerprint images. *Pattern Recognition*, 26(11):1661–1671, 1993.

[15] Jain AK and Farokhnia F. Unsupervised texture segmentation using Gabor filters. *Pattern Recognition*, 24(12):1167–1186, 1991.

[16] Jain AK, Hong L, and Bolle R. On-line fingerprint verification. *IEEE Transactions on Pattern Analysis and Machine Intelligence*, 19(4):302–314, 1997.

[17] Jain AK, Hong L, Pankanti S, and Bolle R. An identity authentication system using fingerprints. *Proceedings of the IEEE*, 85(9):1365–1388, 1997.

[18] Jain AK, Prabhakar S, and Hong L. A multichannel approach to fingerprint classification. *IEEE Transactions on Pattern Analysis and Machine Intelligence*, 21(4):348–359, 1999.

[19] Jain AK, Prabhakar S, Hong L, and Pankanti S. Filterbank-based fingerprint matching. *IEEE Transactions on Image Processing*, 9(5):846–859, 2000.

[20] Jain AK, Prabhakar S, and Pankanti S. Can identical twins be discriminated based on fingerprints? *Pattern Recognition*, 35:2653–2663, 2002.

[21] Jolliffe IT. *Principle Component Analysis*. Springer, Heidelberg, 1986.

[22] Kass M and Witkin A. Analyzing oriented patterns. *Computer Vision, Graphics, and Image Processing*, 37(3)362–385, 1987.

[23] Kuecken MU. *On the Formation of Fingerprints*. PhD thesis (Advisor Prof. A. Newell), Faculty of the Graduate Interdisciplinary Program in Applied Mathematics, University of Arizona, 2004.

[24] Kuecken MU and Newell AC. A model for fingerprint formation. *Europhysics Letters* 68(1):141–146, 2004.

[25] Lee HC and Gaenslen RE. *Advances in Fingerprint Technology*. CRC Press, Boca Raton, FL, 2001.

[26] Lee T. Image representation using 2D Gabor wavelets. *IEEE Transactions on Pattern Analysis and Machine Intelligence*, 18(10):859–971, 1996.

[27] Luo X and Tian J. Knowledge-based fingerprint image enhancement. In *Proceedings of the 15th International Conference on Pattern Recognition*, 4:783–786, 2000.

[28] Maio D and Maltoni D. Direct gray-scale minutiae detection in fingerprints. *IEEE Transactions on Pattern Analysis and Machine Intelligence*, 19(1):27–40, 1997.

[29] Maltoni D, Maio D, Jain AK, and Prabhakar S, Eds. *Handbook of Fingerprint Recognition*, Springer, Heidelberg, 2003.

[30] Maio D, Maltoni D, Cappelli R, Wayman JL, and Jain AK. FVC2000: fingerprint verification competition. *IEEE Transactions on Pattern Analysis and Machine Intelligence*, 24(3):402–411, 2002.

[31] Matsumoto T, Matsumoto H, Yamada K, and Hoshino S. Impact of artificial "gummy" fingers on fingerprint systems. In *Proceedings of SPIE Optical Security and Counterfeit Deterrence Techniques*, vol.4677, 2002.

[32] Mehtre BM. Fingerprint image analysis for automatic identification. *Machine Vision and Application*, 6:124–139, 1993.

[33] O'Gorman L. An overview of fingerprint verification technologies. *Information Security Technical Report*, 3(1):21–32, 1998.

[34] O'Gorman L. Fingerprint verification. In Jain AK, Bolle R, and Pankanti S, Eds., *Biometrics: Personal Identification in Networked Society*, pp. 43–63, Kluwer, Dordrecht, 1999.

[35] Putte T and Keuning J. Biometrical fingerprint recognition: don't get your fingers burned. In *Proceedings of IFIP TC8/WG8.8 4th Working Conference on Smart Card Research and Advanced Applications*, pp. 289–303, Kluwer, Dordrecht, 2000.

[36] Rao AR. *A Taxonomy for Texture Description and Identification.* Springer, Heidelberg, 1990.

[37] Roddy A and Stosz J. Fingerprint feature processing and poroscopy. In Jain LC, Halici U, Hayashi I, Lee SB, and Tsutsui S, Eds., *Intelligent Biometric Techniques in Fingerprint and Face Recognition*, pp. 37–105, CRC Press, Boca Raton, FL, 1999.

[38] Sherlock BG and Monro DM. A model for interpreting fingerprint topology. *Pattern Recognition*, 26(7):1047–1055, 1993.

[39] Sherlock BG, Monro DM, and Millard K. Fingerprint enhancement by directional Fourier filtering. *IEE Proceedings, Image Signal Process*, 141(2):87–94, 1994.

[40] Vizcaya PR and Gerhardt LA. A nonlinear orientation model for global description of fingerprints. *Pattern Recognition*, 29(7):1221–1231, 1996.

[41] Wilson CL, Candela GT, and Watson CI. Neural network fingerprint classification. *Journal for Artificial Neural Networks*, 1(2):203–228, 1994.

[42] Yang J, Liu L, Jiang T, and Fan Y. A modified Gabor filter design method for fingerprint image enhancement. *Pattern Recognition Letters*, 24:1805–1817, 2003.

[43] Yanushkevich SN, Supervisor. Computer aided design of synthetic fingerprints. *Technical Report, Biometric Technologies: Modeling and Simulation Laboratory*, University of Calgary, Canada, 2005.

[44] Yanushkevich SN and Shmerko VP. Biometrics for electrical engineers: design and testing. Chapter 7: CAD of fingerprint analysis and synthesis. *Lecture notes*. Private communication. 2005.

# 5

# Synthetic Faces

The face is a complex and highly mobile object whose detailed topological configuration varies widely between individuals. This chapter discusses the behavioral information carried by facial expressions. It focuses on both the analysis and the synthesis of facial expressions (Figure 5.1). The term "face synthesis" indicates a class of problems aimed at synthesizing a model of a static face. The term "facial synthesis" is defined as a process of modeling facial appearance. Facial topology is a behavioral characteristic called facial expression, the visible result of synthetic emotion indicated in the face.

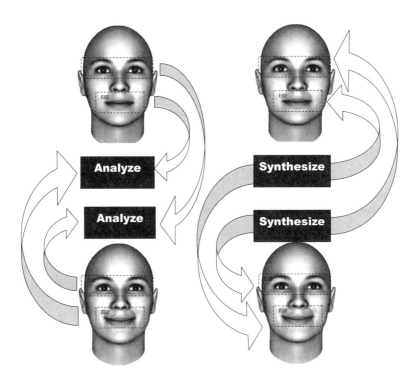

**FIGURE 5.1**
Facial analysis and synthesis.

Physically, the forming of a facial expression involves two steps:

$<$ Emotion\index{Emotion} (brain) $> \Rightarrow <$ Expression (face) $>$

In this chapter, the facial expression model used is described by the following scheme

$<$ Emotion $> \Rightarrow <$ Code $> \Rightarrow <$ Facial muscles $>$

Because of various random conditions, in particular, facial muscles and skin, the appearance of expressions is influenced by certain distortions:

$<$ Facial muscles $> \Rightarrow <$ Distortion $> \Rightarrow <$ Facial topology $>$

According to this model, the analysis and synthesis of expressions is essentially reduced to topological problems and thus, topological representations, transformations, measures, and techniques of manipulation can be used. For example, the smile is a particular topological configuration. This expression can be recognized and distinguished from other expressions directly in topological space or by a polar coordinate representation, by Voronoi diagram or in the frequency domain.

This chapter is organized as follows. Section 5.1 gives a brief introduction to the synthesis of facial expressions. In Section 5.2, the basic properties of facial expressions are introduced, and a face model discussed. In Section 5.3, basics of the measurement and manipulation of facial expression is introduced. Section 5.4 focuses on local manipulations. Some practical applications of facial analysis and synthesis are introduced in Section 5.5. Section 5.6 discusses a new generation of automated systems for supporting experts in the so-called "lie detection problem." A summary is given in Section 5.7 and is followed by some problems (Section 5.8). In this chapter, a software package called FaceGen has been used to generate most of the images.

## 5.1 Introduction to facial expressions design

A face is defined here as a frontal view of the head from the base of the chin to the hair line, and the frontal half of the head from the lateral view. These zones define the mobile areas of the head where facial expressions are formed. In this section, the terminology describing facial data structures, the role of computer aided design of facial expressions, modeling, and applications are briefly introduced.

### 5.1.1 Terminology

Facial modeling includes terminology from psychology, medicine, and the fields of automated analysis and image synthesis. The terminology most often used in this chapter is listed below:

*Topology;* The topology of facial expressions can be generated by deforming a neutral face topology.

*Emotion*; in modeling, this is abstract data that can be assigned to some class of topological configurations of the face, or facial expressions.

*Facial expression*; this is a particular topological configuration obtained by the deformation of a face's topology.

*Emotional and voluntary facial movement* are dynamic (behavioral) characteristics of facial topology. Emotional and voluntary facial expressions have dynamic and topological characteristics.

### 5.1.2 Applications

Computer aided modeling is aimed at the generation of faces for various applications, in particular:

▶ Testing systems for face identification and recognition,

▶ Fast matching during searches in large (global) databases,

▶ Animation,

▶ Automated support of detection of deceit,

▶ Security personnel training,

▶ Human-machine interfaces,

▶ Improvement of human performance, and for

▶ Medical and psychological purposes.

### 5.1.3 Modeling

Each application of facial analysis and synthesis methods requires the development of a particular model. The model will vary depending on requirements and parameters.

> **Example 5.1** *An algorithm for fast face matching can be relatively insensitive to facial expression. However, to support psychologists in decision making, the algorithm must detect the smallest movements of facial muscles.*

Two types of motion can be distinguished:

▶ Global movements as a result of the head movements (rigid body motion) and

▶ Local deformations as a result of various facial expressions (non-rigid body motion).

Detecting small changes is a crucial requirement in the development of tools for facial analysis and synthesis. It can be expected that facial synthesis will tend to the development of a new generation of devices and systems, for example, training systems for medical and psychological applications, non-contact lie detectors, etc.

### 5.1.4   Direct and inverse problems

Facial processing methods can be viewed as a combination of *direct* and *inverse* problems (Figure 5.2):

(*a*)  The group of *direct* problems (Figure 5.2a) is aimed at the analysis of the topological structure of the face, and the result is a set of features that can be used for decision making in various systems.

(*b*)  The group of *inverse* problems (Figure 5.2b) addresses the synthesis of faces; the methods adopted from inverse problems of biometrics provide the rules to manipulate synthetic face primitives to produce a synthetic face.

Figure 5.2c illustrates the postprocessing of synthetic faces: at the final phase, a synthetic image is postprocessed to achieve a realistic image. This is a typical method used in many applications. This is related to the *Analysis-by-synthesis* paradigm where analysis and synthesis are used to verify the perceptual equivalence between the original and synthetic objects.

### 5.1.5   Data structures of facial synthesis

By looking at the faces of people, it is possible to obtain information about their happiness, fear, anger, surprise, disgust/contempt, interest, sadness. It is also possible to interpret the information in terms such as pleasant-to-unpleasant, active-to-passive, and intense-to-controlled. There are various useful classification systems developed by psychologists that aim to describe facial expressions.

A data structure that includes a representation of emotion, a master-expression, and an extension of facial expression is shown in Figure 5.3. This data structure addresses the problem in the following formulation: given a particular facial topology and an emotion, design a master-expression and generate its extension (more details are given in Subsection 5.2.5).

> **Example 5.2** *The extension of the "average" smile (master-smile) can be various facial topologies: closed and open smiles, for example. These smiles should be detected with respect to mouth, eyes and eyebrow dynamic topologies.*

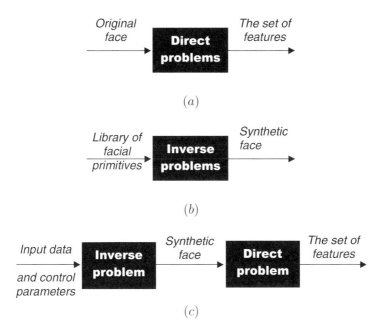

**FIGURE 5.2**

Facial processing: direct problems (a), inverse problems (b), and a scheme for synthetic face processing (c).

**FIGURE 5.3**

Facial expression data structure.

## 5.2 Modeling of facial expressions

In this section, facial models are discussed as generators of synthetic images for various applications. A facial model is a complex dynamic system. These models are developed based on either physical modeling, or topological modeling.

### 5.2.1 Basic properties of facial expressions

There are several unique properties of the face:

▶ The face carries long-term (static) and short-term (dynamic) information. Static information is a topological configuration that changes slowly, for example, progressive changes due to aging and changing state of health. Static information includes also such stable personality characteristic as sex. Dynamic information corresponds to behavioral characteristics (emotions) introduced by changing topological configuration in time (facial expressions).

▶ Unlike, for example, fingerprints, the face cannot be hidden from view, except for in some particular cases, for example, by veils or masks, or partially hidden by sunglasses.

▶ The face includes receivers of information (smell, vision, hearing) and transmitters of information (speech).

▶ The combination of static and dynamic data carries information that can be communicated. For example, this communication plays the role of a verbal (visual) language between a parent and a child of early age.

To analyze and synthesize facial expressions, an appropriate model must be developed. To simplify this model, several basic assumptions are useful, in particular:

(a) The brain (autonomic nervous system) generates emotions, that encode signals which activate facial muscles which produce facial expressions.

(b) Facial expressions are topological configurations that can be measured. These topological structures include the effect of distortions such as (skin and muscle condition, encoding errors, etc.).

> **Example 5.3** *A model based on the above assumptions is shown in Figure 5.4.*

The above general assumptions can be used to develop a model of communication between two persons using facial expressions. Such a model is useful for working on the lie detection problem described later, and as an analogy for designing human-machine interface.

> **Example 5.4** *A simplified model of face-to-face communication involves the following components (Figure 5.5):*
>
> *(a) Transmitter (face) generates various topological configurations.*
>
> *(b) Receiver (eyes) analyzes and recognizes facial topological configurations*
>
> *(c) The channel consists of various distortions, for example, light conditions or the orientation of the subject, that can influence the interpretation of received information (the facial expression can be interpreted as another expression if the light condition is changed).*

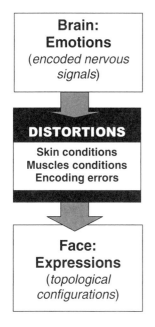

**Brain:
Emotions**
(*encoded nervous
signals*)

**DISTORTIONS**

**Skin conditions
Muscles conditions
Encoding errors**

**Face:
Expressions**
(*topological
configurations*)

- *The face is directly connected to areas of the brain involved in emotion (autonomic nervous system).*
- *When emotion is aroused, muscles on the face begin to fire involuntarily.*
- *It is possible to learn to interfere with these expressions and to conceal them.*
- *A face is usually seen in context, that is, it is integrated with words, voice, body, social setting, etc.*
- *The autonomic nervous system produces behavioral biometric information such as facial expressions, blushing and blanching, and pupil dilation. These changes occur involuntarily when emotion is aroused and are very hard to inhibit, and for that reason can be very reliable clues to deceit.*
- *For example, the polygraph lie detector measures these and other changes such a breathing patterns.*

**FIGURE 5.4**
Model for facial expression generation (Example 5.3).

Additional information, in particular, speech and body movement (for example, the angle of the head to the observer), can be used for supporting communication.

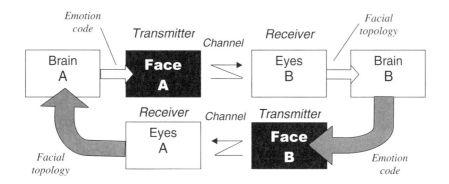

**FIGURE 5.5**
Model of face-to-face communication (Example 5.4).

## 5.2.2    Components of a face model

A face model is a composition of the various sub-models (eyes, nose, etc.).
The level of abstraction in face design depends on the particular application.
Traditionally, at the first phase of computer aided design, a master-face is
constructed. At the next phase, the necessary attributes are added. The
composition of facial sub-models is defined by a global topology and generic
facial parameters. The face model consists of the following facial sub-models:
*eye* (shape, open, closed, blinking, iris size and movement, etc.), *eyebrow*
(texture, shape, dynamics), *mouth* (shape, lip dynamics, teeth and tongue
position, etc.), *nose* (shape, nostril dynamics), and *ear* (shape).

> **Example 5.5** *Figure 5.6 illustrates one of the possible
> schemes for automated facial generation. This includes various
> levels of abstraction and phases:*
>
> The first level *combines libraries of facial primitives.*
> The second level. *The primitives are assembled resulting in a
> master-facial component.*
> The third level. *The master-facial expression is modified by a
> set of parameters.*

## 5.2.3    Facial synthesis based on physical models

Facial expression is the result of the grouped functional synergies of individual
muscles. The modeling of a facial muscle process that is controllable by a
limited number of parameters and is nonspecific to a facial topology is the
main goal of facial expression synthesis methods based on physical modeling.

**Nonrigid synthesis** is based on physical design with materials such as
rubber, cloth, paper, and flexible metals and can simulate the melting of
nonrigid solids into fluids by heat conduction. This modeling is widely used
in facial surgery, archeology, forensic expertise, art, etc.

**Physical facial modeling** is based on:

▶ Muscle activity modeling and
▶ Skin deformation modeling.

Human facial tissue is a deformable object which behaves in accordance with
the laws of Newtowian physics. A combination of both muscle modeling and
skin modeling provides a technique for creating synthetic facial expressions.

The muscles of the face can be grouped according to the orientation of the
individual muscles fibers: linear, parallel, circular, elliptical, etc. The muscles
are attached to the skull at one end and are embedded into the soft tissue of

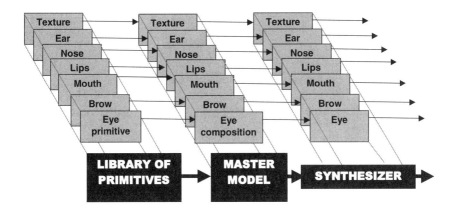

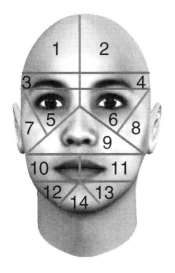

The problem of facial expression synthesis *can be treated as a synthesis of subregions. Thus, the first step is to partition the face into subregions.*

**Example of the face decomposition:**

- *Right pupil* • *Left pupil* • *Nose root* • *Nose tip* • *Upper lip center* • *Lower lip center* • *Right mouth corner* • *Left mouth corner* • *Left outer eye corner* • *Left inner eye corner* • *Right inner eye corner* • *Right outer eye corner* • *Right upper lip* • *Left upper lip* • *Right lower lip* • *Left lower lip* • *Right eyebrow center* • *Left eyebrow center* • *Right nostril* • *Left nostril* • *Right inner eyebrow* • *Left inner eyebrow*

*Synthesis of facial expressions in 2D and 3D can be formulated in terms of inferring feature point motions from a subset, for example, tracking face features along the eye brows, eyes, mouth, and nose.*

**FIGURE 5.6**

Partitioning of the face into regions in the model for facial analysis and synthesis (Example 5.5).

the skin at the other. The skin behaves like a rubber sheet when articulated, i.e., the skin deforms over and around the underlying structures. With the progression of age, the skin loses its elasticity. There are several techniques for modeling the elasticity of skin and its mobility.

In physical facial expression modeling, the skin topology plays an important role. In particular, in modeling aging, skin conditions carry the most valuable information for recognition and generation; in emotions, skin topology significantly contributes in detection of the type of emotion and distinguishing the control and non-controlled facial expressions. Notice that facial expressions closely relates to facial skeleton topology.[*]

> **Example 5.6** *The skull is an enduring immobile structure. The facial skeleton comprises several sections, which uniquely characterize each face. The skull is a long-term biometric characteristic, while the muscle and soft tissues change through life. In particular, the facial skeleton has the following parts: (a) The upper parts of eyeball socket topology and eyebrow ridge that are formed by facial bones, (b) The eye socket topology that formed by a number of bones, and (c) The mouth and nasal cavity topology.*

In physical facial expression modeling, the exact simulation of all face components is not required, i.e., modeling of neurons, muscles, etc. In practice, a physical face model is based on a few dynamic parameters related by a biomechanical description.

> **Example 5.7** *Muscles can be modeled with several control parameters, for example, direction and magnitude.*

**Practical requirements.** In many applications, the synthesis of facial expressions must be performed in real time. This means that manipulating the facial form, the computing muscle and skin activities, and displaying the image must be done at the rate of a dozen frames a second.

### 5.2.4   Facial behavior models

There are two main types of facial behavioral models used in the automatic synthesis of faces:

*Static* models focus mainly on topology, and

*Dynamic* models aimed at the description of a face at the behavioral level.

The type of facial composition depends greatly on whether the model is stable or dynamic. In a dynamic facial model, the facial parameters are distinguished with respect to the *long-term* and *short-term* changes.

---

[*]For example, see paper by Kähler et al. "Reanimating the dead: reconstruction of expressive faces from skull data", ACM Transactions on Graphics, 22(3):554–561, 2003.

**Long-term behavior model,** or age model, captures facial topology and features that change slowly through the life cycle.

> **Example 5.8** *Dynamic long-term face modeling is illustrated by Figure 5.7. Models of degradation are used for facial components.*

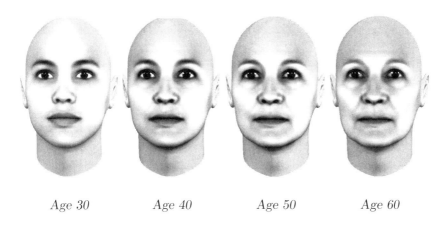

Age 30      Age 40      Age 50      Age 60

**FIGURE 5.7**
Long-term neutral facial expression modeling (Example 5.8).

**Short-term dynamic behavior model.** Facial behavior is usually revealed during speech. Emotion viewed by facial movements may be most evident when a person is speaking. Facial expressions carry information about:

▶ Affective state (fear, anger, enjoyment, surprise, sadness, disgust, and more enduring moods such as euphoria, dysphoria, or irritableness);

▶ Cognitive activity (perplexity, concentration, or boredom);

▶ Temperament and personality (hostility, sociability or shyness);

▶ Truthfulness (leakage of concealed emotions, and clues as to when the information provided in words about plans or actions is false);

▶ Psychopathology (diagnostic information relevant to depression, mania, schizophrenia, and other less severe disorders, and information relevant to monitoring response to treatment).

Movements of the facial muscles pull the skin, temporarily distorting the shape of the eyes, brows, and lips, and the appearance of folds, furrows and bulges in different patches of skin. Facial expressions can cause serious problems in the practice of face identification.[†]

> **Example 5.9** *A person is searched for in the database by a specific photo where this person appears to be serious. The result of the search can be negative if the database includes only images of this person where he has other expressions.*

A possible solution to this problem is to synthesize different expressions for this person and repeat the search in the database.

## 5.2.5   Library

Each topological configuration of facial expression is the result of muscle activity and elastic skin deformation.

**Classification schemes of emotions.** There are various classification schemes of emotions that can be useful for automated analysis and synthesis:

(*a*)  Classification based on 5-8 basic emotions (anger, disgust, fear, etc.). In automated systems, the basic emotions are represented by corresponding basic topological configurations (facial expressions). An extension of the library of basic topological configurations can be achieved by using various techniques, including evolutionary algorithms and artificial neural networks.

(*b*)  Ekman and Friesen have proposed a coding system based on 46 facial movements; each corresponds to units called "action units" (AU). An arbitrary facial expression can be described as a combination of AUs. For example, the happiness expression can be described as follows

$$< \texttt{Happiness 1} > = \underbrace{< \text{AU12+13} >}_{\text{Pulling lip corners}} + \underbrace{< \text{AU25+27} >}_{\text{Mouth opening}}$$

$$+ \underbrace{< \text{AU10} >}_{\text{Upper lip raiser}} + \underbrace{< \text{AU11} >}_{\text{Furrow}}$$

$$< \texttt{Happiness 2} > = \underbrace{< \text{AU6} >}_{\text{Cheek raising up}} + \underbrace{< \text{AU11} >}_{\text{Nasolabial folds}}$$

$$+ \underbrace{< \text{AU12} >}_{\text{Raising of the corners of the lips}}$$

---

[†]Facial expressions are excellent benchmarks for testing biometric systems for facial recognition, identification, and generation.

To apply the above representation and measures for coding, composing and decomposing facial expressions, particular topological representations are needed.

(*c*) Another approach to classification follows from physical modeling: the facial musculature (zygomaticus major, zygomaticus minor, risorius, levator anguli oris, quadratus, mentalis, triangularis) and ear musculature (auricularis superior and anterior).

> **Example 5.10** *The physical basis of the topological configuration classified as a smile is as follows:*
>
> (*a*) *The risorius pulls the mouth corners laterally and mentalis pushes the lower lip out.*
> (*b*) *Five muscles (zygomaticus major and minor, risorius, levator anguli oris, and buccinator) in the lip corner area are able to create movements of the lip corners which are interpretable as smiles.*

In addition, the triangularis causes the down-turning of the mouth corners in the sad-face and crying face.

**The basic topologies of facial expressions** called *facial expression primitives* consists of the smile-topology, frown-topology, furrow-topology, squint-topology, etc. Combinations of these primitives produce more than a thousand different facial topologies (appearances). Decision diagrams are often are used to solve classification problems such as this that are not linearly separable.

**Master expressions.** Facial muscles are sufficiently complex to allow more than a thousand different facial appearances, (corresponding to facial topological descriptions). In automated systems, these topologies must be distinguished, stored, generated, compressed, etc. For automated analysis and synthesis of facial expressions, the data structure used to represent the expression plays a crucial role. For example, the straightforward implementation of known classification systems is not efficient for several reasons, in particular, because of high computational complexity and the high level of abstraction. Therefore, a data structure for representing facial expressions for analysis and synthesis should be easily represented in a computing hierarchy and result in efficient computation.

In the facial expression data structure presented in Figure 5.3, master expressions convey the basic topology of emotion and provides a simplification of the model.

Given an emotion, a master facial expression is the average of a set of topological configurations corresponding to this emotion that also distinguish them from another emotions. In addition, several requirements may be formulated in the problem statement.

> **Example 5.11** *The data structure shown in Figure 5.3 is an abstract representation. The notion of a master-expression is useful because it provides necessary relationships with data structure representation at other levels of the computing hierarchy. For example, in the synthesis of facial expressions, the master facial expression must be convertible to another configuration, convenient for manipulation, etc.*

**Emotion visualization.** Emotions can be visualized in various ways. For example, parameters of emotion after processing (ordering, weighting, and filtering) can be mapped into an appropriate topological structure. Figure 5.8 shows a possible visual representation of facial expressions (color is replaced by gray levels). Well identified positive emotions correspond to intensive colors and negative emotions are inverted compared to positive ones.

**FIGURE 5.8**
Visualization of emotions.

## 5.2.6   Distribution of information sources across an object

The face is a unique source of behavioral and static information. Processing this behavioral information is computationally intensive. In the analysis and synthesis of behavioral information, new paradigms are required. A typical example is the automated support of an expert attempting to distinguish lies from truth in a face-to-face interview.

**Information in speech.** Neurologic disorders, such as Parkinsons disease, ALS, or Huntingtons disease, often manifest themselves in the voice before they are noticed in the rest of the body. The voice is a window to body health and function. We make judgments about age, gender, emotion, personality, physical fitness, and social upbringing from a persons speech. The voice carries emotional information by *intonation*.

> **Example 5.12** *Speech that is characterized by fast pitch changes, large pitch range, and large dynamic amplitude range conveys information that the person may be angry. If pitch changes slowly, its range is small, and the dynamic amplitude range is large, the person is depressed.*

**Information source distributions.** Configurations corresponding to regions such as lips, eyebrows, eyes, etc., contribute to facial expression variably over time. The distribution of this variation indicates the source of information's change over time.

> **Example 5.13** *Facial expressions, speech, and body tracking have different functions of variability in time. For example:*
>
> (a) *Through some time intervals, the area of the eyes provide more information and more accurate information, than other parts of the face.*
>
> (b) *After speaking, the face becomes an intensive transmitter of information.*
>
> (c) *Useful information can be extracted from the face while a liar speaks because he focuses on words, and his control of facial expression is weak.*

There are a lot of problems that can be solved through analyzing information from different regions.

> **Example 5.14** *Given a tested person, the volume of information extracted from various sources that characterize the response of the person to controlled audio/video input information can be discovered through a testing procedure. In particular:*
>
> (a) *Comparison of information extracted from the face compared to other sources of information about emotion,*
>
> (b) *The contribution of context (words, speech characteristics, head and hand tracking, etc.)*
>
> (c) *Relationship of the facial expression and context.*

From a psychological view point, the initial facial expressions caused by aroused emotion are not deliberately chosen, unless they are false. Facial expressions often lie and tell the truth *voluntarily* and *involuntary*, at the same time.

## 5.3 Facial topology transformation and manipulation

In this section, a technique for facial expression synthesis is introduced. The technique consists of various topological transformations.

### 5.3.1   Automated facial measurements

Facial expression or appearance can be formally treated as a topological structure. Measurement characteristics include perimeter, area, max-min radii and eccentricity, corners, roundness, bending energy, holes, Euler number, and symmetry.

> **Example 5.15** *Figure 5.9 illustrates local dynamic measures:*
>
> (a) *Eyebrow configuration, including height, shape, curvature, etc.,*
>
> (b) *Distance $\delta$ between brow and eye, and*
>
> (c) *Frequency $f$ of blink, eyebrow configuration and distance $\delta$ are functions of time.*

All of the above parameters carry psychological information during controlled and non-controlled phases of emotions. This information can be modified with analysis of eyebrow configuration, distance $\delta$, and frequency $f$ of blink together with lip motion. For example, psychologists evaluate the genuineness of a smile and truthfulness of an answer to a question based on information portrayed in these variables.

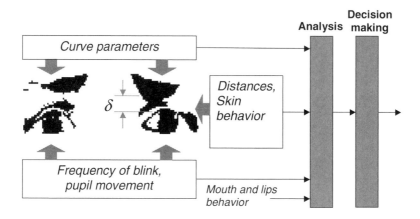

**FIGURE 5.9**
Measures the eye behavior model (Example 5.15).

> **Example 5.16** *A limited number of labeled feature points marked in one face, e.g., the tip of the nose, the eye corner and less prominent points on the cheek, must be located precisely in another face. The number of manually labeled feature points varies from application to application, but usually ranges from 50 to 300.*

In analysis and synthesis techniques, the following measures are used:

▶ Detecting and tracking 3D head position and orientation.
▶ Detecting and tracking eye movements, gaze direction and eye closure.
▶ Detecting and measuring lip movement and mouth opening.
▶ Detecting and measuring facial muscular actions.

> **Example 5.17** *Detected smiles are grouped with eyes, head, and hand tracking as follows:*
>
> *(a) Eyes straight,*
> *(b) Eyes to side,*
> *(c) Eyes down and head down or away,*
> *(d) Eyes down, head away, and face touch by hand.*

## 5.3.2 Morphing technique

A face appearance can be morphed with respect to various parameters. Some typical morphing techniques are:

*Expression morphing* (smile closed and open, anger, fear, disgust, sad, surprise).
*Age morphing* is based on shape and texture modifications.

In addition, the race and gender can be morphed as well as various topological measures, for example, the difference between the average face of the same gender and measure of assymmetry.

## 5.3.3 Facial deformation

Features on a two dimensional face image can be deformed with a number of transformations.

> **Example 5.18** *The first step is to deform the surface grid on which the image is to be mapped (Figure 5.10). The deformation of the initial grid is specified by placing the control points $P_{ij}$ and then moving them. The positions of the deformed grid nodes are found by calculating the two-component bivariate Bernstein polynomial.*

In the next step, the image warping can be performed as a mapping between the initial and the deformed grids.

## 5.3.4 Local and global spatial transformations

Two classes of face transformation are used in facial analysis and synthesis:

*Local* facial transformations and
*Global* facial transformation.

*Initial grid of facial region*          *Control points of facial region*

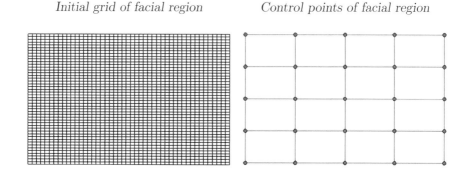

*Displacment of facial expression points   Grid deformation of facial region*

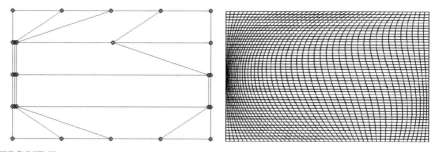

**FIGURE 5.10**
Technique of facial topology deformation of a grid (Example 5.18).

**Local transformations** change local regions of a face. The central idea is based on topological decomposition into parts (regions in Figure 5.6) and the separate modeling of each decomposed part (region). These transformations are based on the set of models used (eye-region, mouth region, etc.) Combinations of local transformations produce different facial characteristics (emotion, age changes, actions, etc.) The effectiveness of the approach to facial synthesis based on local transformations have has been demonstrated in image synthesis and facial animation (see "Further Reading" Section). However, the typical algorithms simply combine various models and control their parameters. In Figure 5.11, most important local facial transformations are illustrated.

**Global transformations** are attractive because a number of control parameters and behavior characteristics can be optimized together. However the complexity of global models is high. Moreover, it is difficult to achieve

▶ Satisfactory observability of a model, and

▶ A modular implementation structure because of the high interrelationship between components.

The state-of-the-art of facial synthesis is based on the first paradigm, the local facial transformation.

> **Example 5.19** *Figure 5.11 illustrates 14 local transforms:*
> $R_1^{20} \Leftrightarrow R_1^{60}$, $R_2^{20} \Leftrightarrow R_2^{60}$, ..., $R_{14}^{20} \Leftrightarrow R_{14}^{60}$

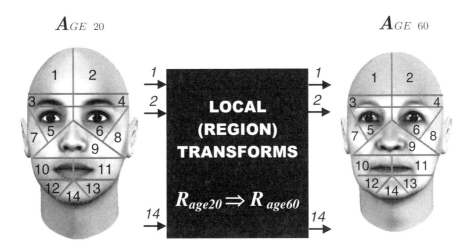

**FIGURE 5.11**
Long-term face behavior design (Example 5.19).

## 5.4   Local facial models

Face components, including the eye, eyebrow, mouth, etc., can be designed with local facial models. In this section, the elements of face component design are introduced.

### 5.4.1   Information carried by eyes, eyebrows, and lips

Information carried by local facial zones (eyes, eyebrows, mouth, and lips) are divided into:

▶ Visual information and

▶ Information in the form of electrophysiological signals.

Visual information is re-processed by image processing algorithms and is used in face identification and recognition, facial expression generation, and the development of various converters of biometric information based on visual effects. The dynamics of local facial zones are the result of facial muscle activity which can be represented in the form of electrical activity. Activated facial muscles including eye-blinks and eye movement[‡] contribute to electrophysiological signals recorded from the scalp. There is a relationship between visual and electrophysiological data structures: to each topology of a local facial zone corresponds an electrophysiological portrait generated by corresponding muscles.

> **Example 5.20** *Electrical activity of frontalis muscle is well detected in the beta- or mu-rhythm frequency range. Eye-blinks can be detected in the theta-rhythm range.*

## 5.4.2   Eye synthesis

The eye is a source of dynamic information. Static images of the eyes carry information about race, emotion, and age. The eyes are considered as channel of non-verbal communication between people. The eyes:

▶ Regulate the conversation,

▶ Control feedback during a conversation, i.e., how the other person follows the flow of conversation,

▶ Look for additional information during a conversation, and

▶ Express emotion.

A sample of images provide information useful for computing the dynamic behavioral characteristics of eyes that is used for medical and psychological study. Irises and retinas provide useful biometric information for identification and medical study including pupil response to light and drugs. As an example, it is reasonable to describe pupil coordinates by fuzzy logic methods when one needs to extract additional information on emotion. Master-eye models chosen from a database can be used to generate eyes with user-defined parameters. A machine learning paradigm can be useful here.

> **Example 5.21** *Figure 5.12 illustrates the result of 30 years of ageing (from 30 to 60 years) of eyes and eye-area.*

---

[‡]see, for example, the result by LaCourse and Hladik [44]

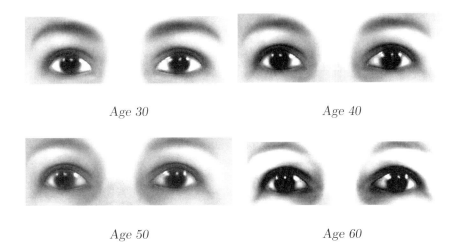

Age 30          Age 40

Age 50          Age 60

**FIGURE 5.12**
A neutral expression showing the effect of aging (long-term eye modeling) (Example 5.21).

### 5.4.3 Eyebrow synthesis

The eyebrow is a good source of information in facial expression analysis and synthesis. The corrugator muscles are responsible for lowering and contracting the eyebrows.

The eyebrow carries information by its

▶ Topological configuration,

▶ Orientation, and

▶ Distance from corresponding eyes.

The eyebrow often is used to increase the reliability of facial behavioral information.

> **Example 5.22** *The simplest method is to detect just eyebrow raising and lowering (Figure 5.13). Eyebrow raising and lowering is related to the speech process, and may also provide information about other cognitive states. Measurements of eyebrow raising and lowering are often combined with the voice detection and face direction.*

Figure 5.14 illustrates fragments of eyebrow primitives which are used in facial expression design.

Brow expressions                    Mouth

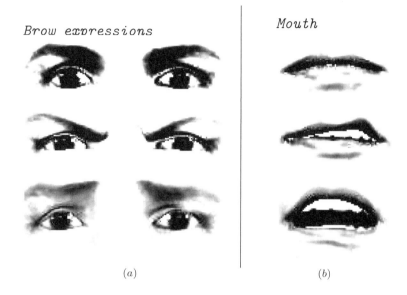

(a)                                      (b)

**FIGURE 5.13**
A sample primitives of eyebrow expressions (a) and mouth expressions (b) from the library of synthetic facial primitives.

## 5.4.4   Mouth and lips synthesis

The main characteristic of the mouth is its flexibility. The mouth and lips provide information about emotion. Sometimes static lip images can be considered as a source of information about the ageing process. Mouth and lip information can be be combined with other sources of information providing useful practical solutions like text-to-lip and lips-to-speech-to-text converters.

Deformation around the mouth region is synthesized just like any other facial expression, except that in this case the movements have to be coordinated with the associated speech signals and brow dynamics. The model parameters necessary to synthesize the lip movements are obtained from the fundamental linguistic speech units. These, in turn, are dependent on the language, age, level of education, etc. of the speaker. Conversely, these are sources of information for emotions identification and social behavioral characteristics useful, for example, in identifying true or false answers to questions.

> **Example 5.23** *Figure    5.13    illustrates    some    possible composition of brow and mouth expressions.*

Visible speech signals can supplement acoustic speech signals, especially in a noisy environment or for the hearing impaired. The face and in particular the region around the lips, contains significant phonemic and articulatory

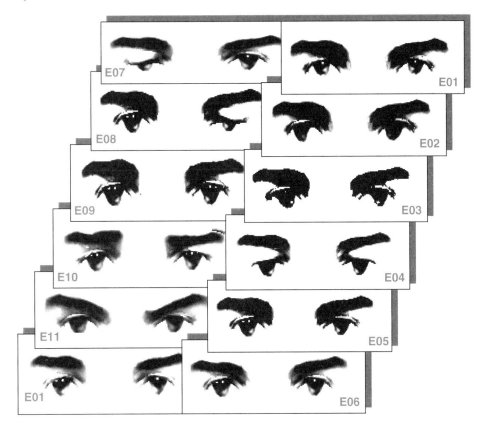

**FIGURE 5.14**

A set of eyebrow topological primitives from the library; each topology is characterized by an electrophysiological portrait.

information. However, the vocal tract is not visible and some phonemes cannot be distinguished from visual observation alone. To synthesize the full appearance of a mouth from a given shape, the first step is to convert its shape parameters into shape-free appearance parameters with a dependence mapping, and the second step is to construct a shape-free appearance, and warp it according to the original shape.

**The features** are defined as height, width, area, spreading (width/height) of the oral cavity, durations (number of frames for vowel articulation and time during vowel articulation). Spreading and the area surrounded by the lips are important features for vowel recognition. There are no absolute fixed and observable positions for the lip and tongue corresponding to specific vowels across different speakers. Lip-readers adjust to different speakers through spatial normalization. That is, constant relative differences among some oral cavity features in the visual perception of the vowel utterances exist for all speakers.

**Lip synchronization.** The inverse problem to the analysis of facial movements in the mouth area (speech recognition) is the synthesis of the facial motions that take place, during spoken utterances.

**Control points** (called also *markers*) and topological measures in synthetic mouth design are shown in in Figure 5.15. Lip shape can be defined by the $x$ and $y$ coordinates of points along its boundary. Usually, all values are normalized to a lip width of 1. Lip topology dynamics can be modeled by a hidden Markov model.

> **Example 5.24** *A smile is defined as a facial expression in which the ends of the mouth curve up or down, often with lips moving apart so that the teeth can be seen (Figure 5.15). A smile usually expresses as a positive feeling such as happiness, pleasure, and amusement of friendliness. In this case, the ends of the mouth curve up slightly. A smile can be recognized as sadness or regret if the ends of the mouth curve down slightly.*

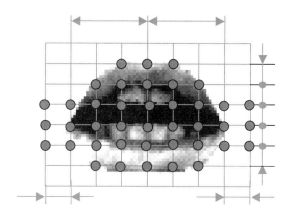

The image of an oral cavity described by:

- Height,
- Width,
- Area surrounded by the lips,
- Spreading (width/height)
- Number of frames for vowel articulation
- Time during vowel articulation
- Lips corners configuration (up, straight, down)

**FIGURE 5.15**

Control points and topological measures in synthetic mouth design (Example 5.24).

**Example 5.25** *Figure 5.16 illustrates a typical texture technique that includes texture and shadow generators with user-defined parameters.*

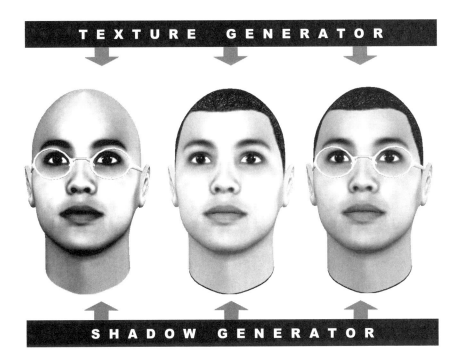

**FIGURE 5.16**
Synthetic faces: texture by hair and glasses (Example 5.25).

### 5.4.5 Facial caricatures

Facial caricatures are widely used in real life. Caricature is the art of making a drawing of a face which makes part of its appearance more noticeable than it really is, and which can make a person look ridiculous.[§] A caricature is a synthetic facial expression, where the distances of some feature points from the corresponding positions in the normal face have been exaggerated (Figure 5.17).

---

[§]Cambridge International Dictionary of English, Cambridge University Press, 1995.

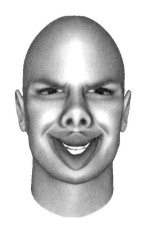

Facial caricatures are characterized by several properties which are useful in facial analysis and synthesis. In particular:

▶ A facial caricature emphasizes the most important facial features and sometimes conveys useful behavioral information.

▶ Facial caricatures can be used in master-expression (average facial expression) design.

▶ Facial caricatures can be used as benchmarks in testing artificial intelligence properties of systems for image processing, pattern recognition, and facial expression synthesis, i.e., the imitation of an art-style.

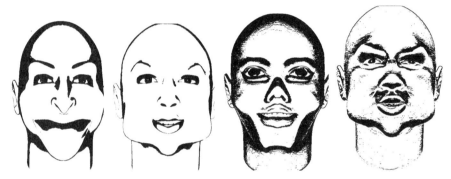

**FIGURE 5.17**

A caricature which has been automatically synthesized given some parameters.

The technique of assigning these distances is called a *caricature style*, i.e., the art-style of the caricaturist. Often the results reported on automatic caricature synthesis implement arbitrary manipulation of distances but without art-style modeling.

The reason why the art-style of the caricaturist is of interest for image analysis, synthesis, and especially of facial expression recognition and synthesis is as follows. Facial caricatures incorporate the most important facial features and a significant set of distorted features. Caricatures emphasize these notable features that make a face different from the average. Some original features (coordinates, corresponding shapes, and the total number of the features) in a caricature are very sensitive to variation, however the rest of the features can be distorted significantly. At their caricatures, people always can recognize themselves. Various benefits are expected in identification, recognition and matching techniques, if the art-style of the caricaturist can be understood.

> **Example 5.26** *Figure 5.18 and Figure 5.19 illustrate designing an artificial caricature by grid deformation. In these Figures, the $4 \times 4$ and $40 \times 40$ grids have been chosen.*

*Original image*
*(4×4 grid)*  *Deformed nose*
*(4×4 deformed grid)*

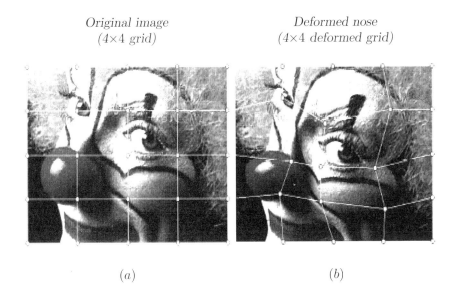

*(a)* *(b)*

**FIGURE 5.18**
Synthetic caricature design: image mapped onto a rectangular grid (a), and image mapped onto the deformed grid (b) creating the so-called elephant-squashed nose (Example 5.26).

## 5.5 Facial synthesizers

In this section, some practical applications of facial models are considered. A speech-to-lips technique synchronizes speech and to lip movement. In the text-to-lips model, technique of text recognition is used. There are a lot of other useful schemes, for example, speech-to-text, lips-to-speech, and lips-to-text. These techniques utilize topological and physical models in the interpretation of facial information and the design of synthetic facial expressions.

Original image                              Deformed face

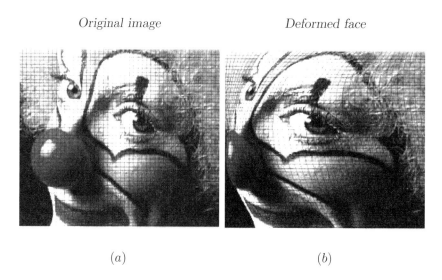

(a)                                                 (b)

**FIGURE 5.19**
Synthetic caricature design: image mapped onto a rectangular grid (a), and
image mapped onto the deformed grid (b) (Example 5.26).

### 5.5.1    Speech-to-lips systems

Speech signals are conventionally treated as acoustical. The visual component
in the production and perception of speech is due to the visible facial
expression (movements of the mouth, face, and head). A facial expression
forms the basis of lip (or speech) reading. Visual aspects of the speech signal
can be relevant to studies in the acoustic domain.

Among various applications, the speech-to-lips technique is especially useful
in noisy conditions. Visual speech recognizers depend upon the capture of
speech gestures from sequences of images of talkers' faces. The image can
be processed to extract essential visual cues or features. A recognition then
proceeds by matching the time-varying templates of features taken from a
test utterance against the stored templates of utterances in the recognition
systems vocabulary.

A possible organization of library of topological primitives can based on
Ekman and Friesen descriptors:

"`Lips pressed`" is a topological configuration where lips are pressed,
   tightened, and narrowed;

"`Jaw drop`" is a topological configuration where lips are parted with a space
   between the teeth;

"`Mouth stretch`" is a topological configuration with an opened mouth;

"Lip corner depressed" is a topological configuration with the corners of
the lips down.

The duration of topological configurations are not critical within scenarios of
dialog. However, this is critical in the prossess of emotion identification and
synthesis of facial expressions.

> **Example 5.27** *Figure 5.20 introduces the schematic structure
> of a speech-to-lips system.*

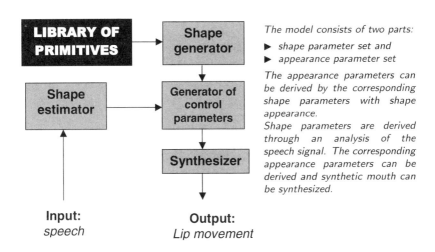

**FIGURE 5.20**
Speech-to-lips system (Example 5.27).

## 5.5.2  Facial transmission

The transmission of a moving face image requires the development of a specific
technique if traffic capacity is limited. The technique of model-based coding
is suitable for this.

The basic idea is as follows: at the encoding side of a visual communication
system, the image is analyzed and the relevant object is identified (Figure
5.21). A model is then adapted to the object, for example, the model is a
wireframe describing the shape of the object.

Instead of transmitting the image pixel-by-pixel, or by describing the
waveform of the image, the image is handled as a 2D projection of 3D objects

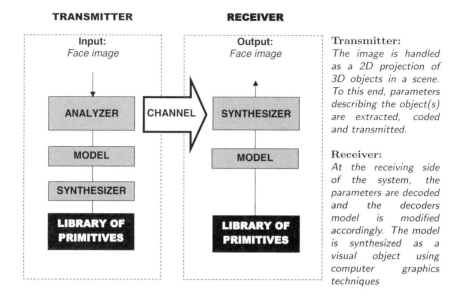

**FIGURE 5.21**
The principle of analysis-by-synthesis coding.

in a scene. Parameters describing the object(s) (size, position, shape, etc.) are extracted, coded and transmitted. To achieve acceptable visual similarity to the original image, the texture of the object is also transmitted. The texture can be compressed with a traditional image coding technique.

The parameters are decoded at the receiver side, and the decoded model is modified accordingly. The image is then synthesized using computer graphics techniques, by shaping a wireframe according to the shape and size parameters, and a texture is mapped onto its surfaces. Thus, the following must be transmitted: parameters describing changes of the model are (information how to rotate and translate the model), and parameters describing the motion of individual vertices of the wireframe, since human face is a non-rigid object. This model-based coder can be implemented as an analysis-by-synthesis coder, that is, the encoder contains a feedback loop and a copy of the decoder, thus enabling it to compare the synthesized image to the original image. The encoder can optimize the transmitted parameters by minimizing the residual image. In this way, the encoder transmits a residual image and improves image quality.

## 5.6 Automated support of deceit detection

The focus of this section is an automated system for the support of an expert estimating deceit. This problem is relevant to the development of a new generation of lie detectors used, for example, in testing job applicants and in immigration control. These generation of biometric systems are characterized by new features, in particular, by the ability to remotely measure, store and process information from various sources (current behavior of person, Internet resources, information from videotape, national and international databases, etc.)

### 5.6.1 Lie detection problem

A lie is defined as something communicated which is not true in order to deceive.[¶] The liar generates behavioral information called *synthetic* information, that does not correspond to the person's natural response to a feeling. For example, pleasure in the form of a smile when something is unpleasant, neutral facial expression when something is offensive, etc. In terms of image processing, this problem is formulated as recognizing information as synthetic or natural. This problem is widely known as the *lie detection* problem.

> **Example 5.28** *A polygraph lie detector is an automated device that is supposed to distinguish synthetic (false) and natural (true) information. A probabilistic decision on information being synthetic (false) or natural (true) is based on measurements of changes in the autonomic nervous system activity including heart rate, blood pressure, skin conductivity, skin temperature, etc., all of which can indicate emotional arousal.*

Lie detectors that are widely used today have several sensors: sensors for measuring changes in the depth and rate of breathing (pneumatic tubes or straps), a blood pressure cuff for measuring cardiac activity, and sensor for measuring minute changes in perspiration. Lie detection is a psychological problem, the role of automated systems is to support an expert in the analysis of a tested person.

---

[¶]In the *Oxford English Dictionary*, the word *lie* is defined as a violent expression of moral reprobation, which in polite conversation tends to be avoided. The synonyms of the word *lie* are *falsehood* and *untruth*.

### 5.6.2    Face as a carrier of false information

Face, body, hand tracking, and speech are all sources of information useful for identifying deceit. However, in contrast to classical identification and recognition problems, specific features must be extracted from this information. These specific features are interpreted based on special rules. Finally, the decision "TRUE" or "FALSE" is made.

A face can misinform by intention or habit. Smiles may be a reliable index of pleasure or happiness, a person may also smile to mask a feeling he wishes to conceal or to present a feeling when he has no emotion at all.

**Mask.** Emotion can be concealed with masking, for example, by covering the face or part of the face with a hand, or by turning away the head. These masks can be easily detected by automated tools. However, the best mask is a false (controlled) emotion.

> **Example 5.29** *The smile is the mask most frequently employed. However, negative emotions are difficult to falsify, even with a smile.*

These masks can be recognized by more complicated algorithms, for example, by the difference of controlled and non-controlled face expressions.

**An information model for detecting deceit** uses various sources of information: the *basic* sources of biometric information include:

(*a*)  facial expression,

(*b*)  Infrared image of the face, and

(*c*)  Speech and voice,

Additional sources of behavioral information include rigid (the orientation of the head, gestures, posture) and non-rigid (motion of facial features) body motion and also respiration, flushing and blanching, sweating, etc.

**Extraction information from various sources.** Facial expressions and speech have different functions of variability in time.

> **Example 5.30** *In Example 5.13, the relationships of various sources of information are given. In the model, sources of information are dynamically combined to produce new sources (Figure 5.22). The relationship of basic and additional sources are represented by a communication tree.*

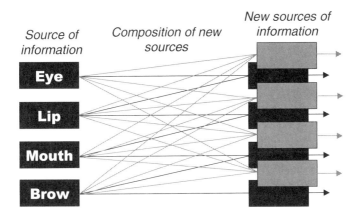

**FIGURE 5.22**
Sources of information are combined to produce new sources (Example 5.30).

**Decision making.** Extracted information is analyzed, classified, separated, and evaluated with respect to reliability, and is the basis for decision making. Decision making is a probabilistic function and can be evaluated by various methods. Interactive analysis can be used to make more reliable decisions, for example, by proposing that the expert ask additional testing questions or demonstrate a video for relaxation, or by indicating reliable features to observe.

> **Example 5.31** *A liar usually controls only two or three sources of information, for example, facial expressions and voice. Hence other sources of information are less-controlled and can be used in decision making.*

The procedure includes:

▶ Parallel processing of all sources from (Figure 5.6),

▶ Grouping and ordering of sources by criterion of reliability, extraction of additional information for decision making, etc.

▶ Determining the intensity of the information conveyed by different sources,

▶ Dynamically overlapping various combinations of sources to produce new sources of information.

In identification, most of these sources of information are considered to be noise. In lie detection, some of this information becomes critical.

### 5.6.3  Models of controlled and non-controlled facial dynamics

Facial expressions are formed by about 50 facial muscles that are controlled by hundreds of parameters. A face model is a complex dynamic system that is not well studied. Psychologists distinguish two kinds of short-time facial expressions:

  *Controlled* facial expressions, and

  *Non-controlled* facial expressions.

Controlled expressions can be fixed in a facial model by generating control parameters, for example, a type of smile. Models of these expressions are well studied. Non-controlled facial expressions are very dynamic and are characterized by short time durations[||]. The difference between controlled and non-controlled facial expressions can be interpreted in various ways. The example below illustrates how to use short-term facial expressions in practice.

> **Example 5.32** *In Figure 5.23, a sample of two images taken two seconds apart shows the response of a person to a question. The first phase of the response is a non-controlled facial expression that is quickly transformed to another facial expression corresponding to the controlled face.* The facial difference *of topological information* Δ *for example, in mouth and eyebrow configuration, can be interpreted by psychologists based on the evaluation of the first image as follows*
>
> $$Mouth = \begin{cases} Irritation; \\ Aggression; \\ Discontent. \end{cases} \qquad Brows = \begin{cases} Unexpectedness; \\ Astonishment; \\ Embarrassment. \end{cases}$$
>
> *Decision making is based on an analysis of the question and the facial response.*

In more detail, the *local facial difference* is calculated for each region of the face that carries short time behavioral information. The local difference is defined as a change of some reliable topological parameter. The sum of weighted local differences is the *global facial difference*.

**Morphology of a genuine smile.** A smile is created by the joint action of the zygomatic major muscle and the lateral part of the orbicularis oculi muscle (the lip corner is pulled up, the cheeks are lifted upward, the eye opening is

---

[||]Visual pattern analysis and classification can be carried out in 100 *msec* and involves a minimum of 10 synaptic stages from the retina to the temporal lobe (see, for example, Rolls ET. Brain mechanisms for invariant visual recognition and learning, *Behavioural Processes, 33:113–138, 1994.*

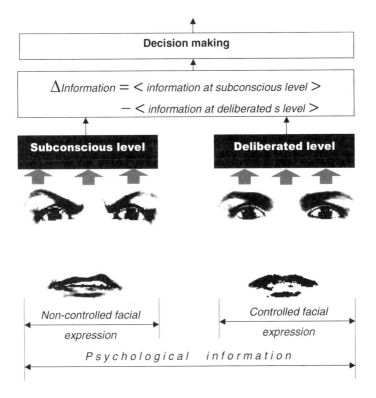

**FIGURE 5.23**

Response of facial model consists of controlled and non-controlled phases of facial expressions (Example 5.32).

narrowed). However, two kinds of smiles are known: (a) an enjoyment smile that is a spontaneous response to emotion, and (b) a false smile consisting of voluntary facial movements.

> **Example 5.33** *False smiles can be recognized by detection of inactivity of the orbicularis orbicularis oculi muscle as follows:*
>
> ▶ *Asymmetry, i.e., voluntary smiles are presented in one half of the face only;*
>
> ▶ *Duration, i.e., voluntary smiles are short or too long;*
>
> ▶ *Asynchronism, i.e., voluntary smiles are asynchronous because, the zygomatic major and orbicularis oculi muscles do not reach their apex at the same time.*
>
> **Example 5.34** *Figure 5.24 illustrates four types of smile.*

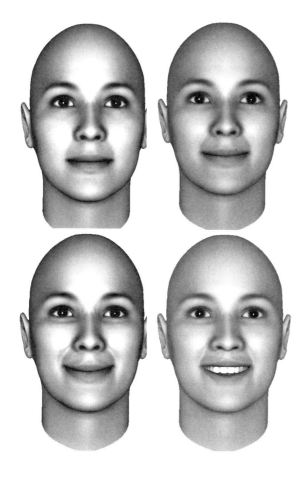

*A smile provides information about emotion.*

*The smile is defined as a facial expression in which the ends of the mouth curve up or down, often with lips moving apart so that the teeth can be seen.*

*A smile usually expresses a positive feeling such as happiness, pleasure, and amusement of friendliness. In this case, the ends of the mouth curve up slightly.*

*A smile can be recognized as sadness or regret if the ends of the mouth curve down slightly.*

**FIGURE 5.24**
Synthetic smile topologies (Example 5.34).

### 5.6.4   New generation of lie detectors

The main new features of the new generation of lie detectors include:

(*a*)  Architectural characteristics (highly parallel configuration),

(*b*)  Artificial intelligence support of decision making, and

(*c*)  New paradigms (non-contact testing scenario, controlled dialog scenarios, flexible sources use, and interactive possibility through an artificial intelligence supported machine-human interface).

**Architecture** of a new generation lie detector includes (Figure 5.25):

▶ Interactive machine-human interface,

▶ Video and infrared cameras, and
▶ Parallel hardware and software tools

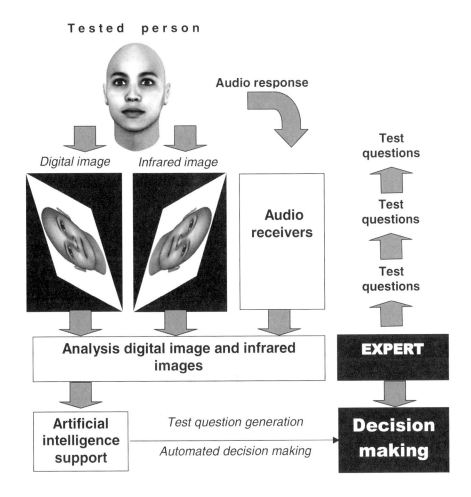

**FIGURE 5.25**
New generation of non-contact lie detector system.

**Information flow processing in non-contact lie detectors.** Three sources of information, i.e., video image, infrared image, and speech, are analyzed by computer. Video image and infrared images provide distance measures (blood pressure, pules, etc.). Expressions in video and infrared images are systemized in time with respect to speech characteristics and body tracking. Prior information can be used to improve decision making.

**Artificial intelligence support** is provided at two levels:

► An *expert system* to support an expert and

► *Machine learning* that adapts to the behavioral models to a tested person.

The expert system works together with the expert augmenting the expert's senses, and his analysis skills. Machine learning is used through the training phase of testing a person. At this phase, the system builds a model of the input (audio/video information) and output (response in the form of behavioral information), i.e., it creates accurate behavioral models from master-models. In other words, through this phase a hypothesis function can be used to predict future data is created. This can be used to play various scenarios and choose appropriate stimuli for the subject that is likely to elicit the desired information.

**Scenario of dialog.** The new generation of lie detectors and expert systems can work in collaboration with an expert. For example, the expert's observation can be introduced through a machine-human interface. Decision making is based on current information. However, prior information can be used too, for example, fragments of speech, facial expressions, etc. This information extracted from situation natural to a given person can be used to pre-adjust a system, increasing the probability of correct decision making, i.e., lie/truth.

**Decision making.** When aiding the expert in his decision, the expert system can use both the body and voice as clues:

*The body* reveals a lot of clues to deception. Most body movements are not directly tied to the areas of the brain involved in emotion. Note that concealment of body movement is easier than concealing facial expressions or voice changes in emotion.

*The voice,* like the face, is also tied to the areas of the brain involved in emotions:

  ► Some of the changes in voice that occur when emotion is aroused can be concealed.
  ► Pauses between words are the most common vocal deception clue.
  ► The length and frequency of pauses carry information too.
  ► Hesitating at the start of speaking, particularly if the hesitation occurs when someone is responding to a question, may arouse suspicion.
  ► So may shorter pauses during the course of speaking if they occur often enough.

▶ The sound of the voice also may reveal deceit. For example, pitch becomes higher when a subject is upset. A raised pitch can be a sign of deceit or of fear or anger.

> **Example 5.35** *Just as a vocal sign of emotion, such as pitch, does not always mark a lie, so the absence of any vocal sign of emotion does not necessarily prove truthfulness. Some people never show emotion, at least not in their voice. And even people who are emotional may not be about a particular lie.*

Known automated devices for speech analysis can extract useful information from speech and voice, however this information is not enough for detecting a lie.

## 5.7   Summary

The human face is a source of various types of information. It carries information about health conditions, current emotions, features for personal identification and recognition, behavioral features, and other features that can be used for predicting personal behavior. These types of information can be classified with respect to various criteria. For example, a face can be analyzed for honesty in a conversation. In general, face modeling can be customized for specific applications.

1. The face can be modelled at topological and physical levels. Topological modeling is based on the statement that face and facial expressions can be treated as topological structures. The generation of faces and facial expressions is the manipulation of topological primitives with special rules. Physical modeling is based on descriptions of facial muscles, skin, and their relationships by mathematical equations. Given a muscle model and emotions, the latter can be encoded or incorporated to produce an output facial expression as a topological configuration corresponding to an emotion.

2. The face is a source of physical and behavioral biometric information. Physical information includes face geometry (relationship of its components) and local geometries (geometries of mouth, nose, etc.). The behavioral information is carried by topological relationships. For example, a smile is defined as the topological composition of local geometries. Synthesis can be accomplished by assembling master expressions from the library of basic facial expressions.

3. In modeling, facial expressions are treated as topological configurations. The problem of analysis is defined as processing these topological

structures with the goal to identify or recognize. The problem of synthesis is defined as construction: topological structures are designed which correspond to facial expressions. Various techniques can be used for converting, transforming, and manipulating facial topologies.

4. A combination of local topological transformations and their dynamic relationships are used in many applications. For example, lip tracking can be synchronized with speech (speech-to-lips systems) and vice-versa (lips-to-speech systems). Other sources of information can be integrated in these systems to design biometric devices. For example, synthesis and recognition techniques can be combined with facial modeling methods to produce systems such as text-to-speech, text-to-lips, and text-to-speech-to-lips.

5. The extraction of various social contexts from speech, body movements and facial expressions can be partially automated, i.e., the expert can be supported by an automated system.

6. An expert is still required in the interpretation of the information and decision making as to social meaning. The role of an automated system is to support an expert in collecting, processing, and classifying information. For example, an expert can be supported by artificial intelligence tools in the formulation of extra control questions in a lie detection session. These tools help to improve the reliability of decision making.

7. Computing synthetic behavioral information such as facial expressions can contribute to social security, in particular, in the development of:

   ▶ New decision making systems for immigration services,
   ▶ A new generation of devices for predicting human behavior to help identify possible drivers for a wide range of socially disruptive events and to put mitigating or preventative strategies in place before the fact, and
   ▶ A new generation of non-contact lie detectors.

## 5.8   Problems

To solve Problems 5.1, 5.2, and 5.3, you can use your images. Some data sets of face images can be found at http://www.imm.dtn.dk/~aam/datasets/datasets.html. Problem 5.4 can be considered as a team project. **.

---

**This section was written by S. Yanushkevich, S. Kiselev, and V. Smerko

**Problem 5.1** The following MATLAB code implement image morphing by a random grid deformaton:

```
load clown; image = ind2rgb (X, map);
[X, Y] = meshgrid
(0.0:1.0/4:1.0, 0.0:1.0/4:1.0); surf (X, Y, zeros(size(X)),
'CData', image, 'FaceColor', 'texturemap');
X(2:4, 2:4) = X(2:4,
2:4) + (0.2 * rand (3, 3) - 0.1); Y(2:4, 2:4) = Y(2:4, 2:4) +
(0.2* rand (3, 3) - 0.1);
surf (X, Y, zeros(size(X)), 'CData',
image, 'FaceColor', 'texturemap');
```

Create several new randomly deformed images.

**Problem 5.2** The following MATLAB code warps the initial image so that the resulting image can be interpreted as generating facial expressions:

```
load clown; image = ind2rgb(X,map);
[grid.x, grid.y] = meshgrid (0.0:1.0/47:1.0, 0.0:1.0/47:1.0);
% Surface Grid
mesh (grid.x, grid.y, zeros (size(grid.x)));
[cp.x, cp.y] = meshgrid (0.0:1.0/4:1.0, 0.0:1.0/4:1.0);  % Control Points
mesh (cp.x, cp.y, zeros (size(cp.x)));
cp.x(2, 2) = cp.x(2, 2) -
0.24; cp.x(2, 3) = cp.x(2, 3) - 0.48;
cp.x(3, 2) = cp.x(3, 2) -
0.24; cp.x(3, 3) = cp.x(3, 3) - 0.48;
cp.x(4, 2) = cp.x(4, 2) -
0.24; cp.x(4, 3) = cp.x(4, 3) - 0.48;
cp.x(2, 4) = cp.x(2, 4) -
0.24; cp.x(3, 4) = cp.x(3, 4) + 0.24;
cp.x(4, 4) = cp.x(4, 4) +
0.24; mesh (cp.x, cp.y, zeros (size(cp.x)));
p = xffd (cat (2,
grid.y(:), grid.x(:)), {cp.x cp.y});
grid.x(:) = p(:, 1);
grid.y(:) = p(:, 2);
mesh (grid.x, grid.x, zeros (size(grid.x)));
surf (grid.x, grid.y, zeros (size(grid.x)), 'CData', image,
'FaceColor', 'texturemap');
```

Change the parameters of the given MATLAB code, so that the resulting image can be interpreted as new facial expressions.

**Problem 5.3** Given images of two eyes with equal size (Figure 5.26a,b), the absolute difference of these two images can be calculated. The result can be interpreted as designing a synthetic eye image (Figure 5.26c). The corresponding MATLAB code is given below. In this code, `strFirstImageFile` is the name of the first image file, `strSecondImageFile` is the name of the second image file, for example, `problem53('eye1.jpg','eye2.jpg')`

```
problem53(strFirstImageFile, strSecondImageFile)
    I1_1 = imread(strFirstImageFile);
    I1_2 = imread(strSecondImageFile);
    I2_1 = rgb2gray(I1_1); imshow(I2_ 1);
    I2_2 = rgb2gray(I1_2); imshow(I2_2);
    I_AbsDiff = imabsdiff(I2_1,I2_2);
    imshow(I_AbsDiff);
```

Find the result of the following transformations:
(a) addition of two images (use MATLAB function `imadd`)
(b) subtraction of two images (use MATLAB function `imsubtract`).

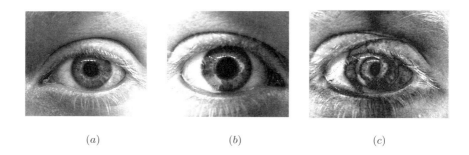

(a)                         (b)                         (c)

**FIGURE 5.26**
Synthetic eye design (Problem 5.3).

**Problem 5.4** The Prewitt method finds edges using the Prewitt approximation to the derivative. It returns edges at those points where the image gradient is maximum. The Canny method finds edges by looking for local maxima of the gradient of image. The gradient is calculated using the derivative of a Gaussian filter. Use the MATLAB function `problem54(strImageFileName)`, where input parameter `strImageFileName` is the name of image file. The MATLAB code is as follows:

```
function problem54(strImageFileName)
    RGB = imread(strImageFileName);
    I = rgb2gray(RGB); imshow(I),figure
    BW1 = edge(I,'prewitt'); BW2 = edge(I,'canny'); imshow(BW1);
    figure, imshow(BW2)
```

Example of the resulting image is given in Figure *5.27;* the MATLAB function `problem59('eye_pair1_1.jpg ')` was applied. Using your images, determine edges by implementing the following transforms available in the MATLAB image processing Toolbox:

(a) Sobel's operator `sobel`
(b) Roberts' operator `roberts`
(c) log operator `log`.

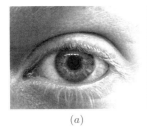 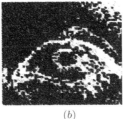 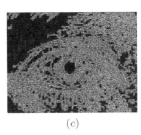

$(a)$ $(b)$ $(c)$

**FIGURE 5.27**
The eye image (a) and edges detecting by Prewitt (b) and Canny (c) methods
(Problem 5.4).

## 5.9 Further reading

**Systems for converting representations of information.** There are various types of automated systems for converting information generated from face attributes:

▶ `Text-To-Speech-To-Face` produces an audio output synchronized with facial motion, from text input,

▶ `Speech-To-Face` produces lips and facial motion, from an audio input,

▶ `Face-To-Speech` produces an audio output, from facial motion,

▶ `Face-To-Speech-To-Text` produce an audio output and text, from facial motions.

These systems are used in telecommunication services, language study, devices for people with disabilities, talking books and toys, vocal monitoring (control of machines, for example), human-machine interfaces, etc. [68, 76, 85, 86].

The *analysis-by-synthesis* paradigm is one underlying theme that this book hopes to illuminate. Accordingly under this paradigm, the utterances are analyzed and resynthesized to verify the perceptual equivalence between the original and the stylized intonation, where the stylized intonation is the output of the model.

Synthetic speech generation is the common component of the above systems. Synthesis of speech by numerical methods has evolved considerably since the first experiments in the 1960s. Intelligibility seems to have reached a sufficiently high level, and new challenges have emerged. Improved audio quality of the synthesis, improved naturalness of speech, synthetic singing, and the rendering of emotions, are among the new targets in speech synthesis technologies.

Several levels of speech description are used: representations in the *acoustic, phonetic, phonological, morphological, syntactic,* and *semantic* domains. Each of these levels can be used to extract behavioral information from speech.

- The acoustic level includes time domain and frequency domain representation. In the acoustic domain, pitch, loudness and timbre are the main parameters. For analysis and synthesis, the most important frequency range is for men from 70 to 200 *Hz*, for women from 150 to 400 *Hz*, and for children from 200 to 600 *Hz*.

- At the phonetic level, the international phonetic alphabet is often used to describe the sounds. The basis of this alphabet are called the *vocal fold* vibrations. Briefly, the respiratory organs provide the energy needed to produce speech sounds. The sound is made by forcing an air flow in the trachea and through the *vocal cords* (or *folds*) which vibrate.

- At the phonological level, speech is described in linguistic units called *phonemes*, the smallest meaningful units in language. The phonological representation is based on three main parameters: duration, intensity, and pitch (also called *prosody*).

- At the morphological level, phonemes combine to form words. Combinations of phonemes at the morphological level are restricted by the rules of syntax that emerge at the syntactic level, ensuring that sequences of words result in a correct sentence.

- At the semantic level, semantic features attach to words giving the sentence meaning.

**Behavioral information in voice.**. The voice carries emotional information by *intonation*. Several models of intonation have been developed for automated speech synthesis. These are grouped into three classes: acoustic, perceptual, and linguistic. For example, d'Alessandro and Mertens [16] have been developing an effective acoustic model. For further reading, an excellent textbook on text-to-speech synthesis by Dutoit [18] can be recommended. More details can be found in [60, 66, 67, 74, 81].

**Lip modeling.** In face-to-face communication, the speech perceived by a person depends not only on acoustic cues, but also on visual cues such as lip movements, and motion of the teeth and tongue is visible. It has been reported that speech intelligibility in noisy environments is improved by combining visual information with the acoustic signal (degraded by noise). Thus a realistic mouth synthesis is a critical part in systems for conveying information with the aid of visual clues. Rendering a convincing mouth requires spatial techniques to produce the large appearance variability between individuals, as well as differences in illumination, mouth opening, and the visibility of teeth and tongue.

There have been many attempts at dynamic lip modeling. For example, Mase and Pentland [50] proposed to track action units with an optical flow instead physical modeling. Kass et al. [41] have used active contour models called snakes as a topological representation of facial structures. In the 3D face model developed by Terzopoulos and Waters [82], linear facial features are used to represent facial expressions. Results by K. Aizawa et al. [1] and Yamamoto et al. [84] contribute to solving the problem of information translation. In lip modeling, the book Walther [88] is useful. Ypsilos et al. [87] reported results on lip modeling in 3D face from given speech sequences.

Bregler and Omohundro [4] developed a lipreading technique using a machine learning paradigm. The model of the lip is based on so-called *manifolds* which are smooth surfaces corresponding to possible lip manipulations. Manifolds describe a changing lip configuration as a point in the lip feature space moving along a smooth curve. The authors used the analysis-by-synthesis approach to test the generated manifolds. For visual speech recognition, the authors used a hidden Markov model (HMM) suitable for recognizing the time sequences in lip images. In this approach, a word utterance corresponds to the feature vector formed by a point moving along a trajectory of the manifold. The domain of the HMM vectors is defined by the manifold. An HMM word model represents a probability distribution over trajectories on the manifold of this word. Li et al. [48] used eigensequences in lip modeling. The eigensequences were computed by training sequences for each spoken letter. Ypsilos et al. [87] reported results on 3D face synthesis from speech.

**Eye movement.** The eyes are considered as channel of non-verbal communication between people. Lee at al. [46] developed an approach for synthesizing the trajectory and statistical distribution of so-called *saccades* eye movements during facial expression. Saccades are rapid movements of both eyes from one gaze position to another. The authors used empirical models of saccades and statistical models of eye-tracking data. The guide by Faigin [29] can be useful in developing eye movement models. Details of psychological studies on eye movements can be found, for example, in [83].

**Automated facial expression synthesis.** Comprehensive literature surveys on facial expression analysis and synthesis are provided in [30]. In particular, Wang and Ahuja [89] define the problem of facial expression synthesis in the formulation: for a given unknown person with known expression, synthesize other expressions. To formalize the problem, the following metric spaces were developed: $U^{person}$, $U^{expression}$, and $U^{feature}$. For example, given a known person with known expression, say, one of happiness, using the above model, all the people in the database can be synthesized with the same expression of happiness.

Synthesis based feedback control systems have also been used to analyze image sequences of facial expressions (Figure 5.28). Facial analysis in Figure 5.28 is formulated as deriving a symbolic description of a real facial image. The aim of facial synthesis is to produce a realistic facial image from a symbolic facial expression model. For example, Koch [42] used a synthesis feedback control system to analyze facial image sequences. The scene is first synthesized with scene parameters such as shape, motion, and texture. The synthesized frame is then fed back to the analyzer, which tunes the scene parameters to produce a better image. In this way the parameters describing the image can be found interactively. Morishima [53] implemented an analysis-by-synthesis paradigm by a 5-layer neural network.

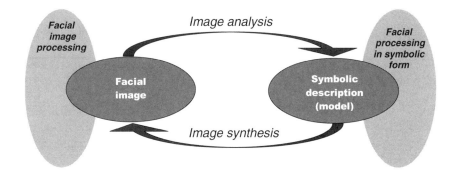

**FIGURE 5.28**

Analysis-by-synthesis approach in facial image.

Essa and Pentland [26] have used physics-based dynamic models of skin and muscles coupled with optimal estimates of optical flow in a feedback controlled framework. They discussed two types of representation of facial motion used to recognize facial expressions: a direct representation based on muscle activations and a representation in the form of spatial motion-energy templates.

Systems for face and facial synthesis are improved by the development of new paradigms. Typically, such a system provides the possibility to design faces in 2D and 3D, to model race, gender and adult age group, control shapes, coloring, etc. In this book, for example, the software package FaceGen was used. More details can be found in [10, 13, 15, 40, 56, 72, 89]. In [85, 86], state-of-the-art facial analysis and synthesis is introduced. The reader can find a very detailed study on automatic facial expression measurement in the paper by Movellan and Bartlett [54].

**Visualization of emotion** is needed in the human-machine interface used in lie detectors to support an expert during a dialog. Morishima [53] used 2D and 3D rectangular coordinates for visualization of basic expressions.

**Facial expressions synthesis and psychology.** The face is capable of displaying a complex array of facial expressions. Historically, Charles Darwin started a systematical study of facial expressions a century ago in the book *The Expression of the Emotions in Man and Animals*. Recent interest in this problem makes it recommend reading. Many researchers use a list of so-called fundamental emotions, for example, [14, 37].

Tian et al. [80] developed system to analyze facial expressions based on both permanent facial features (brows, eyes, mouth) and transient facial features (deepening of facial furrows). This system is based on description of facial expressions in action units (FACS) of Ekman and Friesen facial action coding system [25] and recognizes fine-grained facial changes. According to Bradley and Lang [3] measurable data that gives clues to emotional state can be divided into three groups: (a) functional behavioral sequences (overt acts), (b) emotional language (expressive communication), and (c) physiological reactions (changes in somatic muscles).

Ekman in [22, 25, 24] has developed a data structure and measuring criteria for behavioral information useful in algorithms. His representation of facial information is based on a limited number of expressions (he gives seven basic facial expressions: happiness, sadness, fear, anger, disgust, and surprise) and a limited number of facial feature (Figure 5.29). The facial action coding system provides a useful tool for the creation of grouped functional synergies which makes unnecessary the explicit control of individual muscles. In addition, criteria suggested by Ekman require that a basic emotion must be very brief and be characterized by a specific, interculturally stable facial expression pattern. This data structure representing emotions and facial expressions provides comprehensive coverage of the full spectra of emotion.

More details can be found in [20] (there are more then 20 papers of facial expressions studied). The results of this research provide useful information for the design of automated tools. Also, techniques for measures based on facial action coding systems can be useful for the development of a library of topological primitives corresponding to facial emotions. For example, in Ekman and Friesen measures system, the classifier `AU1` corresponds to the descriptor "`Inner brow raised`" with muscular basis "`Frontalis, pars medialis`". That is the topology of "`Inner brow raised`" can be considered

| *Emotion code* | *Expression code* | *Muscle activity code* |
|:---:|:---:|:---:|
| **Emotion** | **Expression** | **Facial muscles activity** |

**FIGURE 5.29**
Facial expression data structure.

as a topological primitive in a library for master-expression design and automated identification. For modification of this topological primitive and also for physical modeling, information about the muscles that generate this topology is essential. Some other example descriptors include: "`Lip corner pulled`" (AU12), "`Tongue out`"(AU19), "`cheek blow`"(AU33), and "`Eyes closed`"(AU43).

Scherer and Ekman [71] observed more than 7,000 different combinations of action units. A paper by Gosselin at al. [20] contributes to understanding the simulation of emotions by studying the Stanislavsky technique, named after the famous Russian actor at the beginning of the 20th century. This study is useful for

▶ Automatic face-to-face communication aimed to support the recognition of genuine and simulated facial expressions,

▶ Development of controlled dialog scenario,

▶ Testing of algorithms, and

▶ Scenario of attacks on automated tools.

Stanislavsky claimed that the retrieving of past emotional experiences works as an elicitor of genuine emotion and leads to spontaneous-like expressive behavior. His famous exclaim "Hard to believe!" during actor training demonstrates his ability to identify and simulate genuine and voluntary emotions.

Waters [91] used Ekman and Friesen's coding system for a 3D parameterized static and dynamic facial muscle model including biomechanical properties of facial tissue. He has demonstrated basic synthetic facial expressions (surprise, fear, disgust, etc.) and their modifications.

Psychological study combined with advanced techniques of image processing, biometric technology, and artificial intelligence techniques provide new horizons in distance recognition of emotional and voluntary facial movements. In particular, Ekman's definition of lie provides acceptable conditions for formal and algorithmic development and the automated support of decision making [23]. He distinguished situations when the false information can not be classified as a lie (false information is spoken unwittingly). A lie or deceit is a situation that satisfies three requirements: when one person intends to mislead another, doing so deliberately, without prior notification of this purpose, and without having been explicitly asked to do so by the investigator/expert.

Decision making during lie detection is an extremely complicated procedure because a tested person can try:

▶ To *conceal* some information,

▶ To *falsify* information: (a) the liar withholds some true information, or (b) presents false information as if it were true.

Probably, it is impossible to automatically detect the difference between these types of information in all cases because of the limited information measurable by the detector. However, in some situations, it is possible to create improved conditions for detecting the lie, for example, by provoking a person with questions. In this situation, a liar should falsify information that can be detected with reasonable probability.

**Facial muscle models.** Physical (muscle) models have been used by Kalra et al. [39], Platt and Bandler [59], and Waters [90] for character animation. In these models, muscles are simulated taking into account that each has different regions of influence and different ways in which they affect nodes within their regions of influence. Interactions between muscles can be described using well defined models. All muscles are attached to the facial surface at one end, and to the underlying bone structure at the other end. When actuating a muscle, the points of attachment are pulled towards the bone. Each muscle is represented using vector notation, where the tail of the vector is located approximately where the muscle fibers are attached to the skin layer. The surface within the neighborhood of the area where the muscle fibres connect to the skin are under varying degrees of influence from the force due to muscle actuation. Terzopoulos and Waters [82, 91] analyzed facial expressions using a skin-tissue model with simulated muscles based on physical principles. The model is defined in layers with nodes on the surface considered as point masses and the top layers interconnected to the other underlying layers. Zhang et al. [95] studied the skin model using a mass-spring system for synthesizing 3D facial expressions. Zhang et al. [96] developed a deformable multi-layer skin model.

Huang and Yan [36], proposed to control human facial expressions using non-uniform rational B-spline (NURBS) curve. The control points of the NURBS curves are positioned according to facial anatomy. The weights can be changed, or the control points can be repositioned in order to model various facial expressions.

Ezzat et al. [28] developed a machine learning technique for speech animation. They proposed a so-called *multidimensional morphable model* to synthesize mouth configurations from a set of mouth image prototypes. The basic statement underlying the multidimensional morphable model development was that the complete set of mouth images associated with speech lies in a low-dimensional space and mouth appearance can be represented automatically as a set of prototype images extracted from the recorded *corpus*. So, a multidimensional morphable model can be constructed automatically. To represent each image in the video corpus as a set of low-dimensional parameters, a technique of principal component analysis was used. Authors also used a trajectory synthesis technique for automatically training from the recorded video corpus. After training, trajectories of mouth configurations can be synthesized to create desired utterance. Many useful details on the relationship of facial expressions and speech can be found in [58].

Methods of distance field metamorphosis, i.e., given objects, construct intermediate objects [11, 18] can be useful for intermediate facial expression synthesis. Distance fields metamorphosis can be defined as a reconstruction of the surface by interpolating distance values. Cohen-Or et al. [11] described a method for creating a series of objects which form a smooth transition between keyframe objects. Such morphing is an example of distance field metamorphosis.

**Artificial intelligence paradigm in facial expression recognition.** Rosenblum et al. [69] developed a radial basis function neural network that learns the correlation of facial features motion patterns and human expressions (smile and surprise). Zhang [97] investigated various strategies of features extraction from the face using a two-layer neural network.

Lawrence et al. [45] studied the operation of a two-phase recognition strategy for face recognition: at the first phase of processing, a self-organizing map is created for quantizing the image samples into a topological space, and at the second phase, a convolutional network is implemented to provide for partial invariants to translation, rotation, scale, and deformation.[††]

Morishima [53] introduced a human-machine communication system based on synthetic face generation. In modeling emotion, he used a feed-forward neural network architecture. The analysis-by-synthesis paradigm was implemented by 5-layer network containing input and output layers, and hidden layers corresponding to analysis, synthesis, and 3D emotion space. The number of cells in each layer is defined from the number of Ekman and Friesen's AUs: the 3D emotion space (middle layer) corresponds to 3 AUs, input and output layers correspond to 17 AUs, and analysis and synthesis layers correspond to 19 AUs. This architecture assumed that the data patterns distributed in a semicircle feature. The author claims that after training this 5-layer feed-forward neural network, it gives an acceptable model of facial expressions. He reported that basic emotions (anger, disgust, sadness,

---

[††]A self-organizing map, or Kohonen map is defined as unsupervised learning process which learns the distribution of a set of patterns without any class information.

fear, surprise, and happiness) were well recognized and synthesized, and it is possible to detect and synthesize changes between basic emotions. Rowley et al. [70] proposed to use a neural network for face detection.

**Singing voice synthesis.** The voice is considered to be the most flexible, rich and powerful of all musical instruments. A number of successful implementations of singing voice programs exist. One of the recent approaches is to use morphed voice samples for resynthesis as in speech synthesis. A collection of papers listed in [6] introduce the above topic.

Physical models try to imitate the voice generation process by mathematically or physically mimicking the physical laws that reign in the voice organs. Cook [12] has studied sound synthesis related to physical models. A satisfactory real-time modeling of the singing voice based upon the physiology of an individual singer is not yet possible.

In the search for common, fundamental principles of expressive performance, Widmer [92, 93] has used inductive machine learning. The goal is to discover general rules of certain aspects of performance (e.g., expressive timing) that are valid in the sense that they can predict, with a certain level of accuracy, how performers are going to play particular musical passages.

Techniques of emotion synthesis in music performance are similar to synthetic voice synthesis with respect to given emotions, i.e., warping emotion data into a sound data representation. Canazza et al. [8] have developed a system for the real-time control of musical performance synthesis designed to express emotion. Users can change the emotional intent from happy to sad or from a romantic expressive style to a neutral while it is playing. The model is based on the hypothesis, that different expressive intentions can be obtained by the suitable modification of a neutral performance. Authors implemented several innovative ideas, in particular, morphing in a control space (this space reflects how the musical performance might be organized in the listeners mind) and a configurable library to manage rules for musical expression syntheses.

The problem of automated emotional coloring of a music performance is the focus of many authors, in particular, Bresin and Friberg have reported results in [5], and Canazza et al. [9] have contributed to an audio morphing technique. Story and Titze [78] have studied a model for voice synthesis focusing on body movement as the form of emotion representation.

The analysis of facial expressions, vocalic formants generation and the creation of a synthetic computer-generated singers are techniques required for synthesis with realistic expressiveness. They can be used to create sophisticated artistic effects.

**Caricature** is the art of making a drawing of a face. A caricature is a synthetic facial expression, where the distances of some feature points from the corresponding positions in the normal face have been exaggerated. Some researchers developed the *exaggeration rules* for the proportions of eyes, mouth, lips, nose based on analysis of the different art styles. For example, Luo et al. [49] developed a method of exaggerating facial features using several empiric rules. A caricature should only *exaggerate* features without destroying the whole ratio of the face. Fujiwara et al. [32] developed a method to design caricatures and compared these results with the drawings of a professional caricaturist. They used a "mean face hypothesis" that is defined as an average of input faces. Features are extracted from this "mean face" and deformed with respect to several constraints. Gooch et al. [35] present an interactive technique for creating black-and-white caricatures from photographs of human faces. The effectiveness of the resulting image was evaluated through psychophysical studies and speed in both recognition and learning tasks. It has been observed that the caricatures are as effective as photographs in recognition tasks and are learnt one and a half times faster than photographs. Stollnitz et al. [77] described properties of wavelets useful for caricatures. Methods of metamorphosis (given objects, construct intermediate objects) [11, 18] can be useful to caricature design.

**A history of the lie detector.** Origins of modern polygraph techniques begin in 1733 with an invention by S. Hales, an English physiologist and chemist [73]. He proposed a method of measuring arterial blood pressure. The present method of measuring blood pressure was invented in 1896 by S. R. Rocci, an Italian physiologist. Other measurements that are recorded in today's polygraph are pulse and respiration. The precursor of the pneumatic chest tube, an instrument which records respiration, was invented in the latter half of the nineteenth century by J. Hutchinson. The concept of using a machine to judge truth and deception emerged in 1895 by C. Lombroso. He was the first who demonstrated how to use two sources of response information, blood pressure and pulse changes, to help determine the truth of various statements of criminal suspects. In 1914, V. Benussi proposed to use the third source of information, respiration, to improve the accuracy of lie detection. He used a pneumatic chest tube to record changes in respiration; he discovered that the ratio of the length of inspiration to the length of expiration decreased after telling the truth and increased after lying. A machine, which continuously recorded information from three sources of information, blood pressure, pulse, and respiration, was created in 1921 by J.A. Larson, an American police officer. This machine, a polygraph, stimulated researchers to improve the accuracy of lie detection by developing scenarios dialog to provoke a measurable response and by seeking new sources of information. Skin conductance has been found useful to measure arousal. According to recent studies, skin conductance increases linearly as ratings of arousal increases, regardless of emotional valence. This fourth source of information, electrodermal information, is used in modern polygraph detectors.

**Infrared imaging.** Today, facial expression is analyzed predominantly in the visible spectrum, which is only a small band of the electro-magnetic spectrum. Wilder et al. [94] compared the effectiveness of visible and infrared imagery for detecting and recognizing faces in areas where personnel identification is critical. It was found that periorbital blood flow rate carries significant useful information. This information was used to define feature vectors and to create a classifier. A brief review of infrared identification techniques and potential applications can be found in the paper by Prokoski [64].

Sugimoto et al. [79] developed a method for detecting the transition of emotional states which exploits thermal image processing and facial expression synthesis. Facial muscle movement may be intentional. For detecting the face-temperature-change produced from the true expression of emotional states and/or physiological changes, it may be effective to eliminate the influence of facial muscle movement on the faces thermal distribution. For this reason, a test face and a neutral-expression face are re-formed to a frontal view neutral-expression face. The thermal difference between these two re-formed faces is used for analyzing the facial temperature change to detect transitions of emotional states.

Pavlidis and Levine [57] proposed to use thermal images as an additional source of information for improving lie detectors. The measurement such as blood flow rate used in lie detector can be evaluated using thermal images, as redistribution of blood flow in blood vessels causes abrupt changes in local skin temperature. It was confirmed in [17] through experimental study that thermal images contain useful information for human face recognition. Srivastava and Liu [75] used a Bayesian approach to identify faces from infrared facial images in the presence of nuisance variables such as pose, facial expression, and thermal state. Eveland et al. [27] proposed a system for tracking human heads using thermal emission from human skin. Ypsilos et al. [87] developed a synchronized multiple camera system with infrared stereo capture for shape reconstruction and visible capture for color appearance. Ju et al. [38] introduced techniques based on a 3D thermogram scanner that combine a 3D and a thermal camera (the infrared picture is mapped onto the 3D geometry). Kong at al. [43] provide an up-to-date review of research efforts in face recognition techniques based on two-dimensional (2D) images in the visual and infrared spectra.

**The lie detector problem** is a unique area for testing various state-of-the-art methodologies from different scientific research areas, in particular:

▶ Psychology,
▶ Neuroscience,
▶ Criminology and justice,
▶ Ethics,

and technical fields such as:

▶ Modeling of complex systems,

▶ Image processing and pattern recognition,

▶ Artificial intelligence, and

▶ Information theory.

The objective of polygraph testing is to ascertain if the subject under investigation truthfully or deceitfully answers questions in a real time dialog [51, 63, 73]. Lie detectors record not lies but physiological changes. When using a lie detector, the idea is to maximize elicitation by questioning, i.e., emotions might be provoked by questions using a dialog scenario. In today's lie detectors, four physiological parameters are monitored during the testing:

▶ Blood pressure,

▶ Pulse changes,

▶ Respiratory changes, and

▶ Skin conductivity.

This information is recorded and classified, and used to aid an expert in making a decision about the truthfulness of the subject response.

The polygraph is a classical example of a simplest model of a complex system. The effect of this simplification can be observed by the following experiment. The detected physiological changes which are associated with deception are also associated with various conditions other than deception. This means that the extracted features cannot be distinguished contributing to decision making errors. This is a typical situation in modeling complex systems. A possible reliable solution is based on exploiting additional sources of information that provide distinguishing features. For example, in signature and face identification additional sources of information represented as a third dimension. Approaches for enhancing polygraph technique can be divided into two groups:

▶ Intelligent methods to aid an investigator in shaping an interview to get the truth from a subject. This can be in the form of training an investigator, or of real time feedback and analysis obtained during the interview, and

▶ Methods based on using additional sources of information.

**Improving lie detectors.** One of the possible approaches to improving lie detectors is based on the extraction of additional information from the electromagnetic spectrum generated by a human. Some of these problems are discussed by Polonnikov [61]. His extensions and experiments are based on the general statement that a person's electromagnetic field carries valuable information about the processes in his body. This means, in particular, that the individual and their psychological state can be remotely identified and estimated by measuring this electromagnetic field. Polonnikov demonstrated, by experiments, various sources of information about a person's physiological state. In [62], some aspects of this approach are demonstrated by the study of electroencephalograms.[‡‡]

Bradley and Lang [3] reviewed several studies on regional brain imaging that assess regional neural activity indirectly through variations in blood flow, including tomography and functional magnetic resonance imaging [7]. Researchers observed differences in patterns of brain activity when subjects were viewing things that made them emotional, compared to their neutral state. Nomura et al. [55] studied the neural substrates involved in the recognition of ambiguous facial expressions using functional magnetic resonance imaging.

In [63], a collective study of the lie detection problem by several researchers from various scientific fields is given. Alternative techniques and technologies are discussed in four directions: improving the polygraph itself, observation of brain functions, observation of specific behaviors of examinees (voice, facial expressions, choice of words, etc.), and overt investigations (questionnaires, paper-and-pencil tests, etc.).

**New generation of lie detectors.** The main goal of developing a new generation of lie detectors is to improve the rate of success of today's lie detectors. This goal can be approached by using additional sources of information, such as that observed in invisible part of the electromagnetic spectrum. There are several other features which distinguish the new generation of lie detectors from the existing ones, in particular,

▶ Ultimately noninvasive and non-contact (distance) measures.

▶ An increasingly limited role for the examiner; artificial intelligence will be used to support various levels of decision making.

In Figure 5.30, an architecture of a new generation of lie detector is given. This architecture differs from today polygraph in several ways:

▶ Noninvasive collaboration; scanning information is based on distant principles.

▶ Examiner is virtual; its image and voice can be synthesized, and behavior characteristics can be controlled during the testing with respect to the response and scenario.

---

[‡‡]In more detail, Polonnikov has developed the so-called *fractal dynamic* approach to the analysis and synthesis of the human electromagnetic field. He distinguishes three data structures in this field: *regular brain activity* (biorhythms), *chaotic activity*, and *fractal fluctuation*. These three sub-structures compose a *multi-fractal* structure. The hosts of this multi-fractal data are electromagnetic waves. It should be noted that there are many studies worldwide to develop models of electroencephalograms, including fractal structures which suppose the presence of self-similarity in the process. Polonnikov understood fractal structures as an information network between all hierarchical levels of a human's instance including the molecular level. Control of this network is implemented by signals of a semantic nature. Polonnikov likened the problem of learning to recognize this semantic system to learning the natural "language" of the human body.

▶ Multibiometric paradigm of processing information from sensors;

▶ Optimization the response of examinee via feedback analysis;

▶ High-parallel processing; information from sources and between sources are processed in parallel.

In Figure 5.30, overlapping processing of information from eight sources is implemented by a processing systems. The sources of information to be combined in an automated coherent processing system are given in Table 5.1.

**TABLE 5.1**
Sources of information in a new generation of lie detectors.

| Level | Biometric | Technique |
|:-----:|-----------|-----------|
| 1 | Respiration, heart rate, blood pressure, and electrodermal response | Physiological parameters such as these are used in a classical polygraph |
| 2 | Facial expression | Visible response if sufficiently processed is useful |
| 3 | Thermal facial expression | Shows the temperature of the skin, and can give clues to blood flow rate and muscle tension |
| 4 | Acoustic | Voice stress analysis is useful in automated coherent processing with another sources of information, |
| 5 | Brain activity | This can be detected via position emission tomography and magnetic resonance imaging |
| 6 | Other electromagnetic body radiation | Invisible response |
| 7 | Graphological analysis | Detail of personality and emotional state can be obtained through analysing handwriting [2] |
| 8 | Linguistic analysis | This information becomes useful when combined, for example, with facial expressions (in visible and invisible spectrum) and body movement |

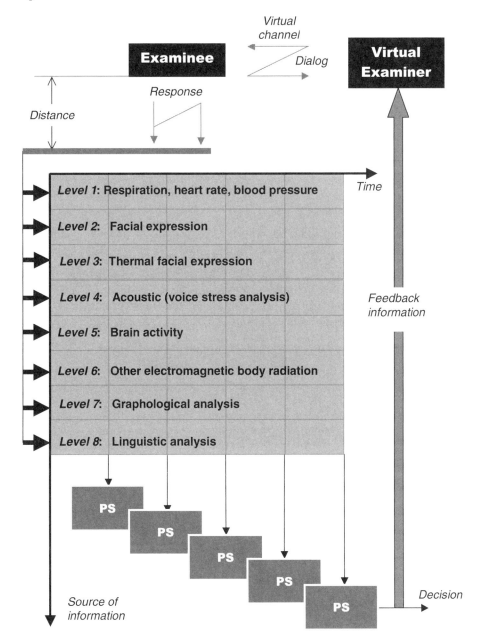

**FIGURE 5.30**
Architecture of a new generation of lie detector.

## 5.10   References

[1] Aizawa K, Choi CS, Harashima H, and Huang TS. Human facial motion analysis and synthesis with applications to model-based coding. In Sezan MI and Lagendijk RL, Eds., *Motion Analysis and Image Sequence Processing*, pp. 317–348, Kluwer, Dordrecht, 1993.

[2] Ben-Shakhar G and Elaad E. The guilty knowledge test (GKT) as an application of psychophysiology: future prospects and obstacles. In Kleiner M, Ed., *Handbook of Polygraph Testing*, pp. 87–102, Academic Press, San Diego, 2002.

[3] Bradley MM and Lang PJ. Measuring emotion: behaviour, feeling, and physiology. In Lane RD and Nadel L, Eds., *Cognitive Neuroscience of Emotion*, Oxford University Press, Oxford, 2002.

[4] Bregler C and Omohundro S. Learning visual models for lipreading. In Shah M and Jain R, Eds. *Motion Based Recognition*, pp. 301–320, Kluwer, Dordrecht, 1997.

[5] Bresin R and Friberg A. Emotional colouring of computer controlled music performance. *Computer Music Journal*, 24(4):44–62, 2000.

[6] Bresin R, Ed., *Proceedings of the Stockholm Music Acoustics Conference*, 2003.

[7] Buxton RB. *Inroduction to Functional Magnetic Resonance Imaging: Principles and Techniques.* Cambridge University Press, Cambridge, 2002.

[8] Canazza S, Rodà A, Zanon P, and Friberg A. Expressive director: a system for the real-time control of music performance synthesis. In *Proceedings of the Stockholm Music Acoustics Conference*, Stockholm, Sweden, pp. 521–524, 2003.

[9] Canazza S, De Poli G, Drioli C, Rodà A, and Vidolin A. Audio morphing different expressive intentions for multimedia systems. *IEEE Transactions on Multimedia*, (3)7:79–83, 2000.

[10] Chellappa R, Wilson C, and Sirohey S. Human and machine recognition of faces: a survey. *Proceedings of the IEEE*, 83(5):705–740, 1995.

[11] Cohen-Or D, Levin D, and Solomovici A. Three-dimensional distance field metamorphosis. *ACM Transactions on Graphics*, 17(2):116–141, 1998.

[12] Cook PR. *Real Sound Synthesis for Interactive Applications*, A K Peters, Natick, MA, 2002.

[13] Cootes TF, Edwards GJ, and Taylor CJ. Active appearance models, *IEEE Transactions on Pattern Analysis and Machine Intelligence* 23:681–684, 2001.

[14] Cowie R, Douglas-Cowie E, Tsuputsoulis G, Votsis G, Kollias S, Fellenz W, and Taylor JG. Emotion recognition in human-computer interaction. *IEEE Signal Processing Magazine,* Jan., 2001.

[15] Craw I, Ellis H, and Sishman J. Automatic extraction of face features. *Pattern Recognition Letters*, 5:183–187, 1987.

[16] d'Alessandro C and Mertens P. Automatic pitch countour stylization using a model of tonal perception. *Computer Speech and Language*, 9:257–288, 1995.

[17] Dowdall J, Pavlidis I, and Bebis G. Face detection in the near-IR spectrum. *Image and Vision Computing*, 21(7):565–578, 2003.

[18] Dutoit T. *An Introduction to TTS Synthesis.* Kluwer, Dordrecht, 1994.

[19] Du Y and Lin X. Realistic mouth synthesis based on shape appearance dependence mapping, *Pattern Recognition Letters*, 23:1875–1885, 2002.

[20] Ekman P and Rosenberg EL, Eds., *What the Face Reveals: Basic and Applied Studues of Spontaneouse Expression Using the Facial Action Coding System (FACS)* Oxford University Press, Oxford, 1997.

[21] Ekman P. Basic emotions. In Dalgleish T and Power T, Eds., *The Handbook of Cognition and Emotion*, Wiley, New York, NY, 1999.

[22] Ekman P and Friesen WV. *Unmasking the Face. A Guide to Recognizing Emotions From Facial Clues.* Prentice-Hall, Englewood Cliffs, NJ, 1975.

[23] Ekman P. *Telling Lies: Clues to Deceit in the Marketplace, Politics, and Marriage.* W.W. Norton & Company, New York, 2001.

[24] Ekman P. *Emotion in the Human Face*, Cambridge University Press, 1982.

[25] Ekman P and Friesen WF. *The Facial Action Coding System: a Technique for the Measurement of Facial Movement.* Consulting Psychologists Press, San Francisco, CA, 1978.

[26] Essa I and Pentland A. Facial expression recognition using image motion. In Shah M and Jain R, Eds., *Motion Based Recognition*, pp. 271–298, Kluwer, Dordrecht, 1997.

[27] Eveland CK, Socolinsky DA, and Wolff LB Tracking human faces in infrared video. *Image and Vision Computing*, 21:579–590, 2003.

[28] Ezzat T, Geiger G, and Poggio T. Trainable videorealistic speech animation. In *Proceedings of ACM SIGGRAPH*, pp. 388–398, 2002.

[29] Faigin G. *The Artists Complete Guide to Facial Expression.* Watson-Guptill Publications, New York, 1990.

[30] Fasel B and Luettin J. Automatic facial expression analysis: a survay *Pattern Recognition,* 36(1):259–275. 2003.

[31] Fröhlich M, Michaelis D, and Strube HW. SIMsimultaneous inverse filtering and matching of a glottal flow model for acoustic speech signals. *Journal of the Acoustic Society of America,* 110(1):479–488, 2001.

[32] Fujiwara T, Koshimizu H, Fujimura K, Kihara H, Noguchi Y, and Ishikawa N. 3D modeling system of human face and full 3D facial caricaturing. In *Proceedings of the 3rd IEEE International Conference on 3D Digital Imaging and Modeling,* Canada, 2001, pp. 385–392.

[33] Hill DR, Pearce A, and Wyvill B. Animating speech: an automated aproach using speech synthesis by rules. *The Visual Computer,* 3(5):277–289, 1988.

[34] George EB and Smith MJT. Speech analysis/synthesis and modification using an analysis-by synthesis overlap-add sinusoidal model. *IEEE Transactions on Speech and Audio Processing* 5:389–406, 1997.

[35] Gooch B, Reinhard E, and Gooch A. Human facial illustrations: creation and psychophysical evaluation. *ACM Transactions on Graphics,* 23(1):27–44, 2004.

[36] Huanga D and Yana H. NURBS curve controlled modelling for facial animation. *Computers and Graphics,* 27:373–385, 2003.

[37] Izard CE. *Human Emotions.* Academic Press, New York, 1977.

[38] Ju X, Nebel JC, and Siebert JP. 3D Thermography imaging standartization technique for inflammation diagnosis. In *Proceedings of SPIE,* vol. 6, pp. 5640–5646, 2004.

[39] Kalra P, Mangili A, Thalmann NM, and Thalmann D. Simulation of facial muscle actions based on rational free form deformations. In *Proceedings of the Eurographics'92,* vol. 11, pp. C59–C69, Blackwell Publishers, 1992.

[40] Kanade T, Cohn J, and Tian Y. Comprehensive database for facial expression analysis. In *Proceedings of the IEEE International Conference on Automatic Face and Gesture Recognition,* pp. 46–53, 2000.

[41] Kass M, Witkin A, and Terzopoulos D. Snakes: active countour models. *International Journal of Computer Vision,* 1(4):321–331, 1987.

[42] Koch R. Dynamic 3-D scene analysis through synthesis feedback control. *IEEE Transactions on Pattern Analysis and Machine Intelligence,* 15(6):556–568, 1993.

[43] Kong SG, Heo J, Abidi BR, Paik J, and Abidi MA. Recent advances in visual and infrared face recognition a review. *Computer Vision and Image Understanding*, 97:103–135, 2005.

[44] LaCourse JR and Hladik FC. An eye movement communication-control system for the disabled. *IEEE Transactions on Biomedical Engineering*, 37(12): 1215–1220, 1990.

[45] Lawrence S, Giles CL, Tsoi AC, and Back AD. Face recognition: a convolutional neural network approach. *IEEE Transactions on Neural Networks*, 8(1):98–113, 1997.

[46] Lee SP, Badler JB, and Badler NI. Eyes alive. In *Proceedings of the ACM SIGGRAPH*, pp. 637–644, 2002.

[47] Lee S-Y, Wolberg G, Chwa K-Y, and Shin SY. Image metamorphosis with scattered feature constraints. *IEEE Transactions on Visualization and Computer Graphics*, 2(4):337–354, 1996.

[48] Li N, Dettmer S, and Shah M. Visually recognizing speech using eigensequences. In Shah M and Jain R, Eds., *Motion Based Recognition*, pp. 345–371, Kluwer, Dordrecht, 1997.

[49] Luo WC, Liu PC, and Ouhyoung M. Exaggeration of facial features in caricaturing. In *Proceedings of the International Computer Symposium*, China, 2002.

[50] Mase K and Pentland A. Lipreading by optical flow. *Systems and Computers*, 22(6):581–591, 1991.

[51] Matte JA. *Forensic Psychophysiology Using the Polygraph - Scientific Truth Verification - Lie Detection*. Williamsville, New York, J.A.M. Publications, 1996.

[52] Milliron T, Jensen RJ, Barzel R. and Finkelstein A. A framework for geometric warps and deformations. *ACM Transactions on Graphics*, 21(1): 20–51, 2002.

[53] Morishima S. Modeling of facial expression and emotion for human communication system. *Displays*, 17:15–25, 1995.

[54] Movellan JR and Bartlett MS. The next generation of automatic facial expression measurement. In Ekman P, Ed., *What the Face Reveals*. Oxford University Press, Oxford, 2003.

[55] Nomura M, Iidaka T, Kakehi K, Tsukiura T, Hasegawa T, Maeda Y, and Matsue Y. Frontal lobe networks for effective processing of ambiguously expressed emotions in humans. *Neuroscience Letters* 348:113–116, 2003.

[56] Pantic M and Rothkrantz LJM. Automatic analysis of facial expressions: the state of the art. *IEEE Transactions on Pattern Analysis and Machine Intelligence*, 22(12):1424–1445, 2000.

[57] Pavlidis I and Levine J. Thermal image analysis for polygraph testing. *IEEE Engineering in Medicine and Biology Magazine*, 21(6):56–64, 2002.

[58] Pelachaud C, Badler N, and Steedman M. Generating facial expressions for speech. *Cognitive Science*, 20(1):1–46, 1996.

[59] Platt SM and Bandler N. Animation facial expressions. *SIGGRAPH'81*, 15(3):245–252, 1981.

[60] Plumpe MD, Quatieri TF, and Reynolds DA. Modeling of the glottal flow derivative waveform with application to speaker identification. *IEEE Transactions on Speech and Audio Processing*, 7(5):569–586, 1999.

[61] Polonnikov RI. *Quasi-Metaphysical Problems*. Russian Academy of Sciences, Institute of Informatics and Automatization, Publishing House "Anatolya", Saint Petesburg, 2003.

[62] Polonnikov RI, Wasserman EL, and Kartashev NK. Regular developmental changes in EEG multifractal characteristics. *International Journal Neuroscience*, 113:1615–1639, 2003.

[63] *The Polygraph and Lie Detection*. The National Academies Press, Washington, DC, 2003.

[64] Prokoski FJ. History, current status, and future of infrared identification. In *Proceedings IEEE Workshop on Computer Vision Beyond the Visible Spectrum: Methods and Applications*, 2000, pp. 5–15.

[65] Provine JA. *3D Model Based Coding*. PhD thesis, Department of Electrical and Computer Engineering, University of Calgary, Calgary, Alberta, Canada, 1997.

[66] Rabiner LR and Schafer RW. *Digital Processing of Speech Signals*. Prentice-Hall, Englewood Cliffs, New York, 1978.

[67] Rao RR and Chen T. Audio-to-visual conversion for multimedia communication. *IEEE Transactions Indust. Electron.* 45(1):15–22, 1998.

[68] Roco MC and Bainbridge WS, Eds., *Converging Technologies for Improving Human Performance: Nanotechnology, Biotechnology, Information Technology and Cognitive Science*. Kluwer, Dordrecht, 2003.

[69] Rosenblum M, Yacoob Y, and Davis LS. Human expression recognition from motion using a radial basis function network architecture. *IEEE Transactions on Neural Networks*, 7(5):1121–1138, 1996.

[70] Rowley HA, Baluja S, and Kanade T. Neural network-based face detection. *IEEE Transactions on Pattern Analysis and Machine Intelligence*, 20(1):23–38, 1998.

[71] Scherer K and Ekman P. *Handbook of Methods in Nonverbal Behavior Research.* Cambridge University Press, Cambridge, 1982.

[72] Sederberg TW, and Parry SR. Free-form deformation of solid geometric models. *ACM SIGGRAPH Computer Graphics*, 20(4):151–160, 1986.

[73] Shattuck J, Brown P, and Carlson S. *The Lie Detector as a Surveillance Device.* American Civil Liberties Union, New York, 1973.

[74] Sondhi MM and Schroeter J. A hybrid time-frequency domain articulatory speech synthesizer. *IEEE Transactions on Acoustics, Speech and Signal Processing*, 35:955–967, 1987.

[75] Srivastava A and Liu X. Statistical hypothesis pruning for identifying faces from infrared images. *Image and Vision Computing*, 21:651–661, 2003.

[76] Stoica A. Learning eye-arm coordination using neural and fuzzy neural techniques. In Teodorescu HN, Kandel A, and L. Jain L, Eds., *Soft Computing in Human-Related Sciences*, pp. 31–61, CRC Press, Boca Raton, FL, 1999.

[77] Stollnitz EJ, DeRose TD, and Salesin DH. Wavelets for computer graphics: a primer, part 2. *IEEE Transactions on Computer Graphics and Applications*, 15(4):75–85, 1995.

[78] Story B and Titze I. Voice simulation with a body-cover model of the vocal folds. *Journal of the Acoustic Society of America*, 97(2):3416–3413, 1995.

[79] Sugimoto Y, Yoshitomi Y, and Tomita S. A method for detecting transitions of emotional states using a thermal facial image based on a synthesis of facial expressions. *Robotics and Autonomous Systems*, 31:147–160, 2000.

[80] Tian Y-l, Kanade T, and Cohn JF. Recognizing action units for facial expression analysis. *IEEE Transactions on Pattern Analysis and Machine Intelligence*, 23(2):97–115, 2001.

[81] Titze IR. The human vocal cords: a mathematical model. Part I. *Phonetica*, 28:129–170, 1973.

[82] Terzopoulos D and Waters K. Analysis and synthesis of facial image sequences using physical and autómatical models. *IEEE Transactions on Pattern Analysis and Machine Intelligence*, 15(6):569–579, 1993.

[83] Vertegaal R, Slagter R, Der Veer GV, and Nijholt A. Eye gaze patterns in conversations: there is more to conversational agents than meets

the eyes. In *Proceedings of the ACM Conference on Human Factors in Computing Systems*, pp. 301–308, 2001.

[84] Yamamoto E, Nakamura S, and Shikano K. Lip movement synthesis from speech based on hidden Markov models. *Speech Communication* 26(1-2):105–115, 1998.

[85] Yanushkevich SN, Supervisor. Computer aided design of synthetic faces. *Technical Report, Biometric Technologies: Modeling and Simulation Laboratory*, University of Calgary, Canada, 2004.

[86] Yanushkevich SN and Shmerko VP. Biometrics for electrical engineers: design and testing. Chapter 6: CAD of face analysis and synthesis. *Lecture notes*. Private communication. 2005.

[87] Ypsilos I, Hilton A, Turkmani A, and Jackson PJB. Speech-driven face synthesis from 3D video. In *Proceedings of the 2nd International Symposium on 3D Data Processing, Visualization, and Transmission*, Greece, pp. 58–65, 2004.

[88] Walther EF. *Lipreading.* Nelson-Hall, Chicago, IL, 1982.

[89] Wang H and Ahuja N. Facial expression decomposition. In *Proceedings of the IEEE 9th International Conference on Computer Vision,* 2003.

[90] Waters K. A muscle model for animation three-dimensional facial expression. *SIGGRAPH'87*, 21(4):17–24, 1987.

[91] Waters K. Modeling three-dimensional facial expressions. In Bruce V and Burton M, Eds., *Processing Images of Faces*, Ablex Publishing Corporation, Norwood, New Jersey pp. 202–227, 1992.

[92] Widmer G. Machine Discoveries: A few simple, robust local expression principles. *Journal of New Music Research,* 31(1):37–50, 2002.

[93] Widmer G. Large-scale performance studies with intelligent data analysis methods. In *Proceedings of the Stockholm Music Acoustics Conference*, Stockholm, Sweden, pp. 509–512, 2003.

[94] Wilder J, Phillips JP, Jiang C, and Wiener S. Comparison of visible and infra-red imagery for face recognition. In *Proceedings of the 2nd International Conference on Automatic Face and Gesture Recognition*, pp. 182–187, 1996.

[95] Zhang Y, Sung E, and Prakash EC. A physically-based model for real-time facial expression animation. In *Proceedings of the 3nd IEEE International Confirence on 3D Digital Imaging and Modeling*, 2001, pp. 399–406.

[96] Zhang Y, Prakash EC, and Sung E. Face alive. *Journal of Visible Languages and Computing*, 15:125–160, 2004.

[97] Zhang Z. Feature-based facial expression recognition: sensitivity analysis and experiments with multilayer perceptron. *International Journal on Pattern Analysis and Artificial Intelligence*, 13(6):893–911, 1999.

# 6

# Synthetic Irises

This chapter introduces techniques for the generation of synthetic irises. The iris is a part of the human eye which is a photo-receptor converting information in the form of light energy to nerve activity (electrical spikes). These electrical spikes are subsequently relayed to the optic nerve and the brain, where further information processing occurs, resulting in the identification and recognition of incoming visual information. Two components of the eye that are useful for identification are the *iris* and *retina* (Figure 6.1). Individual differences are formed during the development of these structures making them unique in each individual.

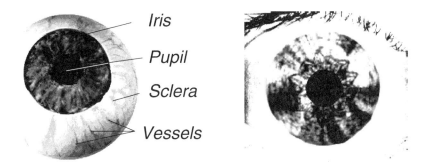

**FIGURE 6.1**
Synthetic eye images.

There are several factors that make the eye a unique carrier of individual information, in particular:

▶ The iris and retina have an extraordinary structure containing abundant textural information. The spatial patterns that are apparent in the iris and retina are unique to each individual.

▶ Compared to other biometrics (such as face, fingerprints, voiceprints, etc.), the iris and retina are more stable and reliable for identification.

▶ An important factor for practical applications is that iris and retina-based personal identification systems are noninvasive to their users because the iris and retina are overt bodies.

The methods of iris processing can be viewed as *analysis-by-synthesis paradigm*, a combination of *direct* and *inverse* problems:

▶ The group of *direct* problems is aimed at analyzing the topological structure of the iris resulting in a set of features that are used in various systems for decision making. These features can also be used to create a set of primitives.

▶ The group of *inverse* problems addresses the synthesis of irises; the methods adopted from the inverse problems give the rules for manipulating synthetic primitives to produce a new, synthetic iris.

The focus of this chapter is approaches to generate a synthetic iris. The state-of-the-art in iris analysis and synthesis is given in Section 6.1. An understanding of the structure of the eye and the functions of its components is necessary for the development of eye models. In section 6.2, the structure of the eye is introduced. The main focus is the iris, the component of the eye where individual information is formed and stored. The eye carries short-term and long-term behavioral information. This information is categorized focusing on the iris. Preprocessing techniques used in synthesis are discussed in Section 6.3. Section 6.5 introduces libraries of iris primitives that are used in synthetic iris design. Then, the simplest generator of synthetic irises is introduced. The warping technique is used to improve this generation technique. After the summary (Section 6.6), problems are given (Section 6.7) and then recommendations for further reading (Section 6.8).

---

## 6.1 State-of-the-art iris synthesis

The direct (analysis) and inverse (synthesis) methods of iris data processing are closely relevant to each other (Figure 6.2).

Both *non-automated* and *automated* design techniques exist for iris synthesis.

### 6.1.1 Non-automated design technique

Long before computerized iris acquisition and processing, oculists used natural techniques to synthesize the iris.

> **Example 6.1** *The oculists' approach to iris synthesis is based on the composition of painted primitives, and utilizes a layering of semi-transparent textures and images from topological and optical models.*

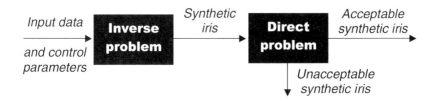

**FIGURE 6.2**
The analysis-by-synthesis paradigm for iris generation.

> **Example 6.2** *Examples of synthetic irises include: photographs, glass eyes, and a fake irises printed on contact lenses.*

## 6.1.2 Automated design technique

**The basic strategies.** In this chapter, the following two strategies for synthetic iris design are distinguished:

Assembling synthetic primitives, i.e.,
$< \text{Synthetic primitives} > \Rightarrow < \text{Synthetic iris} >$

Transformation of original iris into its synthetic equivalent by means of a deformation, i.e.,
$< \text{Original iris} > \Rightarrow < \text{Synthetic iris} >$

**Transformation of an original iris into synthetic equivalent.** A cancellable technique is useful for this approach. Given an original iris, its cancellable analog, a synthetic iris can be created, using an appropriate transform. The new iris does not resemble the original one. The details of the assembling technique are given in section 6.4.

**Assembling technique.** In Figure 6.3, the assembly technique is introduced. The main phases of the design process are as follows:

*Step 1.* A set of random points is generated in the plane defined as the iris region. The coordinates of these points are the centers of primitives and macroprimitives.

*Step 2.* Primitives are taken from the database and placed at these points.

*Step 3.* Based on the appropriate rules, macroprimitives are designed as compositions of initial primitives. The process is stopped when all random points on the iris region are replaced by primitives and macroprimitives.

*Step 4.* An arbitrary color is used from the database of basic colors for primitives and macroprimitives.

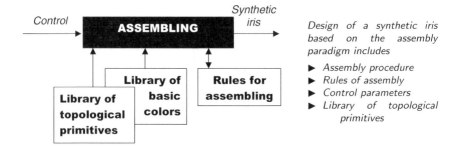

Design of a synthetic iris based on the assembly paradigm includes

▶ Assembly procedure
▶ Rules of assembly
▶ Control parameters
▶ Library of topological primitives

**FIGURE 6.3**
Synthetic iris design based on the assembly paradigm.

> **Example 6.3** *Figure 6.4 illustrates the simplest composition. The resulting macroprimitive is assumed to be colored.*

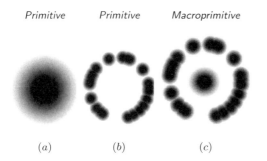

- Two arbitrary topological iris primitives can be composed to produce a macroprimitive.
- The combination of dot and circle primitives results in a new topological structure, a circle with dot, or a macroprimitive

**FIGURE 6.4**
The combination of dot (a) and circle (b) primitives produces a primitive circle with a dot (c) (Example 6.3).

The detail of the assembling technique are given in section 6.5.

## 6.2   Eye model

The aim of this section is to discuss various approaches for the generation of synthetic irises.

## 6.2.1 Structure of eye and models

In this section, the eye and its components are introduced. The focus is the iris as it easier to observe than the retina. Both carry unique biometric data.

It is essential that a model of the eye and its components is developed based on a detailed understanding of the physiological functions of the components of the eye. There are various approaches to eye modeling used for different applications, in particular:

*Optical* models are used in vision systems. For example, certain characteristics (resolution, field of vision, brightness perception, etc.) of the human eye are used in the development of television systems*.

*Iris* and *retina* models are used in the oculist's practice and person identification systems.

A simplified structure of the eye is given in Figure 6.5. The main components of the eye are the *ciliary body, cornea, choroid, iris, lens, sclera,* and *retina.* The retina contains two types of photoreceptors: *rods* and *cones.* The rods, about 100 million in number, are relatively long and thin. They provide *scotopic* vision, which is the visual response at the lower few orders of magnitude of illumination. The cones (about 6.5 million) are shorter and thicker and are less sensitive than the rods. The cones are also responsible for color vision. They provide *photopic* vision, the visual response at the higher 5 to 6 orders of magnitude of illumination. In the intermediate region of illumination, both rods and cones are active and provide *mesopic* vision.

The cones are densely packed in the center of the retina, called the *fovea,* at a density of about 120 cones per degree of the arc subtended in the field of vision. The pupil of the eye acts as an aperture. In bright light it is about 2 *mm* in diameter. These components are included in three basic layers called the *fibrous coat level, uvea or uveal tract,* and *neural layer.* These coats surround the contents of the eye, which are the *lens,* the *aqueous humor* and the *vitreous body.*

## 6.2.2 Structure of iris and models

The human iris, an annular part between the pupil (generally appearing black in an image) and the white sclera, is an extraordinary structure and has many interlacing minute characteristics such as freckles, coronas, stripes, furrows, crypts and so on. These visible characteristics, generally called the *texture of the iris,* are unique to each subject. This uniqueness is the results of individual differences that exist in the development of anatomical structures in the body.

---

*The limit of human eye resolution can be evaluated by the formula $2D \tan \frac{1}{20}$, where $D$ is the viewing distance. For example, for $D = 3$ $m$, the ultimate resolution of the human eye would be about 1 $mm$. The best-size television should have a height of about 500 $mm$ for a viewing distance of 3 $m$.

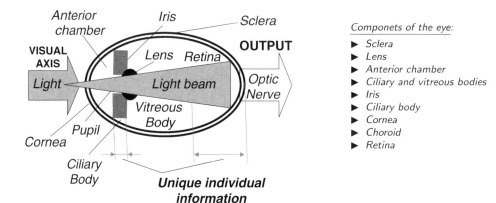

Componets of the eye:
- ▶ Sclera
- ▶ Lens
- ▶ Anterior chamber
- ▶ Ciliary and vitreous bodies
- ▶ Iris
- ▶ Ciliary body
- ▶ Cornea
- ▶ Choroid
- ▶ Retina

*Optical model of the eye.*
The eye contents a lens. The image appears inverted on the retina.

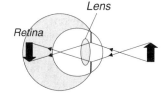

*Light transmission* is provided by three layers of the eye:
- ▶ The sclera,
- ▶ The iris, and
- ▶ The retina

*Dilation of the pupil:*
The pupil is variable in size and under the control of the nervous system. The pupil opens wider, letting more light into the eye in dim light, or if danger threatens. In bright light, the pupil closes down to reduce the amount of light entering the eye and improves its image-forming ability.

**FIGURE 6.5**

Eye structure: input information in the form of light is transmitted to the optic nerve.

The iris is part of the *uveal tract*, which also includes the *ciliary body* and the *choroid*. Its function is to control the amount of light that reaches the retina. Due to its heavy pigmentation, light can only pass through the iris via the pupil, which contracts and dilates according to the amount of available light. Iris dimensions vary slightly between individuals, but on average it is 12 *mm* in diameter. Its shape is conical with the pupillary margin more anterior than the root. The front surface is divided into the *ciliary zone* and the *pupil zone* by a thickened region called the *collarette*.

The iris is composed of four layers, from outside to inside (Figure 6.6[†]):

---

[†]The fourth level of the iris, designed by the model for generation of synthetic irises, will

- ▶ The *anterior border layer* at the front,
- ▶ The *stroma*,
- ▶ The *dilator pupillae muscle*, and
- ▶ The *posterior pigment epithelium*.

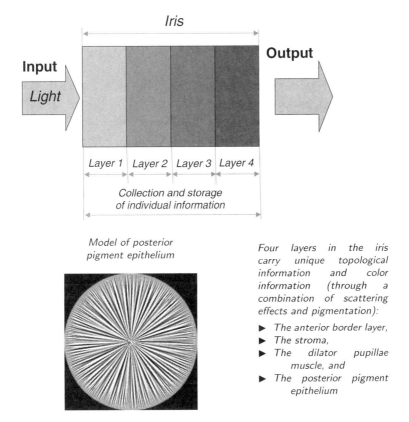

Model of posterior pigment epithelium

Four layers in the iris carry unique topological information and color information (through a combination of scattering effects and pigmentation):

- ▶ The anterior border layer,
- ▶ The stroma,
- ▶ The dilator pupillae muscle, and
- ▶ The posterior pigment epithelium

**FIGURE 6.6**

Layers of the iris collect individual information in the form of topological configurations of primitives, their color, composition in iris, and synthetic topological structure of the fourth layer of the iris.

The first layer, the *anterior border layer*, contains crypts or small pits, and is heavily pigmented in other areas, which appear as naevi. The second layer,

---

be discussed later in this chapter

the *stroma*, is a heavily pigmented layer. The stroma covers nearly all of the iris area and consists of a number of well defined dots, smeared dots or radial smears. The third layer, the *dilator pupillae muscle*, is responsible for dilation (mydriasis), which normally occurs in dark conditions. Its cells are normally lightly pigmented. The fourth layer, the posterior pigment epithelium, prevents light from leaking through the iris. These layers together determine eye color through a combination of scattering effects and pigmentation.

A model of the iris should capture the shape of the iris as well as its patterns and colors.

The model of the iris influences how the synthetic iris is generated. The model of the iris is based on the structure of the eye, the structure of the iris, and the iris color.

### 6.2.3   Measures

In the automated analysis and synthesis of the iris, distance measures are distinguished with respect to geometrical configuration, local and global regions, dimensions, etc. Some measures are defined in the

▶ *Spatial* domain and

▶ *Spectral* (Fourier, Fourier-like, etc.) domain.

> **Example 6.4** *Figure 6.7 illustrates some distance measures that can be used in non-contact detection of the psychological state of a person.*

### 6.2.4   Information carried by the iris

The eye carries information in the form of topological structures in the iris and retina. This section focuses on the information carried by the iris.

## 6.3   Iris image preprocessing

The iris image must be "prepared" for the processing aimed at identification and recognition. This phase is called *preprocessing*. The iris image requires preprocessing before extracting features from the original image because:

▶ It contains not only the iris which is of interest but also some "useless" parts (e.g. eyelid, pupil, reflections, etc.),

▶ The iris image needs to be scaled to the correct size, and

▶ The iris may not be uniformly illuminated.

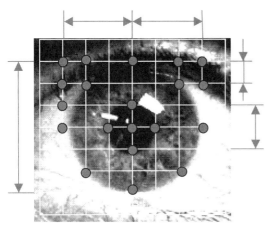

Feature points determine
coordinates for measurements
and transformations.
A limited number of labeled
feature points are marked,
depending the application.
Measures are used to
determine the diameter of
the pupil, parameters of
assymetries, and coordinates
of the iris for the purpose of
finding the psychological state
of a person.

*Normal size*    *Widen iris size*

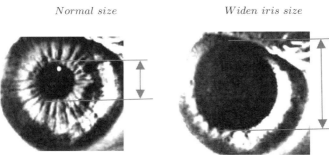

**FIGURE 6.7**
Control points and topological measures (Example 6.4).

Due to the noise and image discrepancies, certain portions of the image can be *recoverable* or *unrecoverable*. A portion of information can also be *redundant* or *insufficient* information. Redundant information needs to be removed from the image before specific features of the iris can be extracted. In particular, the iris must be separated from the background and unwanted features removed.

> **Example 6.5** *Figure 6.8 illustrates noise (like eyelids and lashes) detection in iris processing.*

The basic preprocessing techniques include: localization, normalization, image enhancement and denoising, and feature extraction.

## 6.3.1 Iris localization

Both the inner boundary and the outer boundary of a typical iris can be taken approximately as circles. However, the two circles are usually not perfectly

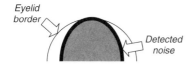

Eyelid border

Detected noise

To detect unwanted features, the threshold function is used.
The threshold function detects a non-white part between the middle of lid and the border

**FIGURE 6.8**
The noise detection (Example 6.5).

co-centered. The iris is localized in two steps:

▶ Determine the approximate region of the iris in an image, and
▶ Calculate the exact parameters of these two circles using edge detection and the Hough transform in a certain region.

> **Example 6.6** *The distortion of the iris caused by pupil dilation can be reduced by counter-clockwise unwrapping of the iris ring onto a rectangular block of texture of a fixed size by piecewise linear mapping. The iris region can be found approximately by projecting the image horizontally, vertically according to the equations* $X_p = \arg\min_x \left( \sum_y I(x,y) \right)$ *and* $Y_p = \arg\min_y \left( \sum_x^y I(x,y) \right)$*, where* $X_p$ *and* $Y_p$ *denote the center coordinates of the pupil in the original* $I(x,y)$*. The exact parameters of these two circles are obtained by using edge detection.*

### 6.3.2  Iris normalization

Irises from different people may be captured at various angles and distances, producing images of different sizes and with distortions. Even for images from the same eye, pupil dilation is influenced by illumination. Such elastic deformation in the iris will influence the results of iris matching. For the purpose of achieving more accurate recognition results, it is necessary to compensate for this deformation.

Locating the iris in the image is performed by finding its inner and outer boundaries. The Cartesian to polar transform produces a rectangular representation of this zone of interest. In this way the stretching of the iris image (pupil changes in size) can be compensated for.

> **Example 6.7** *Figure 6.9 illustrates the normalization and inverse polar transform of the iris.*

### 6.3.3  Iris image enhancement and noise removal

The normalized iris image often has low contrast due to reflections and non-uniform brightness caused by the position of light sources. To obtain a good

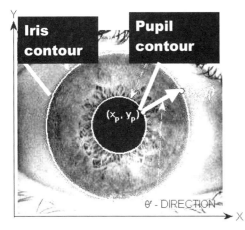

In the iris normalization, given a fixed direction $\theta'$, the image intensities along this direction are given by the equations:

$$I(\rho) = I\left(u(\rho),\ v(\rho)\right)$$
$$u(\rho) = (1 - \rho) \cdot x_P + \rho \cdot x_I$$
$$v(\rho) = (1 - \rho) \cdot y_P + \rho \cdot y_I.$$

The normalized polar image is transformed to the rectangular (cartesian) coordinate system, and further processing (enhancement) is performed.

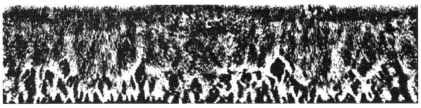

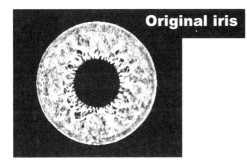

The enhanced iris image is recovered from the polar coordinates by applying the inverse transform:

$$I(x, y) = I\left(x(\rho, \theta),\ y(\rho, \theta)\right)$$
$$x(\rho, \theta) = \rho \cdot \cos \theta$$
$$y(\rho, \theta) = \rho \cdot \sin \theta.$$

**FIGURE 6.9**

Normalization of iris and the inverse polar transform of the normalized iris region: original image with iris region approximated by two circles (a), normalized iris image (b), and recovered iris image (c) (Example 6.7).

image, it is necessary to equalize intensity variations across the whole image.

## 6.3.4 Features in analysis and synthesis

In the task of recognition, image preprocessing is followed by feature extraction. The method is based on the transformation of polar image to the cartesian coordinates. The resulting template (after binarization) can be used for creating a *binary iris code*. This code can be useful data structure in matching process of identification or verification (see "Further Reading" Section).

## 6.4   Iris synthesis by transformation

Transformation of an original image to a deformed, or warped one is also called cancellable technique.

### 6.4.1   Recombination of iris images

A synthetic image can be created by combining segments of real images from a database. Various operators can be applied to deform or warp the original iris image: translation, rotation, rendering etc. The example below considers some simple permutation procedures.

> **Example 6.8** *Figure 6.10 illustrates the generation of a synthetic iris by combining parts of real iris images. The transform applied to the normalized iris image (in cartesian coordinates) is simply a partitioning with random permutation of the parts. The resulting image is transformed to polar coordinates in order to obtain the synthetic iris image.*

### 6.4.2   Iris image warping

Image warping techniques can be used to simulate changes in pupil size, as well as for iris synthesis in general.

> **Example 6.9** *Figure 6.11 illustrates various synthetic irises obtained by deformation of an original iris image. Irises in Figure 6.11a and 6.11b are results of regular function deformation. Random deformation is applied to the iris image in Figure 6.11c.*

## 6.5   Iris synthesis by assembling

This approach assumes usage of a library of iris pattern primitives.

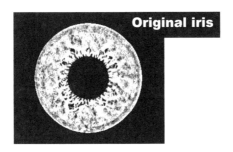

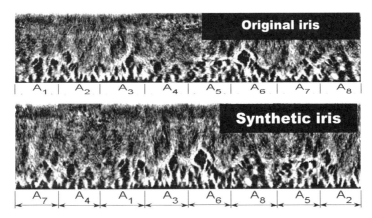

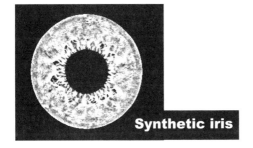

A synthetic iris image is obtained from the original iris image Figure 6.9 as follows:

- The normalized iris image is partitioned into 8 separate regions, $A_1, A_2, \ldots, A_8$.
- Fragments $A_1, A_2, \ldots, A_8$ are randomly permutted.
- The resulting image is transformed by the inverse polar transform into a synthetic iris .

**FIGURE 6.10**

A synthetic iris image is the result of the permutation of separate regions of a normalized iris image and subsequent inverse polar transform of the permutted iris image (Example 6.8).

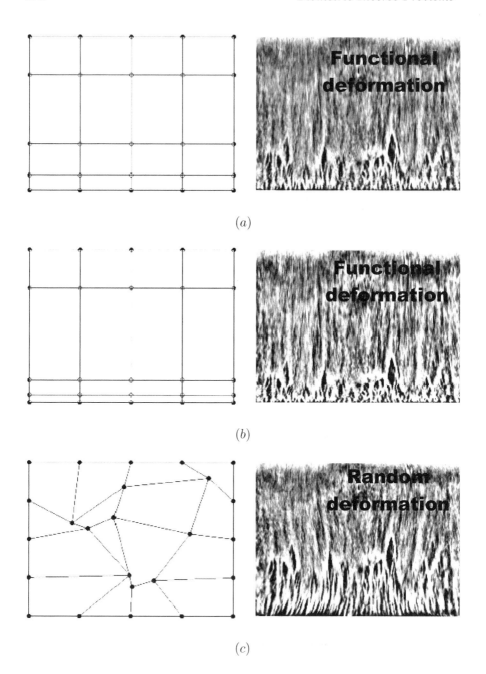

(a)

(b)

(c)

**FIGURE 6.11**
Free-form deformation of iris images (Example 6.9).

### 6.5.1 Assembling

The primitives of different layers (various collarettes, posterior pigment epitheliums, stroma etc.) can be assembled to create a model of an iris pattern. An example of such superposition consists of the following primitives:

► Primitives based on distance transform,
► Primitives using Bezier curves with Fourier masks,
► Primitives based on Voronoi transform, and
► Primitives using wavelet transforms.

> **Example 6.10** *Five various synthetic irises formed by superposition of the iris primitives and a black circle corresponding to the pupil are given in Figure 6.12.*

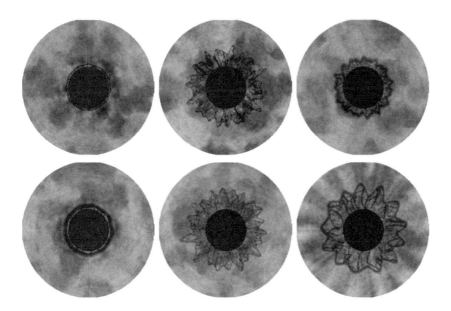

**FIGURE 6.12**
Synthetic irises design by superposition of topological primitives (Example 6.10).

### 6.5.2 Master iris

The above techniques generate a synthetic iris image that is not quite realistic. This image is called a *master iris*. To be realistic, some additional information must be incorporated in a master iris image: noise, acquisition conditions, sensor effects, etc. To implement these effects, models of these effects must be developed and used to simulate them.

The focus of this section is the design of a library of iris colors and topological primitives.

### 6.5.3   Library of iris colors and topological primitives

A library of iris primitives is a collection of synthetic or original elementary patterns of the iris. The amount of detail visible in the iris varies from person to person and is correlated to eye color. For example, brown eyes tend to have less detail and have a smudged or smeared appearance, whereas lighter eye colors show more detail. The number of different colors needed to describe the full spectrum of iris colors is typically less than 10. These base colors are pairwise mixed to form the prevailing color for each layer.

> **Example 6.11** *Ocularists paint artificial eyes using between 30 to 70 different layers of paint, with layers of clear coat between them. The first layer, called the base layer, contains the most dominant colors found in the eye, and is painted on a black opaque iris button. For a grey or blue eye, this layer tends to contain a fair amount of white. For brown eyes, this color may be darker.*

A library of image primitives is used in synthetic iris design. There are various approaches to develop this library. For example, a library can consists of images of painted primitives. Another more formal approach states that a library consists of a minimal number of basis images.

> **Example 6.12** *A dot is a simple topological primitive.*

Simple images can be assembled by a *composer* that implements algorithms of topological design using the rules. A composer produces macroprimitives. As a rule, the primitives or macroprimitives are randomly chosen from the library but the process coloring tools are connected to user-defined parameters.

> **Example 6.13** *In Figure 6.13, a set of painted iris primitives is shown. These primitives are used in lens design with synthetic (fake) iris patterns. Note that additional methods are needed to recognize a synthetic iris, for example, by using the fact that the synthetic iris pattern does not undergo any distortions when the pupil changes in size.*

### 6.5.4   Automated generation of iris primitives

There are various models and strategies for automatically generating iris primitives. A primitive template is defined as a reasonable topology of an

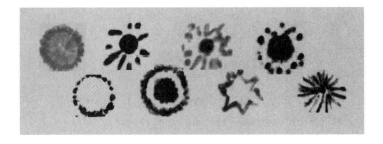

**FIGURE 6.13**
A set of painted iris primitives (Example 6.13).

iris layer pattern to start with in the automated design of the iris primitives. Two different techniques for primitive generation are:

▶ Algebraic-based method, and

▶ Mixed strategy that combines random generation of dots, their distribution in the limited plane, and distance transforms.

### 6.5.5   Primitive template

**Design using distance transform.** The distance transform characterizes the separation of points in space. The distance transform is normally applied to a binary feature map of an image. Given a binary image, the distance transform assigns each pixel a number that is the distance between that pixel and the nearest nonzero pixel of the image (feature pixels are marked with zero). The result is a grey-scale image with real valued pixels.

> **Example 6.14** *Figure 7.19 demonstrates the simplest effect of the distance transform.*
> *The distance transform of the 3 x 3 binary image with the pixel 1 at the center, providing that the distance between neighboring pixels is equal to 1, results in*
>
> $$\begin{bmatrix} 0\ 0\ 0 \\ 0\ 1\ 0 \\ 0\ 0\ 0 \end{bmatrix} \implies \begin{bmatrix} 1.4\ 1.0\ 1.4 \\ 1.0\ 0.0\ 1.0 \\ 1.4\ 1.0\ 1.4 \end{bmatrix}$$

Various effects can be achieved using the distance transform in developing a library of topological primitives for iris design. There are various opportunities to control the effects of images provided by distance transforms.

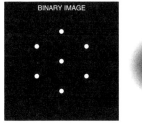
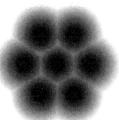

*Distance transform assigns each pixel in a binary source image a number that is the distance to the closest nonzero pixel. The result is a grey-scale distance map with real-valued pixels.*

**FIGURE 6.14**

A binary image consisting of 7 feature points and its distance map (Example 7.22).

**Example 6.15** *Figure 6.15 illustrates various effects in iris design that were achieved using the distance transform. The provided MATLAB code can be used to generate new effects. The distance transform is performed in a rectangular plane, which is then transformed into a disk.*

**Example 6.16** *Figure 6.16a illustrates the design of iris templates by computing functions $f_1 = 2 + \cos \frac{5\theta}{2}$ and $f_2 = e^{\cos \theta} - \cos B\theta + \sin B\frac{\theta}{4}$ in polar coordinates. The templates shown in Figure 6.16b are designed using various methods. The distance transform has been applied to generate iris primitives from the templates. The image has been centered (mapped into a circle).*

**Modeling the collarette.** The area in the iris that is closest to the pupil is called the *collarette*. To model the collarette shape, i.e., to represent the radial lines from the pupil to the iris edge, the Bezier curves can be useful. This representation will require additional manipulation to make the topological configuration more realistic. As a rule, manipulations in the spatial domain are difficult to implement but can be made more easily in the spectral domain.

The Bezier curve is defined by four points (two control points and two end points). It is drawn by interpolation between the two end points using the control points (two in this example) to fix the curve's derivatives:

$$P(t) = \sum_{i=0}^{3} P_i B_i(t) = (1-t)^3 P_0 + 3t(1-t)^2 P_1 + 3t^2(1-t)P_2 + t^3 P_3$$

$$= \left[ (1-t)^3 \ 3t(1-t)^2 \ 3t^2(1-t) \ t^3 \right] \begin{bmatrix} P_0 \\ P_1 \\ P_2 \\ P_3 \end{bmatrix}.$$

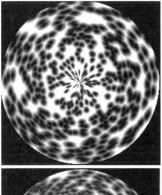

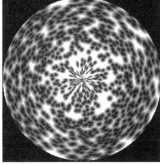

A background layer of an iris can be synthesized from the templates such as the points randomly placed on a black background by the following MATLAB procedure:

```
for i=1:  nbPoint nb=rand(1);
coordY=height/2*rand(1)+height/2;
coordX=lenght*rand(1);
if (nb probability)
coordY=coordY-height/2;
end
matBlack(ceil(coordY),ceil(coordX)) = 1;
end
```

The number of points is equal to 500, 1000, and 3000 for the images shown. Symmetry is added by copying a part of left half-circle to the right. Then, the Euclidean distance transform is implemented with the bwdist function:

```
figAdd=lenght/5;
repeat=matBlack(:,1:figAdd);
matBlack=[matBlack repeat];
[height, lenght]=size(matBlack);
D1=bwdist(matBlack,euclidean);
result=mat2gray(D1);
result(:,lenght-9*figAdd/10:lenght) = [];
result(:,1:round(1*figAdd/10)) = [];
```

The results are the primiives corresponding to the iris layer

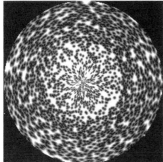

**FIGURE 6.15**

Design of some synthetic iris primitives by using the distance transform (Example 6.15).

Finally

$$P(t) = \begin{bmatrix} 1 & t & t^2 & t^3 \end{bmatrix} \begin{bmatrix} 1 & 0 & 0 & 0 \\ -3 & 3 & 0 & 0 \\ 3 & -6 & 3 & 0 \\ -1 & 3 & -3 & 1 \end{bmatrix} \begin{bmatrix} P_0 \\ P_1 \\ P_2 \\ P_3 \end{bmatrix}$$

The point $P_0$ corresponds to the first end point, points $P_1$ and $P_2$ are the

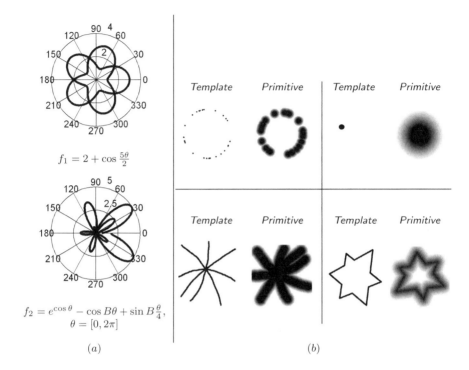

**FIGURE 6.16**
Synthetic iris primitives: algebraically generated templates (a) non-automated templates and the primitives generated from the templates by the distance transform (Example 6.16).

control points, and $P_3$ is the second end point. To create the collarette, the following control points are used:

▶ The *distance* between the two control points is defined by a fixed distance ($\frac{1}{4}$ of the radius of the iris plus a random value (between 0 and $\frac{1}{10}$ of the radius)[‡].

▶ The *range* is a user-defined parameter for computing the coordinates of control points (between the end point and the point at a fixed distance).

---

[‡]In MATLAB `edge=round(height/4)+round(rand(1)*height/10);`

**Example 6.17** *Figure 6.17 illustrates how a synthetic collarette topology has been designed using Bezier curve in a cartesian plane. It is transformed into a disk to form the concentric pattern. The algorithm and MATLAB code are given at the end of the "Problems" Section.*

**FIGURE 6.17**
Synthetic collarette topology modelled by Bezier curves (Example 6.17).

### 6.5.6 Using Fourier transform in collarette design

The collarette's outer boundary edge can be represented by a periodical signal like a Fourier series. Thus, the collarette topology designed using Bezier curves can be combined with a mask formed by a periodic wave. The signal used to construct the mask consists of an offset, a fundamental wave and $n$ harmonics:

$$f(t) = \sum_{i=1}^{k} c_i \ \cos(i\omega t + \varphi_i)$$

where $c_i \in (0, 1)$ is an amplitude and $\varphi_i \in (0, 2\pi)$ is a phase. Note that for the signal to be continuous as a result of this transform, the fundamental wavelength must be equal to the length of the image. The primitive formed with Bezier curves and with the Fourier mask is blurred using the MATLAB `blur` function at the outer boundary of the collarette to make it more real.

**Example 6.18** *Figure 6.18 illustrates forming a Fourier mask and the outer boundary curve of the collarette in cartesian space.*

Next, the obtained image as a rectangle must be transformed to a disk. The result is a primitive of the collarette, as demonstrated in the example below.

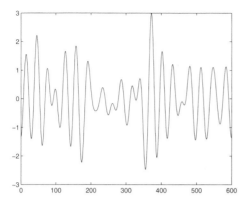

The Fourier series is used as mask applied to a rectangular image formed by Bezier curves. It results in the collarette primitive in cartesian coordinates. All pixels covered by the mask are removed. A blur can be applied to form a blur-alike outer boundary.

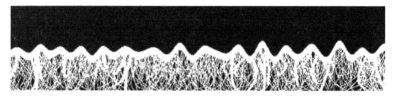

Tranformation back from polar to Cartesian coordinates yield a synthetic collarette pattern

**FIGURE 6.18**

Forming the outer boundary curve of the collarette by Fourier mask and blur (Example 6.18).

**Example 6.19** *Figure 6.19 contains several synthetic collarette primitives. They are obtained by using Bezier curves, Fourier mask and application of blur. The image in cartesian space is then transformed to polar coordinates. Finally, the image is improved by application of the blur. In particular, when designing a blue eye, two primitives are combined: one with many harmonics is placed close to the pupil and a second primitive with large amplitude and few harmonics, is placed further away.*

**FIGURE 6.19**
The resulting synthetic iris primitives (colors are replaced with grey color (Example 6.19).

### 6.5.7 Other transforms for automated design of iris primitives

Among other useful transforms are the Voronoi transform (see more details in Chapter 7) and the wavelet transform. For instance, the Voronoi transform produces patterns similar to the stroma, as shown in the example below.

> **Example 6.20** *Figure 6.20 demonstrates the design of an iris primitive obtained using a Voronoi transform applied to the template such as 100 randomly generated sites within a circle.*

The wavelet transform can be used to form the primitives of the iris ring.

> **Example 6.21** *Design of an iris ring primitive is shown in Figure 6.21.*

Voronoi diagrams are discussed in more detail in Chapter 7.

## 6.6 Summary

The iris is a carrier of unique individual information. The iris is more stable and reliable for identification than other biometrics and provides a possibility for distance identification. In this chapter, the following aspects of synthetic iris design have been discussed:

1. A direct problem is defined as analysis of the iris aimed at extracting unique personal features. An inverse problem is defined as generation of synthetic iris upon user-defined parameters.
2. There are three approaches to synthesizing an iris:

   ▶ *Physical* modeling based on physical processes of iris development through the first months of life,

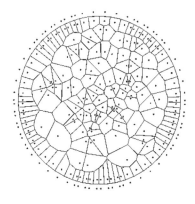

100 randomly generated sites with uniform distribution are placed in a circle. The Voronoi diagram is computed as a set of Voronoi cell.

```
for nbP=1:nbPoints
X=[X x1+width/2];
Y=[Y y1+width/2];
end
```

The sites are assigned with colors, so the colormap cmap of the image is created:

```
voroPrimimage, cmap
colormap(cmap)
h=surface(X,Y,Z,'CData',image,
'FaceColor','texturemap',
'EdgeColor','none','CDataMapping',
'direct');
```

**FIGURE 6.20**

The Voronoi diagram of 100 randomly placed points is the basic topological structure in synthetic iris image design (Example 6.20).

▶ *Topological* modeling based on analysis of topological structures, and

▶ *Topological-physical* modeling that combines both models.

3. Two techniques of topological modeling of synthetic iris can be recommended for further applications:

   ▶ Composition over the library of topological iris primitives,
     < PRIMITIVE >⇒< MASTER IRIS >⇒< SYNTHETIC IRIS >, and

   ▶ Deformation of original iris, also called cancellable technique
     < ORIGINAL IRIS >⇒< DEFORMED IRIS >⇒< SYNTHETIC IRIS >
     .

4. An iris can be easily decomposed to a small set of topological primitives. Due to this property, the problem of iris synthesis is formulated as

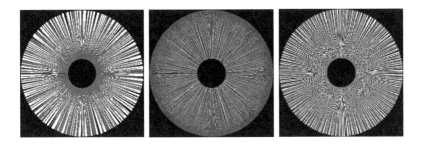

**FIGURE 6.21**
Synthetic iris ring (posterior pigment epitheliums of the iris) (Example 6.21).

follows: given a library of iris primitives, design an arbitrary synthetic
iris topology by random composition of modified primitives over the set
of basic colors.
5. Several practical applications can benefit from using the synthetic iris,
in particular: testing devices for iris-based personal identification,
ocularist's practice, and training.

## 6.7 Problems

Problems 6.1, 6.2, 6.4, 6.5 can be solved using MATLAB and the data given
at the end of this section. Problems 6.3 and 6.6 can be considered as projects.
Problems 6.5 and 6.2 address computer aided design of iris primitives based
on formal (algebraic) representation.

**Problem 6.1** Develop several models of the fourth level of the iris, called
the synthetic posterior pigment epitheliums. This is the basic topology in iris
design as well in synthetic primitive design.
*Hint:* In Figure 6.22, the fourth level of the iris is modeled by wavelet
transform with parameters: (a) $l = 2$, $S = 1$, not inverted; (b) $l = 5$, $S = 3$,
inverted; (c) $l = 2$, $S = 2$, inverted; (d) $l = 2$, $S = 1$, not inverted. In the
design, use the supporting data and MATLAB code given in the end of this
section.

**Problem 6.2** (Continuation of Problem 6.1.) Generate various (as realistic
as possible) topologies of the posterior pigment epitheliums layer of the iris.
*Hint:* The sample topologies in Figure 6.23 are obtained by warping. Use the
supporting data and MATLAB code given in the end of this section.

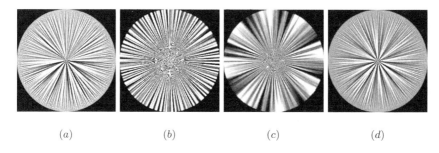

(a)                    (b)                    (c)                    (d)

**FIGURE 6.22**
Synthetic posterior pigment epitheliums of the iris (Problem 6.1).

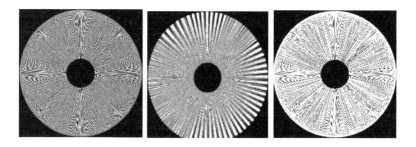

**FIGURE 6.23**
More realistic synthetic posterior pigment epitheliums of the iris (Problem 6.2).

**Problem 6.3** Design synthetic primitives using the distance transform to create
(a) Random points with uniform distribution in a circle.
(b) Randomly generated points of various diameter. Vary the probability of points with large and small diameter, and the probability of being close to the center of circle.
*Hint:* The topologies in Figure 6.24 are obtained by using the distance transform. Use the MATLAB code given in Figure 6.15.

**Problem 6.4** Improve the simplest iris topological primitives given in Figure 6.25 using the MATLAB function `polar`.
*Hint:* Use the supporting data and MATLAB code given in the end of this section.

**Problem 6.5** (Continuation of Problem 6.4.) Using the MATLAB function `polar`, create the simplest iris topological primitives

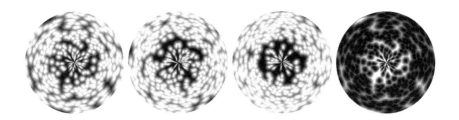

**FIGURE 6.24**
Iris primitives designed by using the distance transform: diameters of points and their probabilities in a circle are different; the last one is an inversion of the first one (Problem 6.3).

(a) $r = A + \cos\frac{5\theta}{2}$ for $A = 2$ and $\theta = [0, 2\pi]$.
(b) $r = e^{\cos\theta} - A\cos(4\theta) + \sin^3(\frac{\theta}{4})$ for $A = 2$ and $\theta = [0, 2\pi]$.
(c) $r = A\sin(3\theta)$ for $A = 2$ and $\theta = [0, 2\pi]$.
(d) $r = 1 - A\sin(\theta)$ for $A = 1$ and $\theta = [0, 2\pi]$.
Try to improve the obtained topological primitives.
*Hint:* In Figure 6.25, the set of graphical models is represented. Use the supporting data and MATLAB code given in the end of this section.

**Problem 6.6** Design synthetic irises similar to those shown in Figure 6.26; see also Figure 6.12.
*Hint:* Use the MATLAB codes given in this Chapter.

**Problem 6.7** Using the MATLAB function $problem6_1$ (see supporting data), create:
(a) An image of iris with `nbPoint` = 1000. For example, $problem6_1(500)$, where `nbPoint` = 500.
(b) An image of iris with `sizeOfBigImage` = [600 1200]. For example, $problem6_1(500, [100\ 200])$, where `nbPoint` = 500, `sizeOfBigImage` = [100 200].
(c) An image of iris with `sizeOfIris` = 250. For example, $problem6_1(500, [100\ 200], 150)$, where `nbPoint` = 500, `sizeOfBigImage` = [100 200], `sizeOfIris` = 250.
(d) An image of iris with `sizeOfPupil` = 1. For example, $problem6_1(500, [100\ 200], 150, 5)$, where `nbPoint` = 500, `sizeOfBigImage` = [100 200], `sizeOfIris` = 250, `sizeOfPupil` = 5.
(e) An image of iris with `sizeOfFinalImage` = [500 500]. For example, $problem6_1(500, [100\ 200], 150, 5, [300\ 300])$, where `nbPoint` = 500, `sizeOfBigImage` = [100 200], `sizeOfIris` = 250, `sizeOfPupil` = 5,

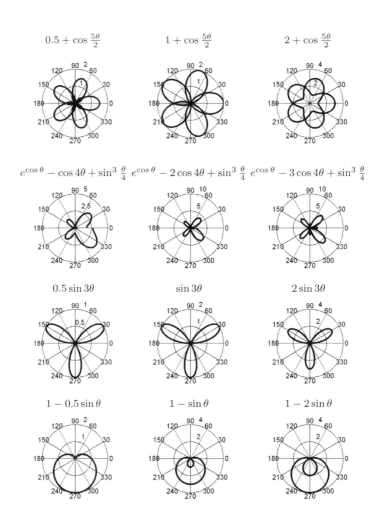

**FIGURE 6.25**

The simplest iris primitives (Problem 6.5).

$\texttt{sizeOfFinalImage} = [300\ 300]$.

(f) An image of iris with $\texttt{Probability} = 0.4$. For example, $\texttt{problem6}_1(500, [100\ 200], 150, 5, [300\ 300], 0.2)$, where $\texttt{nbPoint} = 500$, $\texttt{sizeOfBigImage} = [100\ 200]$, $\texttt{sizeOfIris} = 250$, $\texttt{sizeOfPupil} = 5$, $\texttt{sizeOfFinalImage} = [300\ 300]$, $\texttt{Probability} = 0.2$

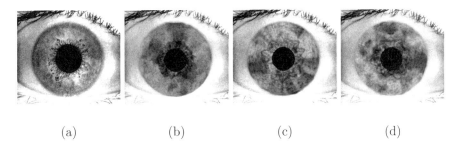

(a)        (b)        (c)        (d)

**FIGURE 6.26**

The real (a) and synthetic irises (b-d) (Problem 6.6).

## Supporting information for problems

The following MATLAB functions can be used for generation of a synthetic images for Problem 6.1. The input parameters are explained in Table 6.1.

MATLAB function: problem6$_1$(nbPoint, sizeOfBigImage, sizeOfIris, ... sizeOfPupil, sizeOfFinalImage, probability)

```
if nargin<6, probability=0.3; end
 if nargin<5,
 sizeOfFinalImage=[256, 256];
    end
    if nargin<4, sizeOfPupil=0;
        end
        if nargin<3, sizeOfIris=floor(sizeOfFinalImage(1,1)/2);
            end
            if nargin<2, sizeOfBigImage=[1000, 2500];
                end
                if nargin<1, nbPoint=1000;
                    end
height=sizeOfBigImage(1); lenght=sizeOfBigImage(2);
matBlack=zeros(height,lenght);
    for i=1:nbPoint
    nb=rand(1);
    coordY=height/2*rand(1)+height/2;
    coordX=lenght*rand(1);
        if (nb>probability)
        coordY=coordY-height/2;
            end
            matBlack(ceil(coordY),ceil(coordX)) = 1;
                end
figAdd=lenght/5; repeat=matBlack(:,1:figAdd); matBlack=[matBlack
repeat]; [height, lenght]=size(matBlack);
D1=bwdist(matBlack,'euclidean'); imshow(D1); pause(2);
result=mat2gray(D1); imshow(result); pause(2);
result(:,lenght-9*figAdd/10:lenght) = [];
```

**TABLE 6.1**

Input parameters.

| Code# | Name | Description, type, default value |
|-------|------|----------------------------------|
| 1 | nbPoint | Number of point to place on iris. Value of this parameter should be integer |
| 2 | sizeOfBigImage | Size of image that we reduce after transformation : [height, length]. Value for this parameter is matrix[2*1] |
| 3 | sizeOfIris | Size of radius of iris on the final imageValue of this parameter is integer. Default value equals sizeOfFinalImage[height]/2 |
| 4 | sizeOfPupil | Size of radius of pupil on the final image. Value of this parameter is integer. Default value is equal 0 |
| 5 | sizeOfFinalImage | Size of final image :[height, length]. It is matrix [2*1] |
| 6 | Probability | Probability that one point will draw between center and height/2: between 0 and 1. Default value is 0.3 |

```
result(:,1:round(1*figAdd/10)) = [];
result=norm2iris(result,[0,0,sizeOfPupil],[0,0,sizeOfIris], . . .
sizeOfFinalImage); imshow(result); pause(2);
```

The material below supports Problems 6.4 and 6.5 (collarette modeling by Bezier curves). The placement of points are implemented by the following algorithm:

Point $P_0$ is placed on the $X$ axis of image in rectangular form at regular intervals.

Point $P_1$ is placed randomly in a square delimited by the range and edge on the right side of the end point.

Point $P_2$ is placed randomly in a square delimited by the range and edge on left side of the end point.

Point $P_3$ is placed at the left or right of the first end point.

To avoid the problem of broken lines when the image is converted to polar coordinates, a modulus function is used. This function continues to draw on the other side, when the Bezier curves overtakes the image's edges. To draw

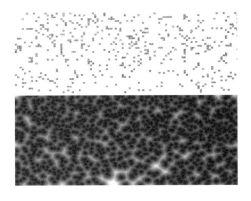 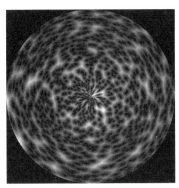

**FIGURE 6.27**
Images of iris (Problem 6.7).

a collarette's large lines, it is possible to specify a proportion and the width
of the lines.

```
for k=1:length(bez)
    Ycoord=ceil(bez(k,2)+0.0001);
    Xcoord=ceil(mod(bez(k,1),lenght+1)+0.0001);
    mat(Ycoord,Xcoord)=color+0.005*k;
    if (randNumber<propBigLines)
        for m=1:bigLineWidthMax
            newXcoord=ceil(mod((Xcoord-m),lenght+1)+0.0001);
            mat(Ycoord,newXcoord)=color+0.005*k;
        end
    end
end
```

## 6.8 Further reading

**Histology of the eye.** The iris is a thin, contractile, pigmented diaphragm
with a central aperture called the *pupil*. The pupil varies in diameter from 1 to
8 *mm*. The iris is composed of elastic connective tissue called the trabecular
meshwork. The prenatal morphogenesis of this meshwork is completed during
the 8th month of gestation. During the first year of life the color of the
trabecular patterns (crypts, rings, furrows, etc. called iris primitives in iris
modeling) often changes but the trabecular pattern itself is stable throughout
the life. The eye color may vary both from one eye to another in the same
person and in different parts of the same iris.

The iris is a pigmented zone of the eye whose pattern can be observed from a distance. The anterior surface is divided into the ciliary zone and the pupil zone by a thickened region called the collarette (Figure 6.5). The ability to dilate and constrict the pupil permits the iris to control the amount of light entering the eye and impinging on the retina. The iris, ciliary body, and choroid together form the middle coat of the eyeball called the uveal tract. Retinal measurements can change in some pathological developments (e.g., diabetic retinopathy).

Further details of anatomy and physiological process in the eye are given in [7, 11, 15, 27]. The book by Leigh and Zee [19] can be useful for development eye movements models.

**Iris analysis and recognition.** Methods of iris recognition have been studied by Daugman [8, 9, 10]. A good introduction to retina-based identification techniques is the paper by Hill [14]. A database of grey-scale iris images is available from the National Laboratory of Pattern Recognition, Chinese Academy of Sciences: http://www.sinobiometrics.com/resources.htm.

**Oculist's approach to iris synthesis.** The basic method in an ocularist's approach to iris synthesis includes a library of iris primitives and library of colors, an algorithm for composition of primitives utilizing a layering of semi-transparent textures, and geometry and optic models. For example, Lefohn et al. [17] has used a geometry model based on approximation by two spheres. A library of primitives consists of so-called *frustumps of right circular cones* or cones. These primitives are chosen over splines because generating the 3-D model using cones is simple. The cones are texture-mapped with semi-transparent textures. Coloring is based on the rule that two of the colors from the library of colors are mixed to produce the dominant color. The pupil is defined implicity by the hole in the center of the iris cones. In this model, a ray incident on a painted layer returns a color and shadow rays that intersect a semi-transparent texture are computed. Results are demonstrated for brown, blue and green eyes.

Oculists distinguish between three different types of eye with the descriptive names: smeared, dit-dot and detail. A smeared eye predominantly consist of smeared versions of the stroma, collarette and sphincter components.

**Computer aided design of synthetic iris.** Various details of the technique for synthetic iris design can be found in [13, 20, 25, 28]. Generally, in iris identification various aspects of the paradigm analysis-by-synthesis is utilized. State-of-the-art computer aided design of the synthetic irises is introduced in [29].

**Feature extraction.** The iris provides abundant texture information. Shapes in the iris (such as freckles, coronas, stripes and furrows etc.) can be considered as elementary components of the iris texture. Thus, shape information, the location of the shape, and the general texture can provide valuable information that can be used for iris identification and synthesis. It is desirable to explore methods which can capture local underlying information in an iris. From the viewpoint of texture analysis, the local spatial patterns in an iris mainly involve frequency information and orientation information. For more details see [3, 6, 18, 30].

**Iris code.** Daugman [8, 9, 10] has developed a method for representing the iris image by *iris codes*. User authentication and identification is achieved by forming a binary code from a processed image of the iris. For example, iris code can be 1024 or 2048 bits in length. The biometric matcher computes the Hamming distance between the input and stored database codes (template representations) and compares it with a threshold to determine whether the two biometric samples are from the same person or not. However the benefits of this simplification should be critically discussed. In particular, the following effect is observed: some percentage of the bits in the code disagree when the enrolled and presented iris patterns are compared because of various sources of corruption (artifacts from contact lenses, glasses, corneal reflections, etc.).

**Non-contact identification.** Unlike fingerprints, it is difficult to classify and localize apparent features in an iris image. From the viewpoint of feature extraction, existing iris recognition methods can be roughly divided into three major categories:

▶ Phase-based methods, for example, [10]
▶ The zero-crossing representation-based method, for example, [3] and
▶ Texture analysis-based methods [20, 21, 28].

An implementation of Daugman's method in MATLAB can be found in work by Masek [22]. In addition, see [5, 12, 23, 24].

**Cancellable iris image.** In [2], a cancellable iris image design is proposed for the problem as follows. The iris image is intentionally distorted to yield a new version.

**The retina** or nervous layer develops from the optic cup and is the internal layer of the eyeball. In this layer the optical image is formed by the eye's optical system. Photochemical transduction occurs so that nerve impulses are created and transmitted along visual pathways to the brain for processing. The retina is a thin, transparent membrane having a purplish-red color. An opthalmoscope permits an observer to view another person's retina with clarity. The retina is illuminated with bright source of light. The arteries are bright red and the veins are darker red. The arteries cross the veins on their superficial or viral surface. Normally the arteries do not compress or nick the veins at the site of crossing. The branching of the vessels is variable. All the above result in a unique topology.

Practical aspects of retina-based identification have been considered, in particular, by Hill [14]. Can et al. [4] studied the retina image registration problem for medical applications: two images taken, for example, of the same eye but at different times, or taken before and after laser surgery, can be registered to track the progress of diseases or to detect locations of scars and burns.

**Iridology** is based on the fact that long-term changes in the appearance of the iris reflecting the state of health. These changes are interpreted with respect to various organs in the body. For this, a specific decomposition of the iris is used, and the obtained regions are interpreted in terms of individual organs of the body behavior. Some causes color changes in the iris occur:

▶ Soon after birth,

▶ Due to special drug treatments, and

▶ Health changes.

For more detail, see papers [1, 16].

## 6.9 References

[1] Berggren L. Iridology: a critical review. *Acta Ophthalmologica*, 1985, 63(1):1–8.

[2] Bolle RM, Connell JH, and Ratha NK. Biometric perils and patches. *Pattern Recognition* 35:2727-2738, 2002.

[3] Boles W and Boashash B. A Human identification technique using images of the iris and wavelet transform. *IEEE Transactions on Signal Processing*, 46(4):1185–1188, 1998.

[4] Can A, Steward CV, Roysam B, and Tanenbaum HL. A feature-based, robust, hierarchical algorithm for registering pairs of images of the curved human retina. *IEEE Transactions on Analysis and Machine Intelligence*, 24(3):347–364, 2002.

[5] Choi SH, Park KS, Sung MW, and Kim KH. Dynamic and quantitative evaluation of eyelid motion using image analysis. *Medical and Biological Engineering and Computing*, 41(2):146–150, 2003.

[6] Clausi D and Jernigan M. Designing Gabor filters for optimal texture separability. *Pattern Recognition*, 33:1835–1849, 2000.

[7] Davision H. *The Eye*. Academic, London, 1962.

[8] Daugman JG. Uncertainty relation for resolution in space, spatial frequency, and orientation optimized by two-dimensional visual cortical filters. *Journal of the Optical Society of America*, 2:1160–1169, 1985.

[9] Daugman JG. High confidence visual recognition of persons by a test of statistical independence. *IEEE Transactions on Pattern Analysis and Machine Intelligence*, 15(11):1148-1161, 1993.

[10] Daugman JG. Statistical richness of visual phase information: update on recognizing persons by iris patterns. *IEEE International Journal Computer Vision*, 45(1):25-38, 2001.

[11] Forrester J, Dick A, Mcmenamin P, and Lee W. *The Eye: Basic Sciences in Practice*. WB Saunders, London 2001.

[12] Fukuda K. Eye blinks: new indices for the detection of deception. *Psychophysiology*, 40(3):239-245, 2001.

[13] Hallinan PW. Recognizing human eyes. In *SPIE Proceedings, Geometric Method in Computer Vision*, vol. 1570, pp. 214–226, 1991.

[14] Hill R. Retina identification. In Jain A, Bolle R, and Pankanti S, Eds., *Biometrics: Personal Identification in Networked Society*, pp. 123–141, Kluwer, Dordrecht, 1999.

[15] Hogan M, Alvarado J, and Weddell J. *Histology of the Human Eye*. WB Saunders, Philadelphia 1971.

[16] Knipschild P. Looking for gall bladder disease in the partient's iris. *British Medical Journal*, 297:1578–1581, 1988.

[17] Lefohn A, Budge B, Shirley P, Caruso R, and Reinhard E. An ocularists approach to human iris synthesis, *Computer Graphics and Applications*, IEEE magazine, 23(6):70–75, 2003.

[18] Lee SY, Chwa KY, Hahn J and Shin SY. Image morphing using deformation techniques. *Journal of Visualization and Computer Animation*, 7(1):3–23, 1996.

[19] Leigh RJ and Zee DS. *The Neurology of Eye Movements*. 3rd Edition, Oxford University Press, Oxford, 1999.

[20] Ma L, Tan T, Wang Y, and Zxang D. Personal identification based on iris texture analysis. *IEEE Transactions on Pattern Analysis and Machine Intelligence*, 25(12):1519–1533, 2003.

[21] Ma L, Tan T, Wang Y, and Zhang D. Local intensity variation analysis for iris recognition. *Pattern Recognition*, 37(6):1287–1298, 2004.

[22] Masek L. *A biometric identification system based on iris patterns*. PhD thesis, The School of Computer Scinece and Software Engineering, The University of Western Australia, 2003.

[23] Moriyama T, Xiao J, Kanade T, and Cohn JF. Meticulously detailed eye model and its application to analysis of facial image. In *Proceedings of the IEEE International Conference on Systems, Man, and Cybernetics*, pp. 629–634, 2004.

[24] Ravyse I, Sahli H, and Cornelis J. Eye activity detection and recognition using morphological scale-space decomposition. In *Proceedings of the IEEE International Conference on Pattern Recognition*, vol. 1, pp. 5080–5083, 2000.

[25] Sanchez-Avila C, and Sanchez-Reillo R. Iris-based biometric recognition using dyadic wavelet transform. *IEEE Aerospace and Electronic Systems Magazine*, pp. 3–6, Oct. 2002.

[26] Seal C, Gifford M, and McCarthney D. Iris recognition for user validation. *British Tellecommunications Engineering Journal*, 16(7):113–117, 1997.
Sederberg TW and Parry SR.

[27] Snell RS and Lemp MA. *Clinical Anatomy of the Eye*. 2nd edition, Blackwell Science, Oxford, 1998.

[28] Wildes R. Iris recognition: an emerging biometric technology. *Proceedings of the IEEE*, 85:1348–1363, 1997.

[29] Yanushkevich SN, Supervisor. Computer aided design of synthetic irises. *Technical Report, Biometric Technologies: Modeling and Simulation Laboratory*, University of Calgary, Canada, 2005.

[30] Zhang J and Tan T. Brief review of invariant texture analysis methods. *Pattern Recognition*, 35(3):735–747, 2002.

# 7

# *Biometric Data Structure Representation by Voronoi Diagrams*

The focus of this chapter* is the application of the technique of *Voronoi diagrams* in the analysis and synthesis of biometric topological structures. Voronoi diagrams are one of the most studied structures in computational geometry. The design of topological structures using Voronoi diagrams concerns the topological characteristics of visual information; topological configurations; and topological primitives and their interconnections. There are several reasons to use Voronoi diagrams in biometric image synthesis:

(a) The Voronoi diagram can be effectively controlled by the appropriate choice of metric space as well as by means of simple parameters such as weights, scale factors, etc. These can be used as fitting parameters in tilings and in other geometrical constraint problems.

(b) The computation of Voronoi diagrams is implemented by various algorithms, and due to its topological compatibility it can be formulated in terms of direct and inverse transforms.

(c) The representation of biometric data with the help of Voronoi diagrams has been found to be very useful. Various manipulations of topological structures due to the properties of Voronoi diagrams can be exploited in assembling and partition of data structures (Figure 7.1).

This chapter is structured as follows. The concept of the Voronoi data structure is introduced in Section 7.1. Basic concepts and principles used in the analysis and synthesis of biometric data structures based on Voronoi diagrams are discussed in Section 7.2. Recovering of topological data structure through topological transformations is often used in biometric data processing. This gives us a reason to discuss the Voronoi diagram technique in terms of direct and inverse Voronoi transforms in Section 7.3. In Section 7.4, the properties of Voronoi diagrams are discussed. Some examples of the concrete

---

*This chapter was written by S. Yanushkevich, O. Boulanov, and V. Shmerko

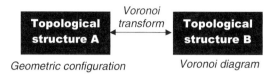

**FIGURE 7.1**
Voronoi diagram technique in direct and inverse problems of biometric data representation and manipulation.

application of the Voronoi diagram technique to the analysis and synthesis of biometric topological structures are given in Section 7.5. Based on the topological compatibility of the Voronoi diagram and the analytical properties of distance functions, an implementation of the Voronoi transform via the distance transform is introduced in Section 7.7. Some applications of this scheme to the computation of Voronoi diagrams for complex topologies are demonstrated in Section 7.8. Unwanted effects of these computations and methods for their suppression are discussed in Section 7.8.1. Finally, after a brief summary (Section 7.9) and problems (Section 7.10), recommendations for further reading are given in Section 7.11.

## 7.1  Voronoi data structure

The *Voronoi data structure* is a representation of data as a Voronoi diagram. The elements of Voronoi diagrams (vertices, polygons or edges) can be assigned certain elements of biometric data. In particular, the information contained in biometric images (iris images, faces or fingerprints) can be classified into three groups: geometrical distribution, color and texture.

For instance, geometric constraint solving consists of finding all possible placements of given geometrical primitives which satisfy a set of constraints between the primitives. Assigning the Voronoi diagram the color and texture primitives will allow us to describe certain geometric constraints between those primitives for effective geometric constraint solving.

### 7.1.1  Topology

The aim of the application of the Voronoi diagram technique in analysis and synthesis of biometric data is to relate the topological characteristic of biometric data to the topology of the Voronoi diagram. Below, some basic concepts used in the calculation of Voronoi diagrams are given.

*Metric space.*  The topology of biometric data can be represented in various metric spaces; the choice of a metric depends on configuration, geometrical construction and the goals of processing.

*Voronoi diagram.*   A given metric space is associated with its distance function.   In a Voronoi diagram, distinct points in a given metric space are called *Voronoi sites*. With each Voronoi site are associated topological structures called *Voronoi regions*. The Voronoi diagram can be defined as a composition of Voronoi regions or as partitioning of metric space into Voronoi regions.

> **Example 7.1** *In the synthesis of biometric data structures, the definition of the Voronoi diagram as composition of Voronoi regions is reasonable.   In the analysis of biometric data structures, the Voronoi diagram can be considered as a partition of the given metric space into Voronoi regions.*

*Topological properties* of the Voronoi diagrams are considered in this chapter with respect to their application to the manipulation of biometric data. For example, a Voronoi diagram can be constructed step by step using the properties of composition, partitioning (decomposition), and symmetry. The scaling property of the Voronoi diagram is associated with the scaling property of image operators.

## 7.1.2   Computing Voronoi diagram

A *Voronoi transform* represents an original topological structure in the form of a Voronoi diagram.   *Computing* a Voronoi diagram is to transform an arbitrary topological data structure into a Voronoi diagram. This topological transformation can be implemented by various algorithms.

> **Example 7.2** *Most of the algorithms for Voronoi diagram design explore particular properties of the input data structure. Typical algorithms for computing Voronoi diagrams include:*
>
> (a) *Algorithms that follow directly the definition of Voronoi diagram. These are not efficient.*
> (b) *An improvement can be made by utilizing the regularity of the initial topology. For example, points are regularly arranged in the form of a lattice.*
> (c) *Using various measures.   For example, to measure the degree of organization in a Voronoi diagram, information-theoretical measures can be used.*

## 7.2    Basics of Voronoi diagram technique

In this section, the basics of Voronoi diagram technique are introduced. The material is focused on using Voronoi diagrams for both representing and manipulating biometric data. Voronoi diagrams are a useful data structure in visual biometric information analysis and synthesis:

*In analysis*, Voronoi diagrams can be used for

> ▶ Partitioning (decomposition) a metric space, and
> ▶ Creating topological measures to extract features for the identification and recognition of images

*In synthesis*, Voronoi diagrams provide possibilities for

> ▶ Designing many classes of topological structures analogous to structures in biometric images; using the incremental properties of the Voronoi diagram, it is possible to manipulate primitives from a library of primitives in order to design biometric images.
> ▶ Colouring metric space via Voronoi regions.

Voronoi transforms can be used in different phases of image analysis and synthesis. Various useful image effects can be achieved by using Voronoi transforms after or before distance transforms.

### 7.2.1    Metric space

In the analysis and synthesis of topological structures, an appropriate metric should be chosen. Sometimes, biometric data is converted several times between various metrics throughout the manipulation.

**Point sets.** A point set is simply a collection of points, called sites, in some space. The topology of the point set carries information about the configuration such as the neighborhood of points, the nearness of two points, the connectivity of subsets of the point set, boundary points, and shape characteristics.

**Euclidean space.** Discrete topology can be defined in an $m$-dimensional Euclidean space $\mathbb{E}^m$. In a topological biometric data representation $m = 1, 2,$ or 3, there is a set of point operations (sum, subtraction, multiplication, Hadamard product, etc). Metric space is the only way of looking at distances between points in space. Formally, a metric space is a set $S$ of arbitrary elements with a single valued, non-negative real function $d$ (called a *distance function*) that, for any two elements $x \in S$ and $y \in S$, gives the distance between them. The classical Euclidean space is a simple and illustrative example.

> **Example 7.3** *In the usual m-dimensional Euclidean space, $\mathbb{E}^m$, every point can be assigned m real numbers, their Cartesian coordinates $(x_1, x_2, \ldots, x_m)$. The distance between any two points*
>
> $$p = (x_1, x_2, \ldots, x_m) \in \mathbb{E}^m \quad and$$
> $$p' = (x'_1, x'_2, \ldots, x'_m) \in \mathbb{E}^m$$
>
> *is given by the expression:*
>
> $$< Distance >= d(p, p') = \sqrt{\sum_{i=1}^{m} (x_i - x'_i)^2}. \quad (7.1)$$
>
> *Figure 7.2 illustrates distance in 2-dimensional Euclidean space, $\mathbb{E}^2$.*

**Other metrics.** However the distance is not limited to the Euclidean distance given by Equation 7.1. Any function can be chosen as long as certain conditions apply. These conditions come about as the generalization of the familiar properties of the Euclidean distance:

▶ Distance is non-negative, i.e., $\forall x \in S \; \forall y \in S : d(x, y) \geq 0$
▶ Distance between two points is equal to zero if and only if the points coincide, i.e., $\forall x \in S \; \forall y \in S : d(x, y) = 0 \Leftrightarrow x = y$;
▶ Distance is symmetric, i.e., $\forall x \in S \; \forall y \in S : d(x, y) = d(y, x)$
▶ Distance satisfies the triangle inequality, i.e., $\forall x \in S \; \forall y \in S \; \forall z \in S : d(x, y) \leq d(x, z) + d(y, z)$.

Biometric data can be analyzed and synthesized in various metric spaces. There are a variety of different distance functions that can be used in the calculation of Voronoi diagrams (City-block, Chessboard distance). The example below demonstrates the *City-block* distance (or *Manhattan metric*).

> **Example 7.4** *The m-dimensional space $\mathbb{R}^m$ with City-block distance is an illustrative example of non-Euclidean metric space. Let $(x_1, x_2, \ldots, x_m)$ and $(x'_1, x'_2, \ldots, x'_m)$ be the Cartesian coordinates of points $p \in \mathbb{R}^m$ and $p' \in \mathbb{R}^m$ respectively. Then the City-block distance is:*
>
> $$< Distance >= d(p, p') = \sum_{i=1}^{m} |x_i - x'_i|.$$
>
> *Figure 7.3 illustrates the City-block distance in $\mathbb{R}^2$.*

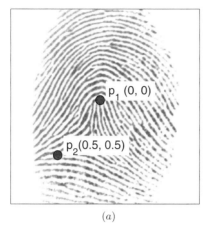

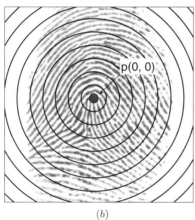

EUCLIDEAN DISTANCE
IN FINGERPRINT MEASURES

In the Euclidian space the distance $d(p_1, p_2)$ between the points $p_1(0,0)$ and $p_2(0.5, 0.5)$ is given by:

$$d(p_1, p_2) = \sqrt{(x_2 - x_1)^2 + (y_2 - y_1)^2}$$
$$= \sqrt{(0.5 - 0)^2 + (0.5 - 0)^2}$$
$$= \sqrt{0.5}$$

(a)

EUCLIDEAN CONTOURS
IN FINGERPRINT MEASURES

The contours of the Euclidian distance from the origin are circles centered at the origin.
The $i$-th contour of the radius $R_i$ $(i = 1, 2, 3, \ldots)$ is described by equation:

$$R_i^2 = x^2 + y^2$$

(b)

**FIGURE 7.2**
Illustration of the Euclidean distance in 2-dimensional space $\mathbb{E}^2$ (a) and the contour lines of the Euclidian metric from the origin for fingerprint image (b) (Example 7.3).

## 7.2.2 Voronoi diagram

**The Voronoi region** is the underlying topological component of a Voronoi data structure. In analysis, the Voronoi region carries useful information for identification and recognition. In synthesis, a Voronoi region can be considered as a geometrical primitive for geometric constraint-solving in the metric space $\mathbb{M}$.

Let $\mathbb{M}$ be a metric space with its associated distance function, $d(x, y)$, and $P = \{p_i\}_{i=1}^{N}$ be a set of $N$ distinct points in $\mathbb{M}$. These points are called

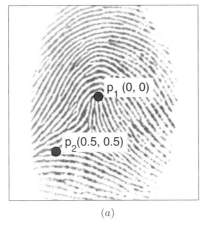

**CITY-BLOCK DISTANCE**
**IN FINGERPRINT MEASURES**

In the real space $\mathbb{R}^2$ the city-block distance between the points $p_1(0,0)$ and $p_2(0.5, 0.5)$ is given by:

$$d(p_1, p_2) = |x_2 - x_1| + |y_2 - y_1|$$
$$= |0.5 - 0| + |0.5 - 0|$$
$$= 1.$$

(a)

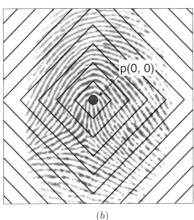

**CITY-BLOCK CONTOURS**
**IN FINGERPRINT MEASURES**

The contours of the city-block distance from the origin are squares centered at the origin. The $k$-th contour ($k = 1, 2, 3, \ldots$) is described by equation:

$$k = |x| + |y|.$$

(b)

**FIGURE 7.3**

Illustration of the city-block distance in 2-dimensional space $\mathbb{R}^2$ (a) and the contour lines of the city-block metric from the origin for fingerprint image (b) (Example 7.4).

*Voronoi sites* or simply *sites*.

Formally, the Voronoi region of the site $p_i \in P$ denoted by $V(p_i)$ is a subset of all points in $\mathbb{M}$ which are closer to the site $p_i$ than to any other site $p_j \in P$, i.e.,

$$V(p_i) = \{x \in \mathbb{M} \ : \ d(x, p_i) \leqslant d(x, p_j), p_j \in P\}. \tag{7.2}$$

It follows from the Equation 7.2 that the Voronoi region carries the following information:

(a) The location of the site $p_i$ with respect to the closer subset of sites, and

(b) The local topology given by a subset of points.

> **Example 7.5** *In 2-dimensional Euclidean space,* $\mathbb{E}^2$, *the Voronoi region of a site presents the nearest neighbor polygonal region surrounding that site. Given the set*
>
> $$P = \{p_1(241, 243),\ p_2(121, 299),\ p_3(250, 80),\ p_4(394, 301)\}$$
>
> *of Voronoi sites in Figure 7.4a, the Voronoi region around the site* $p_1$ *is the region of the plane closer to the site* $p_1$ *than to any other one in* $P$.

**A Voronoi diagram** of the set $P = \{p_i\}_{i=1}^N$ denoted by $V(P)$ is a partitioning of the metric space $\mathbb{M}$ into $N$ Voronoi regions $V(p_i)$, $p_i \in P$, $i = 1, 2, \ldots, N$, such that each Voronoi region $V(p_i)$ contains points that are closer to the site $p_i$ than to any other site in $P$, i.e.,

$$V(P) = \{V(p_i)\}_{i=1}^N.$$

> **Example 7.6** *In Euclidean space* $\mathbb{E}^2$, *the Voronoi diagram consists of the nearest neighbor convex polygons. Given the same set of sites as in Example 7.5, the whole set* $P$ *results in the Voronoi diagram shown in Figure 7.4b.*

**Voronoi edge and Voronoi vertex.** For some points, there is more than one closest generator. Points such as these act as an edge, or boundary, between adjacent Voronoi regions. A point is said to be on a Voronoi edge if every neighborhood of this point contains at least two points from two distinct Voronoi regions.

The Voronoi vertex can be defined by analogy with the Voronoi edge. A point is a Voronoi vertex if every neighborhood of this point contains at least three points from three distinct Voronoi regions. The set of all boundary points forms a network of lines called the *Voronoi skeleton*.

It follows from the above definitions that

▶ The Voronoi diagram is a partitioning of space into convex polygons, one polygon for each point $p_i \in P$.

▶ The Voronoi diagram of $P$ is a collection of Voronoi regions $V(p_i)$. Each region $V(p_i)$ is precisely the region of the plane containing all points closer to the generator $p_i$ than any other generator in $P$.

**Curved boundaries between Voronoi regions.** With points as generators, the Voronoi skeleton consists of straight lines. However the ordinary Voronoi diagram defined for a set of point generators can be generalized in many ways depending on the specific applications. For instance,

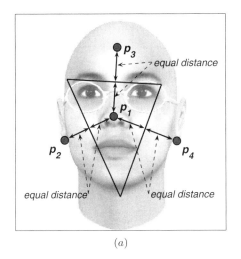

*VORONOI REGION*
*IN FACIAL MEASURES: $V(p_1)$*

The Voronoi region associated with a site is the polygonal region surrounding that site. The dots, $p_1$, $p_2$, $p_3$, and $p_4$, represent the set $P$ of Voronoi sites. $V(p_1)$ is the Voronoi region for the site $p_1$.

- The Voronoi region is the intersection of a finite number of open halfspaces.
- Each edges of the Voronoi region is equidistant from the corresponding two sites.

(a)

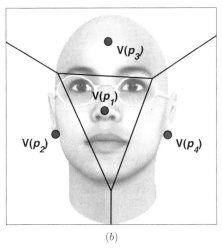

*VORONOI DIAGRAM*
*IN FACIAL MEASURES: $V(P)$*

A Voronoi diagram of a set of points in the plane surrounds each point with the region of the plane closer to that point than any other. The dots, $p_1$, $p_2$, $p_3$, and $p_4$, represent the set $P$.

- Each Voronoi region has a boundary constituting Voronoi edges.
- The unbounded Voronoi regions are the regions corresponding to sites on the convex hull of $P$, the outermos points.

(b)

**FIGURE 7.4**

The Voronoi region (a) of the site $p_1$ in a facial image (Example 7.5) and Voronoi diagram (b) of the set $P$ in a facial image (Example 7.6).

more complex generators may allow us to generate curved boundaries between Voronoi regions. For the purpose of biometric data synthesis it is useful to introduce the Voronoi diagram of *closed connected* area primitives and therefore to extend the ordinary Voronoi diagram for geometrical objects such as line and curve segments, circles, polygons etc.

An *area Voronoi diagram* can be constructed in a quite similar way to that of the ordinary Voronoi diagram. However, before proceeding further, the distance between a point and a set of points needs to be defined.

Given a set, $S$, of points in metric space $\mathbb{M}$, the distance, denoted by $d(x, S)$, from an arbitrary point $x \in \mathbb{M}$ to the set $S$ can be defined as the greatest lower bound of distances between $x$ and $x' \in S$ as $x'$ ranges over all of $S$, i.e.,

$$< \text{Distance} > = d(x, S) = \inf_{x' \in S} d(x, x'), \tag{7.3}$$

where $d(x, x')$ is the distance between two points in $\mathbb{M}$. In Euclidean space this is simply the shortest distance between $x$ and the boundary of $S$. With the help of the distance function $d(x, S)$ the area Voronoi diagram can be defined as follows.

Let $A = \{A_1, A_2, \ldots, A_N\} = \{A_i\}_{i=1}^N$ be a set of $N$ *closed connected* areas in metric space $\mathbb{M}$. The areas in $A$ are assumed not to overlap, i.e.,

$$\forall A_i \in A \ \forall A_j \in A \ : \ (i \neq j) \Rightarrow A_i \cap A_j = \emptyset.$$

**The area Voronoi region** denoted by $V(A_i)$ and associated with the area (also called a generator) $A_i \in A$ is defined as a subset of all points in $\mathbb{M}$ which are closer to $A_i$ (in the sense of the distance function in Equation 7.3) than to any other area $A_j \in A$, i.e.,

$$V(A_i) = \{x \in \mathbb{M} \ : \ d(x, A_i) \leqslant d(x, A_j), A_j \in A\}.$$

**The area Voronoi diagram** generated by the set $A = \{A_i\}_{i=1}^N$ and denoted by $V(A)$ is a partitioning of the metric space $\mathbb{M}$ into Voronoi regions $V(A_i)$, $i = 1, 2, \ldots, N$, such that each Voronoi region $V(A_i)$ contains points that are closer to the area $A_i$ than to any other area in $A$, i.e.,

$$V(A) = \{V(A_i)\}_{i=1}^N.$$

It is important to note that an area generator can be a polygonal figure as well as a line or curve segment, or a point. Therefore the area Voronoi diagram can be interpreted as a generalization of the ordinary Voronoi diagram of point generators.

Curved boundaries between Voronoi regions are useful in biometric topological data manipulation and give, in particular:

▶ The possibility to extract and represent more complex geometric structures from biometric data;

▶ The possibility to create much more realistic biometric data structures;

▶ The possibility to generate more complex geometrical primitives and to define sophisticated constraints between them for geometric constraint solving.

> **Example 7.7** *Given a set of generators* $A = \{p, L\}$ *in* $\mathbb{E}^2$, *where $p$ is a point and $l$ is a horizontal line, the Voronoi diagram $V(A)$ consists of two regions separated by a parabolic arc (Figure 7.5).*

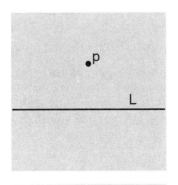

**GENERATORS : POINT AND LINE**

*The ordinary Voronoi generator can be extended from a point to a more complex geometrical object in order to obtain curved boundaries between Voronoi regions. A simple example of such a generalization is a set consisting of a point $p$ and a straight line $L$.*

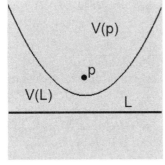

**VORONOI DIAGRAM**

*The generator set of the point $p$ and the line $L$ results in a Voronoi diagram consisting of two regions, $V(p)$ and $V(L)$, separated by a parabolic arc.*

**FIGURE 7.5**
Example of the synthesis of topological structure with non-straight lines (Example 7.7).

## 7.2.3 Information carried by a Voronoi diagram

Voronoi diagrams are useful for the manipulation, evaluation, identification, and recognition of biometric data; the Voronoi diagram can be considered as a useful "extractor" of information from biometric data structures. For instance, a biometric image contains a geometrical distribution of feature points, color information and texture. For the particular purpose of synthesis, the Voronoi diagram can help

▶ To extract from the biometric data information about the geometrical distribution of feature points;

▶ To extract color and texture information from the surrounding area of a feature point;

▶ To create geometrical primitives from the extracted information;

▶ To define the geometric constraints between the primitives for efficient geometric constraint solving.

## 7.3    Direct and inverse Voronoi transform

The focus of this section is the *Voronoi transform*. Manipulation is often required to recover biometric data. For example, data from the spectral domain can be recovered in the time domain by an inverse Fourier transform. The Voronoi diagram is the result of performing a Voronoi transform. The problem is often formulated as recovering initial topology given a Voronoi diagram, i.e.,

$$< \text{Generators} > \Leftrightarrow < \text{Voronoi diagram} > .$$

The Voronoi diagram $V(P)$, can be considered as a *direct Voronoi transform*, denoted by $\mathbf{V}$, and the Voronoi sites are said to be *generators* of the transformation. The concept of the *inverse Voronoi transform*, denoted by $\mathbf{V}^{-1}$, will be introduced as a transform that restores the original distribution of generators $P$ from their Voronoi diagram $V(P)$.

Figure 7.6a illustrates the Voronoi transform $\mathbf{V}(P)$ that turns a given set $P$ into a Voronoi diagram $V(P)$. The inverse transformation that turns the Voronoi diagram $V(P)$ back into a set of generators $P$ is illustrated in Figure 7.6b.

### 7.3.1    Preliminaries

The *function* $f$ is defined as a relation between two sets $A$ and $B$ such that any $a \in A$ is associated with precisely one element $b \in B$ denoted by $f(a)$. The notation $f : a \to f(a)$ specifies that $f$ is a function that for a single element $a \in A$ gives an element $f(a) \in B$. The set A by which a function is defined is called its *domain*. The notation $f : A \to B$ is also used to explicitly specify the domain $A$ and the set $B$ from which the function $f$ takes its values. A slightly different notation is used for the synonymous term *map*. A map $f : A \mapsto B$ is a function $f$ such that for every $a \in A$, there is a unique object $b = f(a) \in B$.

The *image* of $f$, also called the *range* of $A$ under $f$, denoted by $f(A)$ is defined as a set of all values that $f$ takes as its argument ranges over all of $A$, i.e.,

$$f(A) = \{f(a) : a \in A\}.$$

A function $f$ that maps $A \mapsto B$ is said to be a *surjection* (or *surjective map*) if its image $f(A)$ coincides with $B$, i.e., $f(A) = B$. The surjection is also referred as mapping of $A$ *onto* $B$.

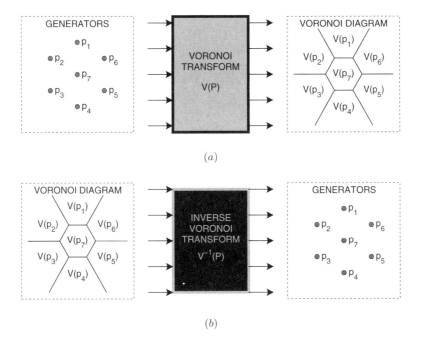

*(a)*

*(b)*

**FIGURE 7.6**

A direct Voronoi transform turns a set of generators $P$ into a Voronoi diagram $V(P)$ (a) and inverse Voronoi transforms turns a Voronoi diagram $V(P)$ into a set of generators $P$ (b).

**Example 7.8** *Let $\mathbb{N}$ be a set of natural numbers. The following examples demonstrate the range of a function, surjective and non-surjective functions:*

(a) *The function $f : \mathbb{N} \to \mathbb{N}$ is defined by $f(n) = 2n$, its range is the set $f(N) = \{2n : n \in \mathbb{N}\}$.*

(b) *The function $f : \mathbb{N} \to \mathbb{N}$, $f(n) = \lfloor \frac{n}{2} \rfloor$, is a surjection from $N$ to $N$.*

(c) *The function $f : \mathbb{N} \to \mathbb{N}$, $f(n) = 2n$, is not a surjection from $N$ to $N$.*

A function $f : A \to B$ is an *injection* (or *injective map*) if distinct arguments to $f$ produce distinct values or, equally, for any $a \in A$ and any $a' \in A$ equality $f(a) = f(a')$ implies $(a = a')$. This property is called *one-to-one* mapping. A function $f : A \to B$ is a *bijection* (or *bijective map*) if it is injective and surjective. This property is called *one-to-one-correspondence*.

**Example 7.9** *Examples of injective and bijective functions are as follows:*

(a) *The function $f(n) = 2n$ is an injection from $N$ to $N$.*

(b) *The function $f(n) = \lfloor \frac{n}{2} \rfloor$ is not an injection.*

(c) *The function $f(n) = (-1)^n \lfloor \frac{n}{2} \rfloor$ is a bijection.*

If a function $f : A \to B$ is a bijection then it is always possible to define its *inverse* $f^{-1} : B \to A$ such that

$$\forall b \in B \; \forall a \in A : a = f^{-1}(b) \Leftrightarrow b = f(a).$$

### 7.3.2   Direct Voronoi transform

The direct Voronoi transform $\mathbf{V} : P \mapsto V(P)$ maps a given set of Voronoi sites $P = \{p_i\}_{i=1}^{N}$ into a Voronoi diagram $V(P) = \{V(p_i)\}_{i=1}^{N}$ in such a way that with every Voronoi site $p_i \in P$ it associates a unique Voronoi region $V(p_i) \in V(P)$:

$$< \texttt{Generators} > \;\; \Rightarrow \;\; < \texttt{Voronoi diagram} > .$$

This can also be written in the form of a function $\mathbf{V} : p_i \to V(p_i)$. The direct Voronoi transform is therefore a one-to-one (or injective) map from $P$ onto $V(P)$:

$$\forall p_i \in P \; \forall p_j \in P : V(p_i) = V(p_j) \Rightarrow (p_i = p_j).$$

As both sets, $P$ and $V(P)$, have exactly the same number of elements, it follows that they are in *one-to-one correspondence*, i.e., the map $P \mapsto V(P)$ is bijective.

### 7.3.3   Inverse Voronoi transform

The inverse Voronoi transform $\mathbf{V}^{-1}$ can be defined as a transform that satisfies

$$P = \mathbf{V}^{-1}\left(\mathbf{V}(P)\right)$$

and therefore constitutes a map $\mathbf{V} : V(P) \mapsto P$ that associates with every Voronoi region $V(p_i) \in V(P)$ its Voronoi generator $p_i \in P$, i.e.,

$$< \texttt{Voronoi diagram} > \;\; \Rightarrow \;\; < \texttt{Generators} > .$$

**Example 7.10** *Given a set of generators in the Euclidean space $\mathbb{E}^2$ and the corresponding Voronoi diagram, denoted by*

$$< \text{Generators} > = P = \{p_1, \ p_2, \ p_3\} \ and$$
$$< \text{Voronoi diagram} > = V(P) = \{V(p_1), \ V(p_2), \ V(p_3)\},$$

*the two sets are in* one-to-one *correspondence which can be represented by the direct* $\mathbf{V}$ *and inverse* $\mathbf{V}^{-1}$ *Voronoi transforms:*

$$\mathbf{V} : p_1 \mapsto V(p_1), \ p_2 \mapsto V(p_2), \ p_3 \mapsto V(p_3),$$
$$\mathbf{V}^{-1} : V(p_1) \mapsto p_1, \ V(p_2) \mapsto p_2, \ V(p_3) \mapsto p_3.$$

*Figure 7.7 gives a graphical representation of the above maps.*

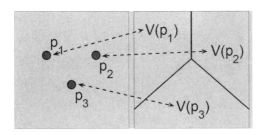

ONE-TO-ONE CORRESPONDENCE:

*Direct Voronoi transform* $\mathbf{V}(P)$ *that turns the set of generators* $P = \{p_1, p_2, p_3\}$ *into the Voronoi diagram* $V(P) = \{V(p_1), V(p_2), V(p_3)\}$ *defines a* one-to-one *correspondence between the sets* $P$ *and* $V(P)$.

**FIGURE 7.7**
Voronoi transform $\mathbf{V} : P \mapsto V(P)$ (Example 7.10).

### 7.3.4 Extension

Let $G$ be a set of all finite subsets $P = \{p_i\}$ in the metric space $\mathbb{M}$. In other words, $G$ can be considered as a collection of all possible sets of Voronoi generators. For each element $P \in G$ the Voronoi transform $\mathbf{V}(P)$ generates a Voronoi diagram $V(P)$. Thus the set $V(G) = \{V(P) : P \in G\}$ is a set of all possible Voronoi diagrams that can be generated in $\mathbb{M}$.

The *Voronoi transform* $\mathbf{V} : G \mapsto V(G)$ maps the set $G$ of all possible sets of Voronoi generators in $\mathbb{M}$ into the set $V(G)$ of all possible Voronoi diagrams in such a way that to every element $P$ in $G$ it associates a Voronoi diagram $V(P)$ in $V(G)$.

In general, the map $\mathbf{V} : G \mapsto V(G)$ is a *many-to-one map*, which means that different generator sets $P \in G$ and $P' \in G$ may result in one and the same Voronoi diagram, i.e, $V(P) = V(P')$. Therefore it is not injective.

**Example 7.11** *Figure 7.8 represents a set of generators*

$$P = \left\{ p_1(-\frac{\sqrt{3}}{2}, \frac{1}{2}), \ p_2(\frac{\sqrt{3}}{2}, \frac{1}{2}), \ p_3(0, -1) \right\}$$

*in $\mathbb{E}^2$ and the corresponding Voronoi diagram*

$$V(P) = \{V(p_1), V(p_2), V(p_3)\}.$$

*Multiplying the generator coordinates by a factor $\alpha$, $p^* = \alpha \cdot p_i$, gives an another distinct set of generators $P^*$ which results in exactly the same Voronoi diagram, i.e., $V(P^*) = V(P)$.*

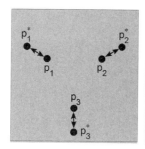 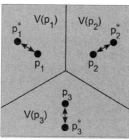

MANY-TO-ONE MAP:

Given a class $G^*$ of generator sets in $\mathbb{E}^2$ that are related to the set $P = \{p_1, p_2, p_3\}$ by the scale factors $\alpha$, the Voronoi transform $\mathbf{V} : G^* \mapsto V(G^*)$ always results in the same Voronoi diagram $V(P) = \{V(p_1), V(p_2), V(p_3)\}$.

**FIGURE 7.8**

Voronoi transform for sets of three generators (Example 7.11).

**Example 7.12** *In contrast to the previous Example 7.11, the Voronoi transform $\mathbf{V} : G \mapsto V(G)$ for a set of four generators in the Euclidean space $\mathbb{E}^2$*

$$P = \left\{ p_1(-\frac{\sqrt{3}}{2}, \frac{1}{2}), \ p_2(\frac{\sqrt{3}}{2}, \frac{1}{2}), \ p_3(-0.2, -1), \ p_4(0,0) \right\}$$

*gives a unique Voronoi diagram*

$$V(P) = \{V(p_1), V(p_2), V(p_3), V(p_4)\},$$

*i.e., the original distribution of the generators can be restored from the Voronoi diagram $V(P)$. Figure 7.9 shows the generator set and the corresponding Voronoi diagram.*

Thus there exist classes of *equivalent* generator sets (similar to Example 7.11) which result in the same Voronoi structures. Generator set reconstruction from such Voronoi diagram is ambiguous and therefore requires some extra constraints. Fortunately in the most practically interesting cases the generators of topological primitives can be unambiguously reconstructed from their Voronoi diagrams.

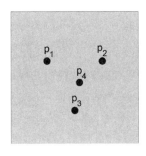
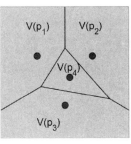

ONE-TO-ONE
CORRESPONDANCE:

*The set of generators*
$P = \{p_1, p_2, p_3, p_4\}$ *in* $\mathbb{E}^2$
*results, by the Voronoi transform*
$\mathbf{V} : P \mapsto V(P)$,
*in the unique Voronoi diagram*
$V(P)$, *i.e., any other distinct set*
*of generators cannot result in the*
*same Voronoi diagram.*

**FIGURE 7.9**
Voronoi transform for four generators (Example 7.12).

## 7.4 Properties

In this section, the basic properties of the Voronoi data structure are given. The focus is the properties that are useful in the manipulation of biometric topological structures, in particular,

(*a*) Assembling (composition) via Voronoi diagrams,

(*b*) Partition (decomposition) via Voronoi diagrams, and

(*c*) Measures via Voronoi diagrams.

In addition, Voronoi properties are useful in identification and recognition. However, these properties are not considered in this chapter. Most of the properties of Voronoi diagrams follow from the fact that the Voronoi polygon around each point is the region of the plane closer to the selected point than to any other point. The basic properties which are used in the generation of topological structures include

▶ *Incremental paradigm,*

▶ *Divide-and-conquer,*

▶ *Locus paradigm,*

▶ *Topological transformation paradigm,*

▶ *Scaling,*

▶ *Rigid motion,*

▶ *Symmetry,* and

▶ *Tilings.*

**The incremental paradigm** provides the possibility of modifying an existing Voronoi diagram by adding new generators one by one. Hence, local properties of a topological structure can be utilized to iteratively expand (or improve) a topological solution.

This method first locates the Voronoi region of the initial Voronoi diagram containing the new generator. Then it updates the Voronoi diagram by calculating the Voronoi region of the new generator and modifying the Voronoi polygons adjacent to this new one.

> **Example 7.13** *Given an initial set of generators*
>
> $$P = \{\ p_1(241, 243), p_2(121, 299), p_3(250, 80),$$
> $$p_4(394, 301), p_5(244, 400)\ \},$$
>
> *the Voronoi diagram $V(P) = \{V(p_1), V(p_2), \ldots, V(p_5)\}$ is shown in Figure 7.10a. A new generator, $p_6(244, 340)$, modifies the Voronoi diagram $V(P)$ as shown in Figure 7.10b,c,d.*

Incremental properties can be utilized by the random order of generators, which is useful in synthetic iris design.

**Divide-and-conquer,** a general paradigm which has been applied at various phases of analysis and synthesis of the topological structures. If the problem instance is small, the solution is found by some direct method. Otherwise, the problem instance is subdivided (decomposed) into two instances. The smaller instances can be solved recursively. The solutions are then *merged* together yielding a solution to the original problem. The merging step is crucial in all divide-and-conquer algorithms. In geometrical divide-and-conquer algorithms, it is often possible to use geometrical aspects of the problem.

**Topological transformation paradigm.** A topological transformation of a given problem can be transformed into another problem whose solution can be transformed back to provide a solution to the original problem. The usefulness of topological transformations to topological structure design has so far been rather limited. However, it has been used to construct some topological data structures.

**Scaling property.** A demonstration of the scaling property can be given in the usual Euclidean space $\mathbb{E}^2$. The simplest modification is to multiply all the coordinates of all generators by a constant factor. This will scale the set of generators in $\mathbb{E}^2$ while preserving their geometrical configuration. The Voronoi diagram of the modified set of generators coincides with the original one which is *multiplied* by the same factor. The multiplication of the Voronoi diagram by a factor means the multiplication of the coordinates of all Voronoi vertices by this factor. This constitutes the scaling property of the Voronoi diagram.

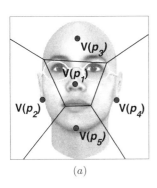

_Phase 1_. The original Voronoi diagram
$$V(P) = \{V(p_1), V(p_2), \ldots, V(p_5)\}$$
shown in (a).

_Phase 2_. The new generator $p_6$ as shown in (b).

_Phase 3_. The Voronoi region $V(p_6)$ of the new generator $p_6$ is presented in (c) superposed over the initial Voronoi diagram.

_Phase 4_. The new Voronoi region $V(p_6)$ updates the initial Voronoi diagram $V(P)$ by modifying the adjacent Voronoi regions $V(p_1)$, $V(p_2)$, $V(p_4)$ and $V(p_5)$ as shown in (d).

(a)

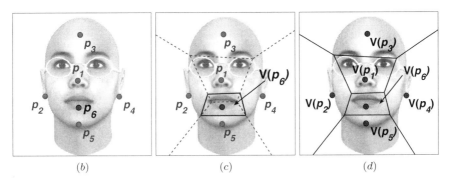

(b)          (c)          (d)

**FIGURE 7.10**
Incremental paradigm (Example 7.13).

**Example 7.14** *The set of generators in the space $\mathbb{E}^2$*

$$P = \{\ p_1(0,0),\ p_2(\frac{\sqrt{3}}{2}, \frac{1}{2}),\ p_3(0,1),\ p_4(-\frac{\sqrt{3}}{2}, \frac{1}{2}),$$
$$p_5(-\frac{\sqrt{3}}{2}, -\frac{1}{2}),\ p_6(0,-1),\ p_7(\frac{\sqrt{3}}{2}, -\frac{1}{2})\ \}$$

*results in the Voronoi diagram shown in Figure 7.11a. The results of scaling by the factors 2 and 2.5 are represented in Figures 7.11b and c respectively.*

The Voronoi transform $\mathbf{V} : G \mapsto V(G)$ that maps a certain set $G$ of finite subsets of distinct points, or Voronoi generators, in metric space $\mathbb{M}$ into the set of all their Voronoi diagrams $V(G)$ is in general a many-to-one map. This fact is closely related to the scaling properties of the Voronoi diagram and can be easily demonstrated by the following illustrative examples.

The first example will concern the *degeneration* of a Voronoi vertex.

Factor 1                    Factor 2                    Factor 2.5

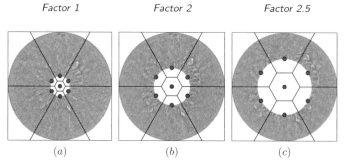

The Voronoi diagram of the modified set of generators coincides with the original one which is multiplied by the same factor.

(a)                         (b)                         (c)

**FIGURE 7.11**
The Voronoi diagram scaling property (Example 7.14).

> **Example 7.15** *Let the sets $P_1$, $P_2$ and $P_3$ of generators in the space $\mathbb{E}^2$ be*
>
> $$P_1 = \{ \ p_1(-1,0), \ p_2(-1,1), \ p_3(1,0), \ p_4(1,-1) \ \},$$
> $$P_2 = \{ \ p_1(-1,-0.4), \ p_2(-1,0.6), \ p_3(1,0.4), \ p_4(1,-0.6) \ \},$$
> $$P_3 = \{ \ p_1(-1,-0.5), \ p_2(-1,0.5), \ p_3(1,0.5), \ p_4(1,-0.5) \ \}.$$
>
> *The corresponding Voronoi diagrams $V(P_1)$, $V(P_2)$ and $V(P_3)$ are shown in Figure 7.12.*

Figure 7.12 illustrates the concept of the *degenerated Vertex*. It shows clearly that as the generators from the set $P_1$ to the set $P_3$ align, the two Voronoi vertices, $V_1$ and $V_2$, move toward each other. When the generators in the set $P_3$ are placed at the corners of a rectangle, the Voronoi vertices $V_1$ and $V_2$ become a single degenerated Voronoi vertex in Figure 7.12c.

The next example is intended to demonstrate the case when the Voronoi diagram contains a single degenerated Voronoi vertex. In this situation, the scaling of the generator set results in the same Voronoi diagram.

> **Example 7.16** *Let the sets $P_1$ and $P_2$ of generators in the space $\mathbb{E}^2$ be*
>
> $$P_1 = \{ \ p_1(-1,-1), \ p_2(-1,0), \ p_3(1,0), \ p_4(1,1) \ \}, \ and$$
> $$P_2 = \{ \ p_1(-1,-1), \ p_2(-1,1), \ p_3(1,-1), \ p_4(1,1) \ \}.$$
>
> *The corresponding Voronoi diagrams, $V(P_1)$ and $V(P_2)$, are shown in Figure 7.13a and c respectively. The scaled sets of generators, $P_1^* = 2 \cdot P_1$ and $P_2^* = 2 \cdot P_2$, result in the scaled Voronoi diagrams $V(P_1^*)$ and $V(P_2^*)$ as shown in Figure 7.13b and d respectively.*

One can observe that the Voronoi diagram $V(P_2)$ contains only one degenerated Voronoi vertex and coincides with its scaled version $V(P_2^*)$.

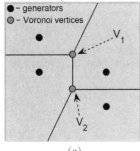

- generators
- Voronoi vertices

The Voronoi diagram $V(P_1)$ of the set $P_1$ in (a) contains two Voronoi vertices $V_1$ and $V_2$.
The generators in the set $P_2$ in (b) differ in their positions from the generators in $P_1$; the left pair of generators is displaced down and the right one up toward the middle line. As the result, two Voronoi vertices, $V_1$ and $V_2$, move toward each other.

(a)

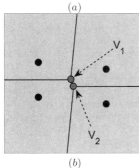

(b)

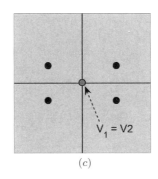

When the generators in the set $P_3$ become aligned so that they are at the corners of a rectangle in (c), the Voronoi vertices $V_1$ and $V_2$ become a single degenerated vertex.

(c)

**FIGURE 7.12**
Degeneration of Voronoi vertex (Example 7.15).

**Rigid motion.** The shape of the Voronoi diagram of a set of generators

▶ Depends only on their relative configuration, and
▶ Does not depend on their absolute position and orientation.

From the above, the following statement can be formulated: if two sets of generators, $P_1$ and $P_2$, can be transformed one into another by some rigid motion $T$ then so can their Voronoi diagrams $V(P_1)$ and $V(P_2)$. In other words, if $T$ is a rigid motion then $V(T(P)) = T(V(P))$.

**Example 7.17** *Given a set of generators*

$$P = \left\{ p_1(-\frac{\sqrt{3}}{2}, \frac{1}{2}),\ p_2(\frac{\sqrt{3}}{2}, \frac{1}{2}),\ p_3(-0.2, -1),\ p_4(0,0) \right\}$$

*the corresponding Voronoi diagram is shown in Figure 7.14a. The new generator set $P^* = \{p_i^*\}$ is obtained from the generators $P = \{p_i\}$ by rotating them around the origin by an angle of $\alpha = 30°$. The corresponding Voronoi diagram $V(P^*) = \{V(p_i^*)\}$ can be obtained by rotating the initial Voronoi diagram $V(P)$ by the same angle $\alpha$ (Figure 7.14b).*

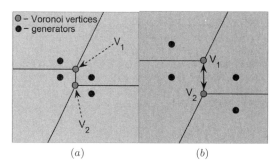

The Voronoi diagram (a) of the initial set of generators $P_1$ differs from the Voronoi diagram (b) of the scaled set of generators $P_1^*$ because the distance between the Voronoi vertices $V_1$ and $V_2$ is scaled as indicated by an arrow in (b).

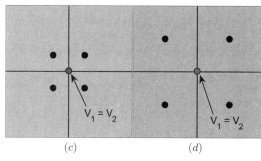

The Voronoi diagram (d) of the modified set of generators $P_2^*$ coincides with the Voronoi diagram (c) of the original set of generators $P_2$. According to the scaling property of the Voronoi diagram, multiplication of the same degenerated Voronoi vertex by the scale factor results in the same diagram.

**FIGURE 7.13**
Scaling of the Voronoi diagrams: one containing the degenerated Voronoi vertex and another which does not contain the degenerated Voronoi vertex (Example 7.16).

**A symmetry** within a set of generators will be manifested as symmetry within their Voronoi diagram (however this property is not invertible).

> **Example 7.18** *The set of generators in Figure 7.15a*
>
> $$P = \left\{ p_1(\frac{\sqrt{3}}{2}, \frac{1}{2}), \ p_2(0, 1), \ p_3(-\frac{\sqrt{3}}{2}, \frac{1}{2}), \right.$$
> $$= \ \left. p_4(-\frac{\sqrt{3}}{2}, -\frac{1}{2}), \ p_5(0, -1), \ p_6(\frac{\sqrt{3}}{2}, -\frac{1}{2}) \right\}$$
>
> *are placed at the corners of a hexagon and have the correspondending symmetry. The Voronoi diagram $V(P)$ in Figure 7.15b also clearly shows the same symmetry.*

Symmetry can be observed within biometric topological structure during various stages of analysis and synthesis, in particular:

▶ Facial components are symmetrical (eyes, brow, nose, mouth, etc.). Algorithms for facial expression recognition and synthesis are based on detection of asymmetries.

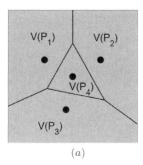

_Phase 1_. *The initial Voronoi $V(P)$ diagram of the set $P$ of generators (a).*

(a)

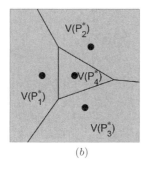

_Phase 2_. *The new generator set $P^* = \{p_i^*\}$ consists of the generators from $P = \{p_i\}$ which are rotated around the origin at an angle of $\alpha = 30°$. The corresponding Voronoi diagram $V(P^*) = \{V(p_i^*)\}$ can be obtained by rotating the initial Voronoi diagram $V(P)$ by the same angle $\alpha$ (b).*

(b)

**FIGURE 7.14**
Example of the rigid motion of a Voronoi diagram (Example 7.17).

▶ Master-iris primitives are designed as symmetric topological configurations.

▶ Most master-fingerprint primitives are locally symmetrical structures.

**Tilings** are useful in biometric structure design. Voronoi diagrams can be interpreted as tilings in the metric space, and inherit any symmetries of the underlying generators. A method for constructing tilings with some set of desired symmetries is as follows: select a set of generators with the desired symmetries and compute the Voronoi diagram.

## 7.5 Voronoi data structure in topological analysis and synthesis

In this section, the Voronoi diagram technique is applied to biometric topological structures, resulting in Voronoi data structures.

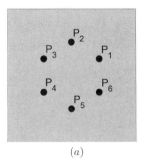

*The set $P$ of generators has the symmetry of a hexagon.*

(a)

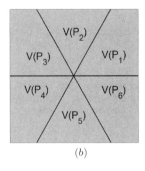

*The symmetry of the set $P$ of generators is also the symmetry of its Voronoi diagram $V(P)$, i.e., the symmetry of a hexagon.*

(b)

**FIGURE 7.15**

Symmetry of Voronoi diagram (Example 7.18).

Topological structuring of biometric data can help discover the spacial relations, or generators, between the elements. This can be used to ensure the integrity of data in models generated during synthesis, by a using a method called geometric constraint solving.

Table 7.1 shows several facial topological elements (primitives) and their corresponding Voronoi diagrams. From observing these primitives it follows that it is reasonable to represent them as a Voronoi configuration for the following reasons:

(a) It seems quite natural to associate the Voronoi generators with the face feature points;

(b) The Voronoi diagram of the facial features reflects their topological structure which can be represented as a Voronoi data structure;

(c) The partitioning of the image into Voronoi regions can help to extract data like color information, texture, etc, relative to the feature points.

Topological synthesis can be started with a collection of several primitives; each primitive consists of a small number of sites which generate a Voronoi diagram. Topological structures can be generated by *superimposing* (combining with interference) Voronoi diagrams or their generating sites.

**TABLE 7.1**
Voronoi diagrams of facial topological primitives.

| Facial primitive | Voronoi diagram |
| --- | --- |
| | |
| | |
| | |
| | |
| | |

## 7.6  Topological compatibility of the Voronoi diagram

*Topological compatibility* is defined as the property of implementing various topological transformations.  The Voronoi diagram is characterized by topological compatibility because it can be computed via various topological transformations.

In this section an approach that computes discrete approximations of Voronoi diagrams (including the Voronoi diagrams of complex areas in different metric spaces) is described.

### 7.6.1  Distance mapping of feature points

To establish the connection between the topological structure of biometric data set to the topology of its Voronoi diagram, it will be useful to introduce *distance maps*.  A distance map is an extension of the distance function in metric space.  Like the usual distance function that characterizes the separation of two points in space, the distance map is intended to characterize the spacial separation of a set of points.  Furthermore, given a set of points, the analytical behavior of the distance map is closely related to the topological structure of its Voronoi diagram.

**Metric space.** By the definition given in Section 7.2, a metric space $\mathbb{M}$ is a space with an associated distance function $f$ which gives the distance between any two points in $\mathbb{M}$. A wide variety of alternate metrics can be used for the calculation of distances. The choice of a concrete distance function depends on the aims of analysis or synthesis as well as on the nature of a particular data set. Below are some of the distance functions commonly used for the calculation of distance $d(p_1, p_2)$ between two points, $p_1(x_1, y_1)$ and $p_2(x_2, y_2)$, in 2-dimensional real space:

$$\text{<Euclidean>} = \sqrt{(x_2 - x_1)^2 + (y_2 - y_1)^2},$$

$$\text{<Chessboard>} = \max\left(|x_2 - x_1|, |y_2 - y_1|\right),$$

$$\text{<City-block>} = |x_2 - x_1| + |y_2 - y_1|,$$

$$\text{<Quasi-Euclidean>} = \begin{cases} |x_2 - x_1| + (\sqrt{2} - 1)|y_2 - y_1|, & |x_2 - x_1| > |y_2 - y_1| \\ (\sqrt{2} - 1)|x_2 - x_1| + |y_2 - y_1|, & |x_2 - x_1| \le |y_2 - y_1|. \end{cases}$$

The following example is to illustrate the usage of the above metrics for the calculation of the distance between two points in 2-dimensional real space.

**Example 7.19** *The results of calculation of the distance between two points, $p_1(0,0)$ and $p_2(1,1)$, in 2-dimensional space for different metrics is given in Figure 7.16.*

$< Euclidean\ distance >$

$$= \sqrt{(x_2 - x_1)^2 + (y_2 - y_1)^2}$$

$$= \sqrt{(1-0)^2 + (1-0)^2} = \sqrt{2}$$

$< Chessboard\ distance >$

$$= \max(|x_2 - x_1|, |y_2 - y_1|)$$

$$= \max(|1-0|, |1-0|) = 1$$

$< City\text{-}block\ distance >$

$$= |x_2 - x_1| + |y_2 - y_1|$$

$$= |1-0| + |1-0| = 2$$

$< Quasi - Euclidean\ distance >$

$$= (\sqrt{2} - 1)|x_2 - x_1| + |y_2 - y_1|$$

$$= (\sqrt{2} - 1)|1-0| + |1-0| = \sqrt{2}$$

**FIGURE 7.16**

Distance between two points, $p_1$ and $p_2$, in 2-dimensional real space (Examples 7.19).

The distance function characterizes the separation of two points in metric space. However an alternative point of view on the distance function will be more suitable for our development and for the definition of a distance map: given a fixed point in a metric space, the distance function can be considered as a function of a single argument so that it gives the distance from any arbitrary point in the space to that fixed point.

## 7.6.2   Distance map

Let $\mathbb{M}$ be a metric space with its distance function $d(p, p')$. The distance from an arbitrary point $p$ to a fixed point $p'$ can be considered as a function of a single argument $p$. Such a function that assigns to each point in $\mathbb{M}$ a value equal to the distance from that point to the fixed point $p'$ will be referred as a *distance map* of a *fixed point* in $\mathbb{M}$.

The distance map for a single fixed point can be easily extended to a *distance map* of a *set* $P = \{p_i\}_{i=1}^{N}$ of N distinct points in $\mathbb{M}$: the distance map, denoted by $d(p, P)$, assigns to each point $p$ in $\mathbb{M}$ the minimal distance $d(p, p_i)$ as $p_i$

ranges over all of $P$, i.e.,

$$d(p, P) = \min_{p_i \in P} d(p, p_i). \tag{7.4}$$

**Euclidean distance map.** Euclidean space $\mathbb{E}^2$ is an excellent metric space to study the analytical properties of the distance map thanks to its nice distance function

$$d\ (p(x, y),\ p'(x', y'))\ =\ \sqrt{(x - x')^2 + (y - y')^2}$$

which is differentiable everywhere in $\mathbb{E}^2$. Therefore the discussion will be restricted to $\mathbb{E}^2$ for a while, solely for the purposes of simplicity, and the necessary generalization will be made later.

The following example shows a contour plot of the distance map $d(p, P)$ for the set $P = \{p_1, p_2, p_3\}$ of three points in $\mathbb{E}^2$.

> **Example 7.20** *Given a set $P$ of three points in $\mathbb{E}^2$,*
>
> $$P = \{p_1(-0.5, -0.25),\ p_2(0.0, 0.7),\ p_3(0.75, -0.5)\},$$
>
> *the distance map $d(x, y) \equiv d(p(x, y), P)$ for the set $P$ is*
>
> $$\begin{aligned} < Distance\ map > \ &=\ \min\ (\ d(p, p_1),\ d(p, p_2),\ d(p, p_3)\ ) \\ &=\ \min\ (\ \sqrt{(x + 0.5)^2 + (y + 0.25)^2}, \\ &\qquad\quad \sqrt{x^2 + (y - 0.7)^2}, \\ &\qquad\quad \sqrt{(x - 0.75)^2 + (y + 0.5)^2}\ ). \end{aligned}$$
>
> *Figure 7.17 represents the contour plot of the above distance map which clearly demonstrates the points and lines where the different contour circles lines meet.*

The plot in Figure 7.17 from the above example makes obvious a noteworthy topological behavior of the Euclidean distance map:

▶ In the areas close to the points $p_1$, $p_2$ and $p_3$ the contours surround each point as entire distinct circles. These actually are the contour lines of the individual distance functions: $d(p, p_1)$, $d(p, p_2)$ and $d(p, p_3)$ respectively. In the areas close to the points, the circles surrounding different points are well separated from each other.

▶ At the same time, the plot clearly demonstrates some contour lines which consist of circles that meet together at common points and lines. At these common points the distance map function becomes non-differentiable.

Thus the Euclidean distance map has a set of *critical points*[†] and these critical points are related to the Voronoi diagram $V(P)$ of the set $P$.

---

[†]Critical points are the points where the function has the gradient equal to zero or where one of its partial derivatives is not defined.

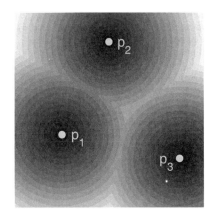

CONTOUR PLOT. The contour plot of the distance map $d(p, P)$ for the set $P = \{p_1, p_2, p_3\}$ clearly demonstrates the points where the different circles in contour lines meet together.

To plot the contours in MATLAB:

```
bw = logical(zeros([512 512]));
bw(192, 129) = 1;
bw(435, 256) = 1;
bw(129, 447) = 1;

dist_map = bwdist(bw);

contour(dist_map, 18);
```

**FIGURE 7.17**
Contour plot of the Euclidean distance function in $\mathbb{E}^2$ (Example 7.20).

**The relation** between the critical points in the distance map and the Voronoi diagram can be expressed in the following statements, which are actually valid in any metric space associated with a monotonous distance function (Euclidean, Chessboard, City-block and Quasi-Euclidean):

▶ The points that lie on the Voronoi edges of the Voronoi diagram are the critical points of the distance map and these points represent the local maximums of the distance map.

▶ If critical points exist inside the Voronoi region, they are not the points of local maximums of the distance map.

The next example is intended to illustrate the analytical properties of the distance map discussed above.

**Example 7.21** *Given the set $P$ of three points in $\mathbb{E}^2$ and the distance map $d(x, y)$ from Example 7.20, Figure 7.18a represents a vector map of the distance map gradient*

$$< Map > \ = \ \nabla d(x, y) = \frac{\partial \ d(x, y)}{\partial x} \cdot \mathbf{e}_x + \frac{\partial \ d(x, y)}{\partial y} \cdot \mathbf{e}_y$$

*and Figure 7.18b – the points where the second derivatives,*

$$\frac{\partial^2 d(x, y)}{\partial x^2} \quad and \quad \frac{\partial^2 d(x, y)}{\partial y^2},$$

*of the distance map are negative.*

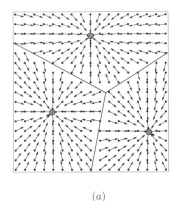

GRADIENT MAP. *The gradient plot of the Euclidean distance map generated by three points.*
*To plot the gradient in MATLAB:*

```
[dx, dy] = gradient (dist_map);
[X, Y] = meshgrid (8 : 31 : 512);
DX = dx (8 : 31 : 512, 8 : 31 : 512);
DY = dy (8 : 31 : 512, 8 : 31 : 512);
quiver (X, Y, DX, DY);
```

(a)

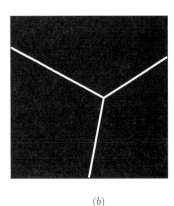

SECOND DERIVATIVES. *The critical points where the second derivatives of the distance map are negative.*
*To plot the points in MATLAB:*

```
diff_1 = diff (dist_map, 2, 1) < 0;
diff_2 = diff (dist_map, 2, 2) < 0;
diff_1 = padarray (diff_1, [1 0], 0);
diff_2 = padarray (diff_2, [0 1], 0);
diff = diff_1 | diff_2;
diff = bwmorph (diff, 'thin', Inf);
imshow (diff);
```

(b)

**FIGURE 7.18**
Analytical properties of the Euclidean distance function in $\mathbb{E}^2$ (Example 7.21).

## 7.7   Implementing the discrete Voronoi transform with a distance transform

Representing biometric data in the form of a Voronoi data structure requires transforming the data into the form of a Voronoi diagram. There are various algorithms to implement this transformation. In choosing an appropriate method of transformation, the following aspects play a critical role

▶ A Voronoi diagram is a topological transformation of a set sites. Hence, input data and output data must be consistent with the topological representation of biometric information.

▶ Ideally other well investigated topological transforms can be used to compute a Voronoi diagram.

In this section, it will be shown how the distance transform, a popular transform in biometric data analysis, can be used to compute a discrete approximation of the Voronoi diagram.

The distance transform is used by many applications in order to map distances from feature points. The distance transform is also used in order to separate the adjoining features in biometric data and to identify primitives for the construction of new topological structures.

**The distance transform** is normally applied to binary images and consists of calculating the distances between feature (colored) pixels and non-feature (blank) pixels in the image; each pixel of the image is assigned a number that is the distance to the closest feature boundary.

> **Example 7.22** *Let the source image be the binary image in Figure 7.19. The distance transform assigns each pixel in the binary source image a number that is the distance to the closest feature (i.e. nonzero) pixel. The result is a grayscale distance map with real valued pixels.*

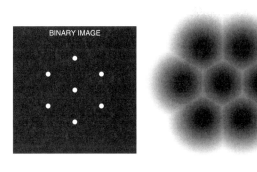

Distance transform assigns each pixel in a binary source image a number that is the distance to the closest feature (i.e., nonzero) pixel.

The result is a grayscale distance map with real valued pixels.

**FIGURE 7.19**

Binary image consisting of seven feature points and its distance map (Example 7.22).

A variety of alternate metrics such as the Euclidean, Chessboard, City-block and Quasi-Euclidean metrics discussed earlier can be used for the calculation of the distance transform.

The following example demonstrates the calculation of the distance transform using these metrics.

**Example 7.23** *Let the source binary image be a* $[3 \times 3]$ *matrix*

$$< Binary\ image >= I_{3\times3} = \begin{bmatrix} I_{11}\ I_{12}\ I_{13} \\ I_{21}\ I_{22}\ I_{23} \\ I_{31}\ I_{32}\ I_{33} \end{bmatrix} = \begin{bmatrix} 0\ 0\ 0 \\ 0\ 1\ 0 \\ 0\ 0\ 0 \end{bmatrix},$$

*where the only feature pixel has been placed at the center of the image. The distance transform replaces the zeros in the image by the distances indicated by the arrows:*

$$\begin{bmatrix} 0 & 0 & 0 \\ 0 \longleftarrow & 1 \longrightarrow & 0 \\ 0 & 0 & 0 \end{bmatrix}$$

A distance transform based on different distance metrics gives the results presented in Table 7.2.

**TABLE 7.2**

Distance transform in Euclidean, Chessboard, City-block and Quasi-Euclidean distance metric (Example 7.23).

| Transform | Computing |
|---|---|
| $\begin{bmatrix} 0\ 0\ 0 \\ 0\ 1\ 0 \\ 0\ 0\ 0 \end{bmatrix} \Rightarrow \begin{bmatrix} I_{11}\ I_{12}\ I_{13} \\ I_{21}\ I_{22}\ I_{23} \\ I_{31}\ I_{32}\ I_{33} \end{bmatrix} = \begin{bmatrix} \sqrt{2}\ 1\ \sqrt{2} \\ 1\ 0\ 1 \\ \sqrt{2}\ 1\ \sqrt{2} \end{bmatrix}$ | EUCLIDEAN METRIC<br>$I_{11} = I_{13} = \sqrt{1^2 + 1^2} = \sqrt{2}$<br>$I_{31} = I_{33} = \sqrt{1^2 + 1^2} = \sqrt{2}$<br>$I_{12} = I_{32} = \sqrt{0^2 + 1^2} = 1$<br>$I_{21} = I_{23} = \sqrt{1^2 + 0} = 1$ |
| $\begin{bmatrix} 0\ 0\ 0 \\ 0\ 1\ 0 \\ 0\ 0\ 0 \end{bmatrix} \Rightarrow \begin{bmatrix} I_{11}\ I_{12}\ I_{13} \\ I_{21}\ I_{22}\ I_{23} \\ I_{31}\ I_{32}\ I_{33} \end{bmatrix} = \begin{bmatrix} 1\ 1\ 1 \\ 1\ 0\ 1 \\ 1\ 1\ 1 \end{bmatrix}$ | CHESSBOARD METRIC<br>$I_{11} = I_{13} = \max(1,1) = 1$<br>$I_{31} = I_{33} = \max(1,1) = 1$<br>$I_{12} = I_{32} = \max(0,1) = 1$<br>$I_{21} = I_{23} = \max(1,0) = 1$ |
| $\begin{bmatrix} 0\ 0\ 0 \\ 0\ 1\ 0 \\ 0\ 0\ 0 \end{bmatrix} \Rightarrow \begin{bmatrix} I_{11}\ I_{12}\ I_{13} \\ I_{21}\ I_{22}\ I_{23} \\ I_{31}\ I_{32}\ I_{33} \end{bmatrix} = \begin{bmatrix} 2\ 1\ 2 \\ 1\ 0\ 1 \\ 2\ 1\ 2 \end{bmatrix}$ | CITY-BLOCK METRIC<br>$I_{11} = I_{13} = 1 + 1 = 2$<br>$I_{31} = I_{33} = 1 + 1 = 2$<br>$I_{12} = I_{32} = 0 + 1 = 1$<br>$I_{21} = I_{23} = 1 + 0 = 1$ |
| $\begin{bmatrix} 0\ 0\ 0 \\ 0\ 1\ 0 \\ 0\ 0\ 0 \end{bmatrix} \Rightarrow \begin{bmatrix} I_{11}\ I_{12}\ I_{13} \\ I_{21}\ I_{22}\ I_{23} \\ I_{31}\ I_{32}\ I_{33} \end{bmatrix} = \begin{bmatrix} \sqrt{2}\ 1\ \sqrt{2} \\ 1\ 0\ 1 \\ \sqrt{2}\ 1\ \sqrt{2} \end{bmatrix}$ | QUASI-EUCLIDEAN METRIC<br>$I_{11} = I_{13} = (\sqrt{2} - 1)1 + 1 = \sqrt{2}$<br>$I_{31} = I_{33} = (\sqrt{2} - 1)1 + 1 = \sqrt{2}$<br>$I_{12} = I_{32} = (\sqrt{2} - 1)0 + 1 = 1$<br>$I_{21} = I_{23} = 1 + (\sqrt{2} - 1)0 = 1$ |

## 7.7.1 Distance mapping of biometric data

As the distance transform is normally applied to binary images, the feature points and boundaries representing biometric data in the image must be determined in advance. The features can be detected by an algorithm for feature extraction, or a set of feature points can be chosen manually. The detected features can also serve for the creation of topological primitives.

In the following example, the distance maps for three images representing different types of biometric data are calculated.

> **Example 7.24** *A face image, an image of a fingerprint and an image of a signature are shown in Figure 7.20a, 7.21a and 7.22a respectively, together with some feature points. The sets of feature points are as binary images shown in Figure 7.20b, 7.21b and 7.22b. The distance maps computed for these images based on the Euclidean distance are presented in Figure 7.20c, 7.21c and 7.22c.*

## 7.7.2 Transformation of the distance map of point generators to a discrete approximation of Voronoi diagrams

When dealing with digital images and when a discrete approximation to the Voronoi diagram of distinct feature points is desired, the problem can be formulated as follows: given a set of generator points, compute a discrete approximation of the Voronoi diagram for these generators. A solution of the problem can be achieved by taking the next steps:

1. *Computing the discrete distance map,* i.e.,
   $<$ `Generators` $> \Rightarrow <$ `Binary image` $> \Rightarrow <$ `Distance map` $>$

2. *Computing second derivatives of the distance map,* i.e.,
   $<$ `Distance map` $> \Rightarrow <$ `Second derivative map` $>$

3. *Computing the Voronoi diagram,* i.e.,
   $<$ `Second derivative map` $> \Rightarrow <$ `Voronoi diagram` $>$

Voronoi edges are the critical points of the distance map where the order differences are negative indicating a local maxima of shade (not all the critical points are on the Voronoi edges). To determine the Voronoi edges, the second order differences of the distance map $d(i,j)$ of an image are calculated:

$< \text{Differences (X-direction)} > = \text{diff}(i,j) = d(i+2,j) - 2d(i+1,j) + d(i,j)$
$< \text{Differences (Y-direction)} > = \text{diff}(i,j) = d(i,j+2) - 2d(i,j+1) + d(i,j).$

Those with negative values are selected as possible Voronoi edges.

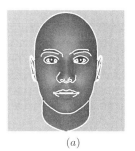

(a)

*Phase 1*. Synthetic face image with a set of feature points. The features can be selected manually or determined by an edge detector.

*To read image into MATLAB:*

```
face = imread ('face.bmp');
```

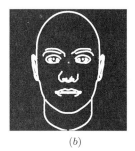

(b)

*Phase 2*. Binary image with a set of feature points extracted by canny edge detector.

*MATLAB command line:*

```
edges = edge (face, 'canny', 0.15);
```

(c)

*Phase 3*. Distance map of the feature points calculated based on the Euclidean distance.

*MATLAB command line:*

```
map = bwdist (edges, 'euclidean');
imshow (mat2gray (map));
```

**FIGURE 7.20**

Illustration of distance mapping of the face features (Examples 7.24).

**Example 7.25** *Let $P$ be a set of feature points of the fingerprint image in Figure 7.21a. These points serve as the generators for the calculation of a discrete approximation of the Voronoi diagram $V(P)$. Figure 7.23 demonstrates the following procedure:*

$$< \text{Set of generators} > \Rightarrow < \text{Distance map} >$$
$$< \text{Distance map} > \Rightarrow < \text{Derivative map} >$$
$$< \text{Derivative map} > \Rightarrow < \text{Voronoi diagram} >$$

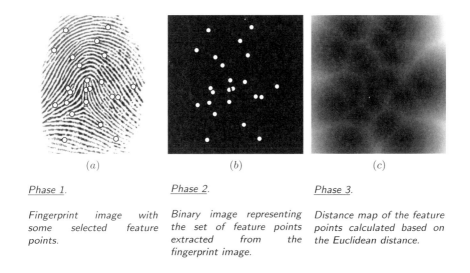

|  (a) | (b) | (c) |
|---|---|---|

*Phase 1.*

Fingerprint image with some selected feature points.

*Phase 2.*

Binary image representing the set of feature points extracted from the fingerprint image.

*Phase 3.*

Distance map of the feature points calculated based on the Euclidean distance.

**FIGURE 7.21**

Illustration of distance mapping of the fingerprint features (Examples 7.24).

## 7.8 Calculating area Voronoi diagrams using nearest-neighbor transform

**The distance map of area generators.** In Section 7.2.2 the distance function $d(p, P)$ from the points $p$ to the set $P$ in the metric space $\mathbb{M}$ was defined. In a similar way as for the set of distinct points, the distance map for a set $A = \{A_i\}_{i=1}^{N}$ of $N$ distinct area generators, denoted by $d(p, A)$, can be introduced; the distance map $d(p, A)$ assigns to each point $p$ in $\mathbb{M}$ the minimal distance $d(p, A_i)$ as $A_i$ ranges over all of $A$, i.e.,

$$d(p, A) = \min_{A_i \in A} d(p, A_i). \qquad (7.5)$$

However, when attempting to use the analytical properties of this distance map in order to calculate Voronoi diagrams some difficulties arise. Two types of possible errors produced in such calculations for the area generators are described in the next section.

### 7.8.1 Effects of approximation to Voronoi diagrams via distance transform

**Nonconvex primitives.** The distance map of a non-convex generator contains some critical points (with negative second derivatives) that do not

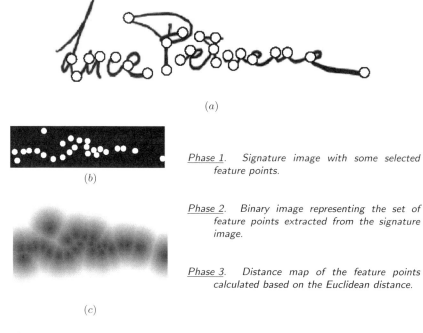

(a)

(b)

*Phase 1.* Signature image with some selected feature points.

*Phase 2.* Binary image representing the set of feature points extracted from the signature image.

*Phase 3.* Distance map of the feature points calculated based on the Euclidean distance.

(c)

**FIGURE 7.22**
Illustration of distance mapping of the signature features (Examples 7.24).

lie on the Voronoi edges. Figure 7.24 illustrates this situation. Figure 7.24a shows a non-convex primitive. Its distance map is presented in Figure 7.24b and its critical points in Figure 7.24c. However, a single generator primitive cannot produce Voronoi edges; such extra features must be removed from the result of the calculation.

**Effect of rasterization.** Voronoi diagrams calculated via distance transforms from digital images have another type of extra feature that needs to be removed from the final results also. Namely, the rasterization of an area object results in stairs (artefacts) on the object borders. These stairs in turn result in a great number of local maximums and minimums in the distance map. Therefore, the distance map contains critical points (with a negative second derivative) that are not related to the Voronoi diagram. Figure 7.25 demonstrates such extra features.

The area object is shown in Figure 7.25a. The corresponding distance map is represented in Figure 7.25b.

As in the above case, a single generator primitive should not produce Voronoi edges. The lines in Figure 7.25c are due to the rasterization of the generator object.

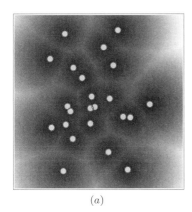

(a)

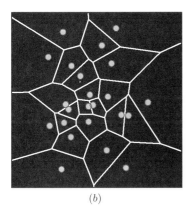

(b)

*Phase 1.* Calculating the distance map for a set of fingerprint feature points:

*MATLAB code:*

```
P = int32(...
[232 321; 180 156; 302 149; 207 186; ...
109 130; 113 333; 236 241; 232 276; ...
246 272; 291 247; 332 301; 354 303; ...
149 461; 287 405; 154 56; 178 366; ...
346 457; 273 96; 316 45; 413 264; ...
162 270; 171 285; 157 324]);

BW = logical (zeros ([512 512]));
for i = 1 : 23 BW(P(i, 2), P(i, 1)) = 1; end;

dist_map = bwdist (BW);
imshow (mat2gray (dist_map));
```

*Phase 2.* Determining the negative second differences of the distance map, determining the Voronoi diagram:

*MATLAB code:*

```
diff_1 = diff (dist_map, 2, 1) < 0;
diff_2 = diff (dist_map, 2, 2) < 0;
diff_1 = padarray (diff_1, [1 0], 0);
fiff_2 = padarray (diff_2, [0 1], 0);
diff = diff_1 | diff_2;
diff = bwmorph (diff, 'thin', Inf);
imshow (diff);
```

**FIGURE 7.23**

Calculation of a discrete Voronoi diagram from a distance map of fingerprint features (Examples 7.25).

## 7.8.2 Nearest-neighbor map

Fortunately, there exists another simple and effective method for computing a discrete approximation to the Voronoi diagram. For instance, the MATLAB bwdist function also computes the so-called *nearest-neighbor transform*. The nearest-neighbor transform is very like the distance transform; it assigns each element of a binary image the index of the nearest feature (i.e., nonzero) pixel. A more detailed description of the function bwdist is given in Problem Section 7.10.

Every pixel in the nearest-neighbor transform by definition contains the index of the nearest feature pixel in the original binary image. So, even for a primitive of the same area, the closest points will be assigned different

Non-convex object          Distance map          Critical points

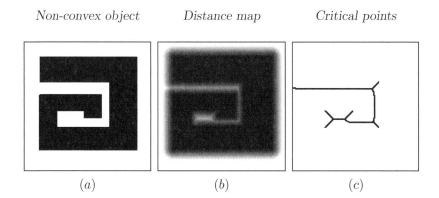

(a)                          (b)                          (c)

**FIGURE 7.24**
Non-convex generator: generator object (a), distance map of the object (b),
and critical points calculated from the distance map (c).

indices. Therefore it is difficult to devise a scheme for transferring directly the
nearest-neighbor transform into the discrete Voronoi diagram. Instead, the
area primitives will be represented by a grayscale image, where each distinct
primitive is assigned a unique gray-level value. After the calculation of the
nearest-neighbor transform, every index is replaced by the value from the pixel
in the grayscale image this index points to. The final image obtained in this
way will be referred as *nearest-neighbor map* of area primitives.

Figure 7.26 gives an illustration of the distinction between the nearest-
neighbor transform and the nearest-neighbor map of area primitives.

### 7.8.3  Transformation of nearest-neighbor map of area primitives into a Voronoi diagram

To produce a discrete approximation to the Voronoi diagram from a set of
area primitives, the following scheme can be used:

*Represent the topological primitives as a binary image,* i.e.,
    < Generators > ⇒ < Binary image >

*Compute the nearest-neighbor map of the binary image,* i.e.,
    < Binary image > ⇒ < Nearest-neighbor map >

*Compute the Voronoi diagram,* i.e.,
    < Nearest-neighbor map > ⇒ < Voronoi diagram >

The next example illustrates the application of this scheme to the
calculation of a discrete approximation to the Voronoi diagram of the features
on a face.

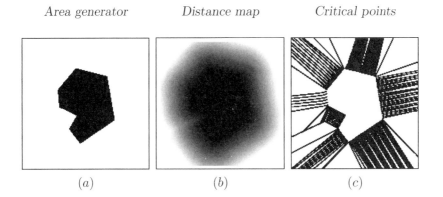

**FIGURE 7.25**
Effect of rasterization: generator object (a), distance map of the object (b), and critical points calculated from the distance map (c).

**Example 7.26** *Given the set of features on the face image in Figure 7.20, the procedure to calculate of a discrete approximation to Voronoi diagram is presented in Figure 7.27:*

$$< \text{Features} > \Rightarrow < \text{Binary image} >$$

$$< \text{Binary image} > \Rightarrow < \text{Nearest-neighbor map} >$$

$$< \text{Nearest-neighbor map} > \Rightarrow < \text{Voronoi diagram} >$$

## 7.9 Summary

Most biometric objects (fingerprints, iris, retina, face, etc.) can be naturally modelled as topological data structures. Processing and manipulating topologies are key to computer aided analysis and synthesis of such structures. Voronoi diagrams play a special role in the topological representation of biometric data:

1. The Voronoi diagram structure can encode certain characteristics of the biometric data object such as geometrical distribution, color and texture information:

    *In analysis*, Voronoi diagrams can be used for partitioning (decomposition) metric space in order to extract features for the recognition of images.

    *In synthesis*, Voronoi diagrams provide possibilities for the design of topology related to the structures of biometric images, the

Area generator          Nearest-neighbor tran.      Nearest-neighbor map

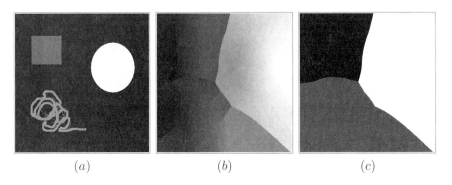

      (*a*)                   (*b*)                   (*c*)

**FIGURE 7.26**

Nearest-neighbor mapping: grayscale image of generator primitives (a), nearest-neighbor transform (b) of the primitives, and nearest-neighbor map (c) calculated from the nearest-neighbor transform.

      manipulation of primitives using incremental properties of Voronoi diagrams, and for coloring the metric space via Voronoi regions.

2. The Voronoi transform can be applied to simple data structures such as separate points on the plane, or to complex topological structures such as area primitives.

3. Some properties of Voronoi diagrams are useful for manipulating topological data structures:

   ▶ Hierarchical and incremental properties of Voronoi diagrams make them compatible with the assembling paradigm of synthetic data structure design.

   ▶ The Divide-and-conquer approach used in Voronoi diagram techniques is useful for decomposing a model into local and global levels. This simplifies the analysis and synthesis of biometric data.

   ▶ Scaling and symmetry properties are useful for geometric constraint solving and in generating tilings.

   ▶ Some topological primitives can be designed by manipulating a Voronoi diagram and then performing an inverse Voronoi transform.

4. The Voronoi transform possesses the property of *topological compatibility*. This provides several benefits for the manipulation of biometric topological structures:

   ▶ The Voronoi transform can be implemented with various different transforms such as distance or nearest-neighbor transforms.

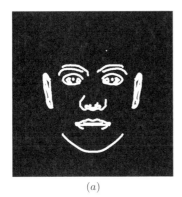

(a)

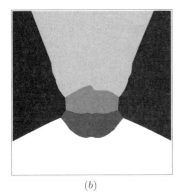

(b)

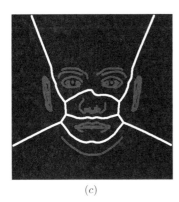

(c)

*Phase 1.* Creating a binary image representing the face features (a):

The generator sets are represented by nonzero pixels in a gray image. The following distinct objects are represented by pixels of different values:

| | |
|---|---|
| ears | – gray value of 55 |
| lips | – gray value of 100 |
| nose | – gray value of 150 |
| eyes | – gray value of 200 |
| jawline | – gray value of 250 |

MATLAB code:

```
generators_map = imread ('features.bmp');
BW = im2bw (generators_map, 0.1);
imshow (BW);
```

*Phase 2.* Calculating the nearest-neighbor map for the face feature sets (b):

MATLAB code:

```
[dist, ind] = bwdist (BW);
neighbor_map = zeros ([512 512]);
neighbor_map(:) = generators_map (ind(:));
imshow (mat2gray (neighbor_map));
```

*Phase 3.* Determining the edges in the nearest-neighbor map (c):

MATLAB code:

```
V = edge (neighbor_map, 'canny', [], 0.04);
```

To display the Voronoi diagram:

```
imshow (V);
```

To display the Voronoi diagram together with the features (c):

```
V = im2double (V);
features = im2double (BW);
imshow (mat2gray (V + 0.4 * features));
```

**FIGURE 7.27**

Calculation of a discrete Voronoi diagram from the nearest-neighbor transform of a facial features (Examples 7.26).

▶ The Voronoi transform can be implemented in various metric spaces. For example, a Voronoi diagram can be created in polar coordinates,

< TOPOLOGY ⇒ < POLAR REPRESENTATION ⇒ < VORONOI DIAGRAM

---

## 7.10   Problems

Problems from 7.1 through 7.7 can be solved using MATLAB functions bwdist, contour and voronoin. Problem 7.8 can be considered as a project.

**Problem 7.1** Using MATLAB functions bwdist and contour, compute contour maps from the origin for the following metrics (see Examples 7.3 and 7.4):

(*a*) Euclidean

$$\rho(\mathbf{p}, \mathbf{p}_o) = \rho(\mathbf{p}, \mathbf{0}) = \sqrt{x^2 + y^2}.$$

(*b*) Cityblock $\rho(\mathbf{p}, \mathbf{p}_o) = \rho(\mathbf{p}, \mathbf{0}) = |x| + |y|$;

(*c*) Chessboard

$$\rho(\mathbf{p}, \mathbf{p}_o) = \rho(\mathbf{p}, \mathbf{0}) = max\ (|x|, |y|).$$

(*d*) Quasi-Euclidean

$$\rho(\mathbf{p}, \mathbf{p}_o) = \rho(\mathbf{p}, \mathbf{0}) = \begin{cases} |x| + (\sqrt{2} - 1)|y|, \text{ if } |x| > |y| \\ (\sqrt{2} - 1)|x| + |y|, \text{ if } |x| \le |y| \end{cases}$$

*Hint:* For example, for (*a*), the point at the origin can be represented by a binary $257 \times 257$ image with the only non-zero pixel at the centre:

```
map = logical (zeros ([257 257]));
map(129, 129) = logical (1);
```

The MATLAB function bwdist can be used to compute the matrix representing the Euclidean distances:

```
map = bwdist (map, 'euclidean');
```

The MATLAB function contour displays the isolines of a matrix:

```
contour (map), axis equal;
```

**Problem 7.2** Given a set of generators

$$P \equiv \{p_1(0, 0),\ p_2(-0.5, -0.25),\ p_3(0, 0.7),\ p_4(0.75, -0.5)\},$$

compute, using MATLAB function voronoin:

(*a*) Voronoi vertices of the Voronoi cell $V(p_1)$ for the generator $p_1$;

(b) Plot the Voronoi cell $V(p_1)$ of the generator $p_1$ using MATLAB functions `convhulln` and `plot`;

(c) Find all bounded Voronoi regions $V(p_i)$, i.e., the cells that do not contain points at infinity;

(d) Find all unbounded Voronoi regions $V(p_i)$, i.e., the cells that contain points at infinity.

*Hint:* For example, for (a), the generators are

```
generators = [0 0; -0.5 -0.25; 0 0.7; 0.75 -0.5];
```

Voronoi vertices and Voronoi cells can be computed by the MATLAB command

```
[vor_vertices, vor_cells] = voronoin (generators);
```

To find the Voronoi vertices of $V(p_1)$ display the elements of `vor_vertices` indexed by the indices in the first cell in `vor_cells`:

```
disp (vor_vertices(vor_cells{1},:));
```

**Problem 7.3** Compute Voronoi diagrams for sets of generators representing the facial feature points similar to those given in Table 7.1:

(a) Generators are pupils.

(b) Generators are pupils and nose.

(c) Generators are pupils and the two points of the mouth.

(d) Generators are seven points: pupils, nose, the two points of the mouth, and two extra points on the face. Follow Example 7.28.

*Hint:* Use the following MATLAB functions: `voronoi` and `plot`.

**Problem 7.4** Using the MATLAB function `bwdist`, compute the distance transform of $256 \times 256$ binary images representing the following geometrical primitives:

(a) isolated points at $p_1(64, 64)$, $p_2(192, 64)$, and $p_3(128, 192)$.

(b) a horizontal line from $p_1(64, 64)$ to $p_2(192, 64)$ and an isolated point at $p_3(128, 192)$.

(c) a point at $p_1(160, 160)$ and a square with top left corner at $p_2(32, 32)$ and bottom right corner at $p_3(64, 64)$.

(d) a point at $p_1(160, 160)$, a line from $p_2(160, 32)$ to $p_3(160, 128)$ and a square with top left corner at $p_4(32, 32)$ and bottom right corner at $p_5(64, 64)$. Follow Example 7.27.

*Note:* The MATLAB function `bwdist` takes as its input argument a binary image with colour depth equal to 1, i.e., of `logical` type.

**Problem 7.5** Using MATLAB function `diff`, compute the second differences of the distance maps calculated in Problem 7.4 and determine the negative one. Follow Example 7.25 or 7.21.

**Problem 7.6** Using the MATLAB function `bwdist`, compute the nearest-neighbor maps for the primitives given in Problem 7.4.

Follow Example 7.26.

*Note:* To calculate the nearest-neighbor map, represent the primitives by the grayscale image rather than by a binary one. Then apply a threshold to the grayscale image with the function `im2bw` in order to obtain a binary one. Use the grayscale image to translate the nearest-neighbor transform returned by the function `bwdist` into a nearest-neighbor map.

**Problem 7.7** Using the MATLAB function `edge`, compute the Voronoi diagrams from the nearest-neighbor maps calculated in Problem 7.6. Follow Example 7.26.

**Problem 7.8** This problem can be considered as a project and it concerns some phases of iris primitive synthesis.
(*a*) Repeat the experiment given in Figure 7.28a: A Voronoi diagram is computed for a set of sites uniformly distributed on the concentric circles.
(*b*) Repeat the experiment given in Figure 7.28b: A Voronoi diagram is computed for a set of sites composed of the points randomly distributed over the ring in between two circles.
(*c*) Repeat the experiment given in Figure 7.28c: A Voronoi diagram is computed for two sets of sites superposed together to form the composite set.

---

## Supporting information for problems

**MATLAB function** `bwdist`.

```
D = bwdist (BW)
D = bwdist (BW, metric)
[D, L] = bwdist (BW)
[D, L] = bwdist (BW, metric)
```

The function `bwdist` computes the distance transform of the binary image BW and returns it in D, i.e., it assigns each pixel in the input binary image BW a number that is the distance between that pixel and the nearest nonzero pixel. `bwdist` uses the following distance metric: euclidean, chessboard, cityblock and quasi-euclidean. The metric can be specified by the optional argument `metric` which can take the following values: `'euclidean'` (default), `'chessboard'`, `'cityblock'` and `'quasi-euclidean'`.

If the parameter L is specified, each element of the nearest-neighbor transform L contains the linear index of the nearest nonzero pixel of the binary image BW.

For the detailed description of the function `bwdist`, type `help bwdist` in the MATLAB command window or consult the MATLAB Help.

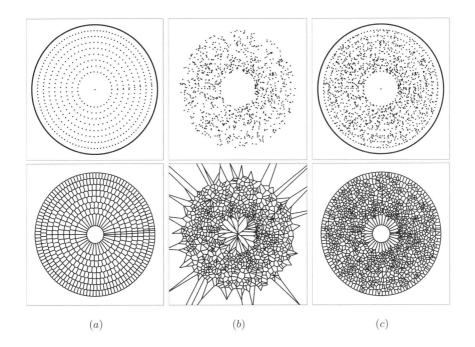

(a)                          (b)                          (c)

**FIGURE 7.28**

Topological structures for Problem 7.8.

> **Example 7.27** *Let the binary image* BW *of the size* $512 \times 512$
> *contain three nonzero feature points:*
>
> $$< \text{Points} > = \{p_1(128, 128),\ p_2(384, 192),\ p_2(192, 384)\}.$$
>
> *The distance transform and the nearest-neighbor transform of
> the image* BW *are shown in Figure 7.29a and b respectively.*

**MATLAB function** voronoi *and* voronoin:

```
[VX, VY] = voronoi (X, Y)
[V, C] = voronoin (P)
```

The function voronoi returns the finite vertices of the Voronoi edges[‡] for the
point set whose coordinates passed to the function in the arrays X and Y. The
Voronoi vertices are returned in the arrays VX and VY.

---

[‡]The edges of unbounded Voronoi cells can be calculated by adding some extra generators
to the generator set so that these additional points produce the absent edges. For an
illustration see Phase 3 in Figure 7.30.

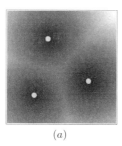

(a)

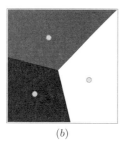

(b)

Phase 1. Create an empty binary image BW:

```
BW = logical(zeros([512 512]));
```

Draw the feature points from Example 7.27 on the binary image BW:

```
BW(128, 128) = logical(1);
BW(192, 384) = logical(1);
BW(384, 192) = logical(1);
```

Phase 2. Calculate the distance and the nearest-neighbor transforms of the binary image BW:

```
[D, L] = bwdist(BW);
```

Phase 3. Display the distance transform D (a) and the nearest-neighbor transform L (b):

```
figure, imshow(mat2gray(D));
figure, imshow(mat2gray(L));
```

**FIGURE 7.29**

The distance (a) and the nearest-neighbor (b) transforms (Example 7.27).

The function `voronoin` can be used to obtain the vertices of each Voronoi cell. `voronoin` returns the Voronoi vertices in V and the Voronoi cells in C. Each row in V represents a Voronoi vertex and contains its coordinates. C is a cell array where each element contains the indices into V of the vertices of the corresponding Voronoi cell.

For a detailed description of the function `voronoi` (`voronoin`), type `help voronoi` (`help voronoin`) in the MATLAB command window or consult the MATLAB Help.

**Example 7.28** *Let the set of generators be*

$$P = \left\{ p_1(-\frac{\sqrt{3}}{2}, \frac{1}{2}), \ p_2(\frac{\sqrt{3}}{2}, \frac{1}{2}), \ p_3(-0.2, -1), \ p_4(0, 0) \right\}.$$

*Figure 7.30a represents the topology of the Voronoi cell $V(p_4)$. The Voronoi diagram $V(P)$ is shown in Figure 7.30b.*

**MATLAB function** `contour`.

```
contour (Z)
contour (Z, N)
```

The function `contour` displays isolines of the matrix Z, i.e., draws a contour plot of the matrix Z. Z is interpreted as heights with respect to the $x$-$y$ plane.

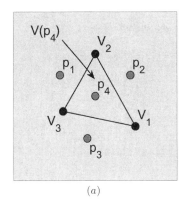

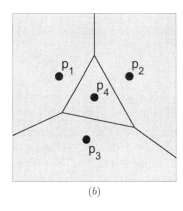

(a)

(b)

*Phase 1*. Create a set of generators $P$:

```
t = sqrt(3)/2;
P = [t 0.5; t 0.5; −0.2 − 1; 0 0];
```

*Plot the generator points $P$*:

```
plot (P(:, 1), P(:, 2), 'ok');
hold on;
```

*Phase 2*. Calculate and display the Voronoi vertices $V_1$, $V_2$ and $V_3$ of the Voronoi cell $V(p_4)$ (a):

```
[VP, CP] = voronoin (P);
V1 = VP(CP{4}(1), :);
V2 = VP(CP{4}(2), :);
V3 = VP(CP{4}(3), :);

plot (V1(1), V1(2), 'ok');
plot (V2(1), V2(2), 'ok');
plot (V3(1), V3(2), 'ok');
```

*Phase 3*.  Calculate and display the Voronoi diagram $V(P)$ (b):

```
t = 10000;
P = cat (1, P, [−t t; t t; t − t; −t − t]);
[VX, VY] = voronoi (P(:, 1), P(:, 2) );

plot (VX, VY, ' − k');
```

**FIGURE 7.30**
The topology of Voronoi cell (a) and Voronoi diagram (b) (Example 7.28).

Function argument N, if specified, indicates the number of contour levels to be displayed.

For a detailed description of the function `contour`, type `help contour` in the MATLAB command window or consult the MATLAB Help.

> **Example 7.29** *The binary image* BW *of the size* $512 \times 512$ *contains three nonzero feature points:*
>
> $<\text{Points}> = \{p_1(128, 128), \ p_2(384, 192), \ p_2(192, 384)\}.$
>
> *The contour plot of the distance transform of the image* BW *is shown in Figure 7.31.*

**MATLAB function** `diff`:

```
D = diff (A)
```

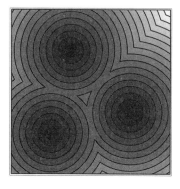

*Phase 1.* Create a binary image BW and calculate the distance transform of BW:

```
BW = logical (zeros ([512 512]));

BW(128, 128) = 1; BW(192, 384) = 1;
BW(384, 192) = 1;

D = bwdist (BW);
```

*Phase 2.* Display the contour plot of the distance transform D of the image BW:

```
figure, contour (mat2gray (D), 18, ' − k');
```

**FIGURE 7.31**

Contour plot of the distance transform (Example 7.29).

```
D = diff (A, N)
D = diff (A, N, dir)
```

The function `diff` calculates differences between adjacent elements of A along the first non-singleton dimension of A. The size of the returned matrix D along the processed dimension is one element shorter than A. If the argument N is specified, `diff` is applied recursively N times. The argument `dir`, if specified, indicates the dimension in which the differences are calculated.

For a detailed description of the function `contour`, type `help contour` in the MATLAB command window or consult the MATLAB Help.

Notice that the MATLAB function `gradient` can also be used to calculate the differences. The following gives an example of the `diff` function usage:

```
A = [1 2 3 4 5];
D = diff (A);
```

The resultant vector of differences is: D = [1 1 1 1].

## 7.11    Further reading

**Computational geometry and the computer-aided design of spacial forms** is concerned with the design and analysis of algorithms for [solving] geometric problems. Topological structure design problems are concerned with the design and analysis of topologies such as images or networks. Topological structures are composed of topological primitives, their nodes, arcs, relationships, and the topological and assembling primitives. Topological methods can be used to analyze or characterize a system. Topological structure design problems arise in applications where points and edges are embedded in some metric space.

The classical *closed-point problem* is formulated as follows. Let $S$ be a set of $n$ points in the plane, called *sites*. Construct a data structure for $S$ so that the Euclidian distance between two closely spaced points can be found quickly. State-of-the-art methods for topological structure representation and manipulation can be found in [4]. The fundamentals of computing geometry are introduced in [11, 7, 16, 17, 18] and [22].

**The Delaunay triangulation** is the triangulation of the plane containing a set $P$ of distinct points $p_i \in P$, such that the triangle's edges connect neighbouring points and do not overlap. The Delaunay triangulation and Voronoi diagram are dual to each other. By extension, space or hyperspace may also be tessellated. A Voronoi diagram is sometimes also known as a *Dirichlet tessellation*. The cells are called *Dirichlet regions*, *Thiessen polytopes*, or *Voronoi polygons*. The word *tessellate* means to arrange small pieces in a mosaic pattern in such a way as to leave no region uncovered. A brief introduction to Voronoi diagram and Delaunay triangulation techniques is given in by Fortune [8]. The fundamentals of 2D and 3D Voronoi diagram and Delaunay triangulations and their applications in pattern recognition, communication topology, and for manipulating various topological structures are considered by Okabe et al. [15].

**Technique of computing.** Algorithms for incrementaly computing Delaunay triangulations are presented, in particularly, by Fortune [9], Guibas and Sibson [10], and Lischinski [14]. Applications and algorithms for computing Voronoi tessellations are introduced by Du et al. [6], Besag [2], Bowyer [3], and Watson [24].

**Synthetic biometric data design.** Voronoi diagrams and Delaunay triangulations are useful methods for data representation and manipulation, and for the analysis and synthesis of topological configuration. Yanushkevich et al. [25, 28] used Voronoi diagrams in the computer aided design of synthetic biometric data: iris, signatures, fingerprints, etc. Tsalakanidou et al. [23] used 2D Delaunay refinement algorithm by Ruppert [19] in coloring 3D face images.

**Using in image mesh models.** Voronoi diagrams and Delaunay tessellations are used in reconstruction of the mesh as follows. The selected feature points are located at random. To construct mesh, the Voronoi region is determined of the feature points, then Voronoi diagram is transformed to a Delaunay tessellation. The resulting topological net is a mesh model that can be mapped into 3D space.

**Using in watermarking techniques.** Bas et al. [1] have used Delaunay tessellations in the design of a watermarking algorithm, namely, to decompose the image into a set of disjoint triangles and embedding the mark in each triangle.[§] In the embedding phase, a Delaunay tessellation can be recreated because feature points of the image. A signature is embedded in each triangle of the tessellation. In the detection phase, the Delaunay tessellation can be performed because feature points can be detected.

**Other applications.** Many other applications of Voronoi diagrams and Delaunay triangulations can be found in recent publications. For example, Yanushkevich [27] utilizes the topological properties of Voronoi diagrams to represent hypercube-like topology in nanoscale logic circuits. The relationship of Voronoi diagrams and trees is the focus of many recent studies, in particular, Schroeder and Shephard [20], Liotta and Meijer [13]. Fractal structures are useful for data structure representation. The Voronoi diagram is an effective tool for generating fractals [21], texture synthesis and analysis using Markov random fields [5, 12].

## 7.12   References

[1] Bas P, Chassery J-M, and Macq B. Image watermarking: an evolution to content based approaches. *Pattern Recognition*, 35:545–561, 2002.

[2] Besag J. Statistical analysis of non-lattice data. *The Statistician*, 24(3):179–195, 1975.

[3] Bowyer A. Computing Dirichlet tessellations. *Computer Journal*, 24:162–166, 1981.

[4] Cormen TH, Leiserson CE, Riverst RL, and Stein C. *Introduction to Algorithms*. The MIT Press, Cambridge, MA, 2001.

[5] Cross GC and Jain AK. Markov random field texture models. *IEEE Transactions on Pattern Analysis and Machine Intelligence*, 5(1):25–39, 1983.

[6] Du Q, Faber V, and Gunzburger M. Centroidal Voronoi tessellations: applications and algorithms. *SIAM Review* 41(4):637–676, 1999.

[7] Farin GE. *Curves and Surfaces for Computer Aided Geometric Design: A Practical Guide*, 2nd edition. Academic Press, Boston, MA, 1990.

---

[§]Watermarking consists of two phases: embedding the mark in an initial image and detecting the mark watermarked image from possible distorted.

[8] Fortune S. Voronoi diagrams and Delaunay triangulations. In Du DZ and Hwang F, Eds., *Computing in Euclidean Geometry*, pp. 193–233, World Scientific, 1992.

[9] Fortune S. A sweepline algorithm for Voronoi diagrams. In *Proceedings of the 2nd Annual Symposium on Computational Geometry*, New York, pp. 313–322, 1986.

[10] Guibas L and Sibson R. Primitives for the manipulation of general subdivisions and the computation of Voronoi diagrams. *ACM Transactions on Graphics*, 4(2):74–123, 1985.

[11] Hofmann CM and Vermeer PJ. Computing in Euclidian geometry. In *Geometric Constraint solving in R2 and R3*, pp. 266–298, World Scientific, 2nd edition, 1995.

[12] Li SZ. *Markov Random Field Modeling in Image Analysis*. Springer, Heidelberg, 2001.

[13] Liotta G and Meijer H. Voronoi drawings of trees. *Computational Geometry*, 24(3):147–178, 2003.

[14] Lischinski D. Incremental Delaunay triangulations. In Heckbert PL, Ed., *Graphics Gems*, Academic Press/Morgan Kaufmann, New York, pp. 47–59, 1994.

[15] Okabe A, Boots B, and Sugihara K. *Spatial Tessellations. Concept and Applications of Voronoi Diagrams*. Wiley, New York, NY, 1992.

[16] Penna MA and Patterson RR. *Projective Geometry and its Applications to Computer Graphcis*. Prentice-Hall, Englewood Cliffs, NJ, 1986.

[17] Preparata FP and Shamos MI. *Computational Geometry: An Introduction*. Springer, Heidelberg, 1985.

[18] Rogers DF and Adams JA. *Mathematical Elements for Computer Graphics*. McGraw-Hill, Boston, MA, 1990.

[19] Ruppert J. A Delaunay refinement algorithm for quality 2D mesh generation. *Journal of Algorithms*, 8(13):548–585, 1995.

[20] Schroeder WJ and Shephard MS. A combine octree/Delaunay method for fully automatic 3-D mesh gencration. *International Journal for Numerical Methods in Engineering*, 29:37–55, 1990.

[21] Shirrif K. Generating fractals from Voronoi diagrams. *Computers and Graphics*, 17(2):165–167, 1993.

[22] Strang G. *Linear Algebra and its Applications*, 2nd ed. Academic Press, New York, 1980.

[23] Tsalakanidou F, Malassiotis S, and Strintzis MG. Exploitation of 3D images for face authentication under pose and illumination variations. In

*Proceedings of the 2nd International Symposium on 3D Data Processing, Visualization, and Transmission*, Thessaloniki, Greece, Sept. 2004.

[24] Watson DF. Computing the $n$-dimentional Delaunay tessellation with application to Voronoi polytopes. *Computer Journal*, 24:167–172, 1981.

[25] Yanushkevich SN., Supervisor. Computer aided design of synthetic data based on Voronoi diagrams. *Technical Report, Biometric Technologies: Modeling and Simulation Laboratory*, University of Calgary, Canada, 2004.

[26] Yanushkevich SN and Shmerko VP. Biometrics for electrical engineers: design and testing. Chapter 9: Voronoi diagrams and Delaunay triangulations in analysis and synthesis of biometric data. *Lecture notes*. Private communication, 2005.

[27] Yanushkevich SN. Nanostructures representation by Voronoi diagrams. In *Advanced Logic Design of Electronic and Nanoelectronic Devices*, Lecture notes, Department of Electrical and Computer Engineering, University of Calgary, Canada, 2004.

[28] Yanushkevich SN, Boulanov OR, and Shmerko VP. Analysis and synthesis of biometric data: Voronoi diagrams and Delaunay triangulation in MATLAB, *Unpublished manuscript*. Private communication. 2005.

# 8

# *Synthetic DNA*

This chapter focuses on studying synthetic sequences of amino acids like DNA (deoxyribonucleic acids) and proteins as one of the future targets of inverse biometric research. Existing biometric technologies which deal with objects like fingerprints, iris scans, face scans, voice patterns, signatures and combined biometrics can be effected by various disturbances (both environmental and physical) which reduce accuracy in personal identification systems. The environmental disturbances include bright light for faces and irises, dry skin for fingerprint, or background noise for voice recognition. The physical disturbances include voice changes due to illness or other factors, and signature and handwriting changes due to aging and mood. DNA is not affected by environmental and physical disturbances. Historically, DNA signatures used in forensics have proven that people can be identified with high accuracy. The main problem with DNA is the enormous time and resources required for sequencing and processing (alignment, matching, storing, etc.). However, future technology will ensure real time DNA sequencing and processing. Another problem is a privacy concern which promotes work on synthetic DNA instead of actual sequences taken from humans.

This chapter outlines some concepts of synthetic biometrics and methods principles for generating DNA sequences based on training techniques. A proposed algorithm is based on interpreting sequences of amino acids acquired by sampling the existing database of DNAs or proteins. Current genomic data are accessible through databases which contain billions of nucleotides. Both the meaning of DNA sequences and their statistical characteristics must be considered to ensure variability and separatability for artificially generated amino acids. Experimental results show that the generation method and the developed algorithm give a quick and reliable way to synthesize sequences while preserving distinctiveness within the existing database of DNAs/proteins.

Techniques associated with biological sequence processing can be viewed as a combination of *direct* and *inverse* transformations (Figure 8.1):

▶ Synthesis sequences are generated by applying (i) direct transformations, as a preprocessing step, to create libraries of primitives and features obtained by analyzing statistical dependencies derived from DNA and proteins; and (ii) inverse transformations to generate a collection of synthetic sequences (see Figure 8.1a).

▶ Synthesis sequences are verified by applying (i) inverse transformations in order to generate a collection of synthetic DNAs or proteins; and (ii) direct transformations, as a postprocessing step, to check the validity of newly synthesized sequences by performing an analysis of their biological importance and statistical relevance (see Figure 8.1b).

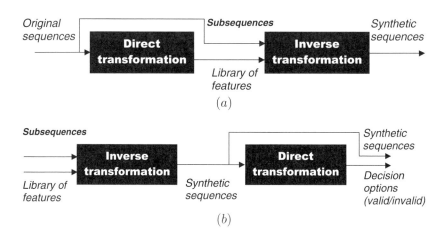

**FIGURE 8.1**

Sequence processing techniques and methods: (a) two stage direct and inverse transformations to generate synthetic DNAs or proteins, (b) two stage inverse and direct transformations to synthesize sequences and check their validity.

The chapter is structured as follows. An overview of some current trends in DNA biometrics is given in Section 8.1. The background of this study is given in Section 8.2. Concepts and methodological principles are introduced in Section 8.3, where statistically meaningful synthesis based on such characteristics as the probabilities of given nucleotides, characteristics of Markov models, etc. A more detailed study on postprocessing is presented in Section 8.5. This section describes syntheses which satisfy the above requirements and ensure compliance with the genomic structure. Examples of Markov models are introduced in Section 8.4. The algorithm is given in Section 8.6. After the summary (Section 8.7), a set of problems and recommendations for "Further Reading" are given in Sections 8.8 and 8.9 respectively.

## 8.1  Introduction

In this section, the DNA characteristics, identification properties, and privacy issues related to DNA processing are considered.

### 8.1.1  DNA characteristics

Each individual inherits his/her genomic DNA from both parents, thereby producing a composite of DNA sequences that are unique to that individual (except in the case of identical twins). DNA composition of cells within an individual remains constant throughout one's lifetime*. Appropriately, DNA is considered to be a special type of biometric having so called *digital signature* properties. The following characteristics make DNA a special biometric type:

- ▶ An actual physical sample of DNA should be larger than samples from any other biometric type in order to draw meaningful conclusions.
- ▶ Semi-automated sequencing, sampling and decision making techniques are necessary to perform DNA matching.
- ▶ Actual DNA sequences should be used without any feature extraction, scaling or prototyping in order to compare sequences.
- ▶ Processing should take advantage of digital nature of DNA information which is suitable for digital computations.

The current process of using DNA to verify identity or to determine identity of an unknown person is a time-consuming process. Biological samples must be processed in multiple stages in a laboratory setting in order to extract DNA for testing purposes. DNA samples must be available in sufficient quantities, and actual analysis can take one to two days. Therefore, the inability to test the DNA on a real-time basis remains the main limitation of DNA's biometric applications. Because of these reasons and the advanced technology required, it is not the most cost-efficient biometric solution, but when a positive identification is needed, it is the most reliable.

### 8.1.2  Ideal identification properties

Despite the enormous time and complexity characteristics of DNA sequencing and matching, a genetic code is unique to each individual, making it an ideal biometric. Minor differences in each person's DNA sequences can be used as identification markers, and researchers are continually developing additional

---

*Even though mutations in DNA sequences occur because of environmental agents (ultraviolet light, nuclear radiation or certain chemicals) or mistakes that occur when a cell copies its DNA in preparation for cell division, these mutations are insignificant and not considered in DNA biometrics.

markers to identify individual genetic sequences. Thus, a unified system for analysis is employed by forensic laboratories which utilize the same 13 regions of DNA (DNA loci) to assess a match between DNA evidence and a suspect.

### 8.1.3   Privacy issues

Regardless of DNA's unique properties as an identifier, there are many privacy concerns. DNA biometrics requires taking tissue or fluid physically as opposed to other biometrics. That is why it is considered to be more intrusive, breaching one's rights to privacy and medical information. Understanding DNA testing requires a fundamental knowledge of biology by the general public in order to assess the true impact on privacy issues of this type of testing.

DNA based identification for forensic purposes is one of the premier technologies (see the work by Rudin [16]).

> **Example 8.1** *Since 1987, hundreds of cases have been decided with the assistance of DNA fingerprint evidence. Another important use of DNA fingerprints in the court system is to establish paternity in custody and child support litigation. Every organ or tissue of an individual contains the same DNA fingerprint and they can be collected. The DNA method is superior to using dental records and blood types.*

With the impending increase in the use of DNA signatures in security applications, there has been a constantly growing demand for benchmark data in the form of DNA sequences to perform an extensive testing of biometric applications on large quantities of data while avoiding concerns of privacy advocates. Since the acquisition of large anonymous quantities of DNA in many cases is problematic due to personal concerns of confidentiality, a method to automatically generate DNA sequences is needed. Creating DNA sequences is another example of an inverse problem in biometrics.

The necessary requirements for DNA and protein synthesis from genomic databases are as follows:

*Requirement 1* Sequences of amino acids should be genetically meaningful.

*Requirement 2* Synthesis should incorporate mutations within a class of sequences.

*Requirement 3* Sequences of amino acids should have meaningful statistical distributions.

## 8.2 Basics of DNA biometrics

The characteristics of all living organisms, including humans, are essentially determined by information contained within DNA that they inherit from their parents. Such a statement gives an answer to a question asked by biometrics "What are the differences among living things?" The units that govern the differences between living organisms are called genes.

### 8.2.1 DNA code

Genes contain information in the form of a specific sequence of nucleotides that are found in DNA molecules. Only four different bases are used in DNA molecules: guanine, adenine, thymine and cytosine (G, A, T and C). For example, the sequence ACGCT represents different information than the sequence AGTCC or the sequence TCCAG even though they use the same letters. The characteristics of a human being are the result of information contained in the DNA code. Protein structures are described by sequences containing 20 amino acids: A, C, D, E, F, G, H, I, K, L, M, N, P, Q, R, S, T, V, W, and Y.

DNA is a double-stranded molecule. While the information content on each strand of a double-stranded DNA molecule is redundant, it is not exactly the same – it is complementary. For every G on the strand, a C is found on its complementary strand and vice versa. Similarly, for every A on the strand, a T is found on its complementary strand and vice versa. The chemical interaction between the two different kinds of base pairs is actually so stable and energetically favorable that it alone is responsible for holding the two complementary strands together.

Within almost every cell in the human body there exist two types of DNA: (i) genomic DNA found in a membrane sack called the nucleus, and (ii) mitochondrial DNA (mtDNA) found in multiple membrane-bound energy factories. Both forms of DNA offer information for human identification and have been routinely used in forensic work and for developing DNA databases. Both types of DNA are present in almost every cell in the human body and, as a result of scientific advances made over the last 20 years, a single hair or a licked stamp can be used for identification.

A person's genomic DNA can be thought of as consisting of about 70,000 genes and many repetitive sequences that represent seemingly nonessential or noncoding DNA. The noncoding portion of the genome can withstand alterations in its composition over generations without compromising the viability of the cell or the human being as a whole. Largely because of the variability in the composition of noncoding DNA, one person's DNA sequence will differ from another person's.

Proteins are large molecules composed of one or more chains of amino acids in a specific order; the order is determined by the base sequence of nucleotides in the gene coding the protein. Proteins are required for the structure, function, and regulation of the body's cells, tissues, and organs. Each protein has its own unique functions.

## 8.2.2   Terminology

***Amino acids:*** A group of 20 different kinds of small molecules that link together in long chains to form proteins. They are often referred to as the "building blocks" of proteins.

***Cell:*** The basic unit of any living organism. It is a small, watery, compartment filled with chemicals and a complete copy of the organism's genome.

***Chromosome:*** One of the threadlike "packages" of genes and other DNA in the nucleus of a cell. Different kinds of organisms have different numbers of chromosomes. Humans have 23 pairs of chromosomes, 46 in all: 44 autosomes and two sex chromosomes. Each parent contributes one chromosome to each pair, so children get half of their chromosomes from their mothers and half from their fathers.

***DNA:*** Deoxyribonucleic acid. The chemical inside the nucleus of a cell that carries the genetic instructions for making living organisms.

***DNA sequencing:*** Determining the exact order of the base pairs in a segment of DNA.

***Gene:*** The functional and physical unit of heredity passed from parent to offspring. Genes are pieces of DNA, and most genes contain the information for making a specific protein.

***Genome:*** All the DNA contained in an organism or a cell, which includes both the chromosomes within the nucleus and the DNA in mitochondria.

***mtDNA:*** Mitochondrial DNA. The genetic material of the mitochondria, the organelles that generate energy for the cell.

***Nucleotide:*** One of the structural components, or building blocks, of DNA and RNA. A nucleotide consists of a base (one of four chemicals: adenine, thymine, guanine, and cytosine) plus a molecule of sugar and one of phosphoric acid.

***Protein:*** A large complex molecule made up of one or more chains of amino acids. Proteins perform a wide variety of activities in the cell.

Information about available databases is given in "Further Reading" Section.

**Example 8.2** *An example of the collection of DNA sequences given in the FASTA format is illustrated below (the sequence is given in Figure 8.2):*

>1

```
ATCAACAACCGCTATGTATTTCGTACATTACTGCCAGCCACCATGAATAT
TGTACGGTACCATAAATACTTGACCACCTGTAGTACATAAAAACCCAATC
TTCTTTCATGGGGAAGCAGATTTGGGTACCACCCAAGTATTGACTCACCC
CACATCAAAACCCCCCCCCCCATGCTTACAAGCAAGTACAACAATCAACCT
```

>2

```
ATCAACAACCGCTATGTATTTCGTACATTACTGCCAGCCACCATGAATAT
TGTACGGTACCATAAATACTTGACCACCTGTAGTACATAAAAACCCAATC
TTCTTTCATGGGGAAGCAGATTTGGGTACCACCCAAGTATTGACTCACCC
CACATCAAAACCCCCTCCCCATGCTTACAAGCAAGTACAGCAATCAACCT
```

>3

```
ATCAACAACCGCTATGTACTTCGTACATTACTGCCAGTCACCATGAATAT
TGTACAGTACCATAAATACTTGACCACCTGTAGTACATAAAAACCCAATC
TTCTTTCATGGGGAAGCAGATTTGGGTACCACCCAAGTATTGACTCACCC
CACATCAAAACCCCCTCCCCATGCTTACAAGCAAGTACAGCAATCAACCT
```

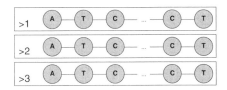

**FIGURE 8.2**
Multiple DNA sequences (Example 8.2).

**Example 8.3** *A set of protein sequences in the FASTA format is given below (the sequence is given in Figure 8.3):*

>1

```
GQPHGGGWGQPHGGGWGQPHGGGWGQGGGTHNQWHKPSKPKTSMKHMAGA
AAAGAVVGGLGGYMLGSAMSRPLIHFGNDYEDRYYRENMYRYPNQVYYRP
MLVLFVATWSDLGLCKKRPKPGGWNTGGSRYPGQGSPGGNRYPPQGGGGW
VDQYSNQNNFVHDCVNITIKQHTVTTTTKGENFTETDVKMMERVVEQMCI
```

>2

```
GQPHGGGWGQPHGGGWGQPHGGGWGQGGGTHNQWHKPSKPKTSMKHMAGA
AAAGAVVGGLGGYMLGSAMSRPLIHFGNDYEDRYYRENMYRYPNQVYYRP
MLVLFVATWSDLGLCKKRPKPGGWNTGGSRYPGQGSPGGNRYPPQGGGGW
VDQYSNQNNFVHDCVNITIKQHTVTTTTKGENFTETDVKMMERVVEQMCI
```

>3

```
QPHGGGWGQPHGGGWGQPHGGGWGQPHGGGWGQGGGTHNQWNKPSKPKTN
MKHMAGAAAAGAVVGGLGGYMLGSAMSRPLIHFGNDYEDRYYRENMYRYP
MLVLFVATWSDLGLCKKRPKPGGWNTGGSRYPGQSSPGGNRYPPQSGGWG
NQVYYRPVDQYSNQNNFVHDCVNITIKQHTVTTTTKGENFTETDVKIMER
```

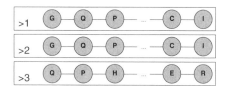

**FIGURE 8.3**
Multiple protein sequences (Example 8.3).

## 8.3    DNA/protein synthesis techniques

One can consider the following techniques for generating synthetic DNA/protein sequences:

*Mutation technique* Applying irreversible transformations
       < Original sequence > ⇒ < Synthetic sequence >

*Recombination technique* Assembling primitives such as amino acids
       < Structural primitives > ⇒ < Synthetic sequence >

### 8.3.1    Mutation technique

Figure 8.4 illustrates the design of a sequence synthesis system based on the mutation technique. In nature and in artificial design, DNA/protein sequences can be affected by three operations [15]: mutation of individual amino acids, insertion (increasing the length of a sequence) and deletion (decreasing the length of a sequence). The main phases of the design process are the following:

*Phase 1.* Preprocess the original collection of DNA/protein sequences.

*Phase 2.* Choose appropriate mutation rules and control parameters like mutation speed, number of insertions and deletions.

*Phase 3.* Apply mutation, insertion and deletion operations to selected positions of the sequence transforming a sequence from the original collection into a mutated equivalent.

*Phase 4.* Use the synthesized sequence for matching.

Some results of the mutation technique are given in Example 8.4.

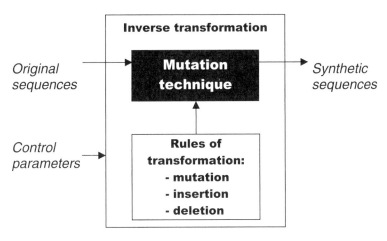

**FIGURE 8.4**

Design of synthetic DNA/protein sequences based on the mutation technique

> **Example 8.4** *Let us consider a DNA subsequence from Example 8.2: ACCGCTATGTATT. Applying three permissible operations separately, the subsequence can be transformed to:*
>
> *mutation: ACGGGTAAGTACT;*
>
> *insertion: ACGCAGCTATTGTATTC;*
>
> *deletion: ACC‖CT‖TGTAT‖.*

## 8.3.2 Recombination technique

Figure 8.5 depicts the design of a sequence synthesis system based on the recombination technique. The main phases of the design process are the following:

*Phase 1.* An arbitrary set of two structural primitives (e.g., subsequences) is taken from the database of primitives.

*Phase 2.* Based on appropriate recombination rules, a macroprimitive is created by merging two initial primitives.

*Phase 3.* The next topological primitive is taken, and then added to the macroprimitive by applying appropriate recombination rules.

*Phase 4.* Control parameters direct the process of "growing" the macroprimitive. This process is stopped by a predefined criterion from the set of control parameters.

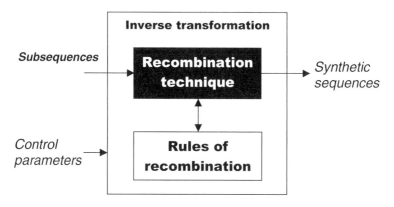

**FIGURE 8.5**

Design of synthetic DNA/protein sequences based on the recombination technique.

> **Example 8.5** *Let us consider two DNA subsequences extracted from Example 8.2:*
> *ACCGCTATGTATT and AGTACCATAAATA.*
> *A simple recombination rule can concatenate two sequences resulting in*
> *ACCGCTATGTATTAGTACCATAAATA.*
> *Another recombination rule can place amino acids from the first subsequence in odd positions, and amino acids from the second subsequence in even positions:*
> *AACGCTGACCTCAATTGATAAATTTA.*

In order to generate sequences with a high percentage of original parts, we introduce a method based on repositioning sequences of sub-parts randomly selected from the pool of training sequences. Since the structure is described in sequential form, repositioning implies vector recombination. This method can be outlined in the following series of steps:

*Step 1:* The sequential representation of original sequences is treated as the pool of sequences.

*Step 2:* Randomly select the start point within the pool of sequences and select as many nucleotides as the length of the current sequence or subsequence.

*Step 3* Repeat Step 2 for the given number of subsequences.

Experimental results demonstrate that the method based on vector recombination is fast and accurate, but it can produce biologically meaningless data (see the work by Khudyakov and Fields [12] on artificial DNA sequences).

### 8.3.3 Statistically-driven synthesis

The first step in a statistically meaningful synthesis is *learning*. Learning from the pre-existing collection of sequences (or DNA databases) allows us to accumulate a priori knowledge about sequence structures, detailed sequence composition, and other important characteristics of a statistical nature. The existing databases of DNA and protein sequences from GenBank (see Section 8.9 for details) have been used to generate statistical data.

**Preprocessing: learning.** The learning stage results in multiple distributions with a distribution function $\Phi(z)$ which give the probability that a standard normal variate assumes a value from $z$. The following statistically meaningful characteristics can be acquired through the learning process from the pre-existing database:

*Distribution of nucleotides* $\Phi(N)$. A simple observation is that various nucleotides have different probability distributions within sequences.

*Transition matrix* $P$. The entries of this matrix are transition probabilities from one state to another. The transition matrix $P$ is used to interpret sequence generation as a Markov system.

> **Example 8.6** *Figures 8.6 and 8.7 show the distribution of nucleotides for a database of DNA and protein sequences respectively. A series of experiments with this distribution revealed that the original composition of the database has a significant impact on the distribution. Hence, the sequence generation technique should be trained on a selected subset of database sequences.*

**Sequence generation based on Markov chains.** The main phases of generating sequences based on various distributions are:

*Step 1:* Generate lengths of subsequences so that the total length is the given length of the sequence.

*Step 2:* For each subsequence of length $s$, generate the set of nucleotides based on their statistical distribution.

In real DNA molecules, the nucleotide bases are highly dependent on their neighbors. The sequential nature of DNA allows us to design a model which encompasses the likelihood of possible combinations of values and their transitions from one to another accommodating genetic relevance in nucleotide generation (Step 2). Thus, in our study we used Markov systems to simulate sequential behavior of DNA/protein sequences. Models based on Markov chains proved to be efficient in terms of storage and sequence synthesis runtime. Figure 8.8 illustrates a Markov model for nucleotide transitions in DNA sequences.

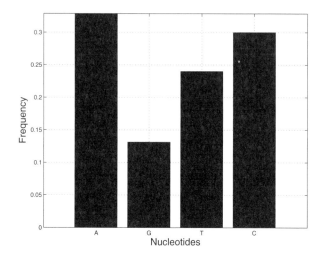

**FIGURE 8.6**

Distribution of nucleotides in DNA sequences for the pre-existing database.

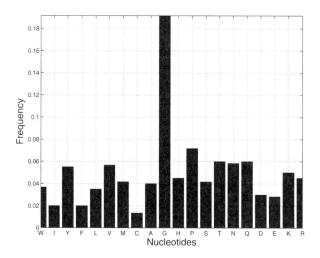

**FIGURE 8.7**

Distribution of nucleotides in protein sequences for the pre-existing database.

Consider a discrete time finite-state Markov chain $\{N_t, t = 0, 1, 2, \ldots\}$ with stationary transition probabilities $p(N_{t+1} = i | N_t = j) = p_{ij}$. Let $P = (p_{ij})$ denote the matrix of transition probabilities. The transition probabilities between $N_t$ and $N_{t+n}$ are denoted $p_{ij}^{(n)}$ and the transition matrix $P = P^{(n)}$. In the context of DNA sequence analysis, each nucleotide base may be considered

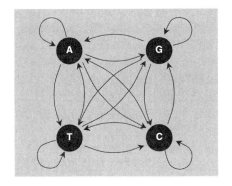

A discrete time finite-state Markov chain

$$\{N_t, t = 0, 1, 2, \ldots\}$$

with stationary transition probabilities

$$p(N_{t+1} = i | N_t = j) = p_{ij}$$

A random variable $N_t$ takes on four possible values A, C, G, and T

**FIGURE 8.8**
Markov model for nucleotide transitions.

a random variable $N_t$ which takes on four possible values A, C, G, and T. The time index $t$ is actually the position of the base in the DNA sequence. The transition probability $p_{ij}$ is the probability that the base at the next position is $j$ given that the base at the present position is $i$. It is also possible that a nucleotide base of DNA is not only dependent on its immediately preceding base, but rather on a few, say $n$, preceding bases. This is referred to as an $n$th order Markov nucleotide sequence.

## 8.4    Examples of Markov models

In this section, three Markov models of various orders are discussed:

▶ 0th order Markov model,

▶ 1st order Markov model, and

▶ 2nd order Markov model.

The distribution of nucleotides in DNA and protein sequences for the Markov chains are analyzed.

### 8.4.1    0th order Markov model

In a 0th order Markov chain, bases do not depend on preceding bases, so the generation procedure relies on the distribution of nucleotides $\Phi(N)$ (see Figures 8.6 and 8.7).

**Example 8.7** *(Continuation of Example 8.2) The following sequences have been generated based on the 0th order Markov model:*

>1
```
AATTTTTGAAACCGCACATCTGGCCCACTGCCCAGACAAAGTATCAAACG
CCAATTCCCTATGACTCTCTTTACGCTCTCTAAACTACACGGTCCAACAC
CTTAGAACCAAAGTCAGAGATATTGACCTAACTATAGATCCTCACACCAC
TGCAAGTCTACGGATAATTGAGCATCCCTAAAAGAGAAAGGCCAAATTCC
```
>2
```
TAGTCCACTGGGTAAGGCTATCAATCATTACCAATAACTCGGAGGTACAC
CCAGATCACTATCGTTTTTCGTATTATCACTGTTTGGCTTTGCGATCCTA
TCCGACTTGCTCCAGGGTCATAAATTTTCTCGTCGAATCATCTGAACATT
AAATGTTTAGCTCGCCCCAATAACAATATTAACTGCTAGCTAAGAGCGAC
```
>3
```
CAACTGCATTTACCCTCCCACCCATGAAACATATTGAATAGAAAAATAAT
AATCAAAACACAATTCAAGATACTAAGCGAGACTCAGGAATCGCCTCCCC
AAAACGCACGTGGACCCAGTCTCATAATGCGACCCCGCACAAAAGATTTG
TCCCACATCAAAAGCCATCAACGAACATATTCTAGTCAAACACCGTTACC
```

**Example 8.8** *(Continuation of Example 8.3) The following protein sequences have been generated based on the 0th order Markov model:*

>1
```
PNPEPRGGGYGHYIIHKHIHGNSNYRYSYWGGEYWVYYYQSNNNACRKPW
CAQGYRWHAQGGYGGVPGHFLNVCGKFTNTIKTVRGKPQTTNTKLNTGEN
GPWRVYGNHNHSPGNHFKQGWHKLKYTGGEQDGVEPGPVPQGQYRKGAWN
PVGGGMINGSGNGTGPTPDIGNNKTPVDVYKYDNTGNKWVAVEDQSDGML
```
>2
```
GQGGMQGQILNNSDSKECEDGRVGGKNDGLVYLGMGHGHTYDHGVAPIMP
KRPIQTWGKGEGGNHGDANPMRWMHAGRIYYGPMYRQNGKRGGEGMLSDM
GIAKTTQGMNRLTEPCGKTNFTYQSMMFWTGGGEDGGLRIPFSGMGGKYK
PYGYCQSGEGPLSGWFGATGMTLTKTTTQGNLNQCKHRLEWQSYEGQKQH
```
>3
```
HQAGGGPLWQTHICCQTSGYCLGGKPEGDCTPPQGGNFKPGYWGYGCGLH
GNPNRSPNRWQKPLGNQEFGMWSGKGWYGQGGQTMDVPEHRVPQRGSNNQ
NVIGDTQYGRIDYAGPGGVPMPVGYQHVGGMYTCSGLPHGTSGYFGQGGH
LWGGGVLNYMVGDTHKEGGHGELWQNPWGSGGHKHNAVMGYESGVYGYWP
```

## 8.4.2   1st order Markov model

In a 1st order Markov chain, each base depends on the preceding base improving the quality of results. Figures 8.9 and 8.10 illustrate the probability of transition for the 1st order Markov model. For instance, the probability of the DNA base A followed by A is 0.29, the probability of the protein amino acid W followed by G is 0.59 (all probabilities are evaluated in a group).

**Example 8.9** *(Continuation of Example 8.2) The following DNA sequences have been generated based on the 1st order Markov model.*

>1
```
ACAATGCCTAGCCCAATACCACGCTTTTATATATTGCATCCAATCGAACA
ACTGCCGAACACCTCAAATGAGCCATACTAAGAATCACAAATGGGCCGGA
AACCACAAACTCAGCCCTAACCCCCCGCTACGTTGTCGCGGTAATCACGT
CTACCCAACAGCCCAATAAGCAGCGTATTACAATGGGTTTTATCACCCAC
```
>2
```
AGATAAAACTCCCCCCATATGAAAATAAAGTACGCCACCGGCAACAAACA
ATTTTTCTCAACCCGTCACTAAAGATCAAAAACAAACGTATATGCACCCC
AGGATTCGAAATATGATGACTCACCTTACGCTCAATAGATAACACATCAT
CATAACGCGACTATACTAAGTCAAGAGTATTTCCACCCAGAACCACCTCC
```
>3
```
ACAGCTGGCACAACCCGATGTGCCATTACGTCTCCTCGCAAATCTCGCTT
ATAACCCCAACATACAAAAAACACCATAGATAGTGACATAACACGCTAAC
TCCTCCGTTCGTCTCAAAGTAATCCAAGTCCTGCCAAGCTAAGCTTGCGT
TAAAGCATCATACAAACTTTACGCACAACCGCTATATTCATGCAACTGCA
```

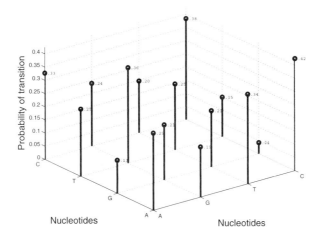

**FIGURE 8.9**
Probability of transition in DNA sequences for the 1st order Markov chains
$< \Phi(N_{t-1}) \ \Phi(N_t) >$ (Example 8.2).

**Example 8.10** *(Continuation of Example 8.3). The following protein sequences have been generated based on the 1st order Markov model.*

```
>1
        GSYEPLGKWPGRGYGGTRGGLGKAHFGAGEPLRFHGQHGTPGKYGMQGSS
        GEGMQMYLKWKSMQQPRYQVERAHGQWEFKVMGPPRSKVYGHDGTTRYGH
        EDYKGIKVMHTYYSKKNFKNEDYLMPPHNKPMIYEVRFFAQTPQGMKDEA
        GPYWPRTQKGTQASGPQQTGMYGKNGGGEHLTRYWGVGTGDCMGNRVARK
>2
        RDNPNEHEGTWPENHNNVTGYVGGEVSANMYVDQAGVYYSLNRHPSQYKP
        ELGSIGYTPPMINLRNNYDKWLSGHGRGLCGDWGGAKWLLPWYRGWSEYG
        GNIWGGDTKSGSSLDGMYVMKGLPPTHGEPFIGHFWGWENGEGFGWKWRA
        SEPGRKGWGHSKQTPDGMGYGAERTIGGAGHRKWKCMCQTGGLTHRMDLQ
>3
        HGHEPAQSLPSKPWRVDNKMVDQNPPLVGRSVPTRGGPGNQPGTHGLYLT
        THFLLGNGNPFNVSGGVYGGGMGWKDGRMNPYSGGNKQAGGNGYCQQDGT
        NQGLKVSASVRNSWPGRTRLGSVKQHSLTGGIGRKWEGRGGYDSLGRGQH
        MEELGNGYRIVLSMNPGTTKQPCSQRDHYNMGMGGGAMGGYGDHKMPWLWL
```

## 8.4.3   2nd order Markov model

In a 2nd order Markov chain, each base depends on two preceding bases, further improving the quality of results. Figures 8.11 and 8.12 illustrate the probability of transition for the 2nd order Markov model. For instance, the probability of the DNA base A followed by AA is 0.31, the probability of the protein amino acid W followed by NK is 0.25 (all probabilities are evaluated in a group).

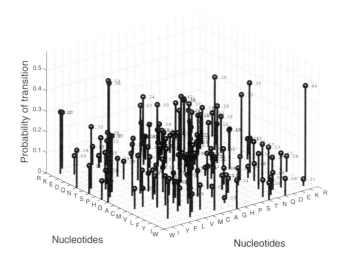

**FIGURE 8.10**

Probability of transition in protein sequences for the 1st order Markov chains $< \Phi(N_{t-1}) \ \Phi(N_t) >$ (Example 8.3).

**Example 8.11** *(Continuation of Example 8.2) The following DNA sequences have been generated based on the 2nd order Markov model.*

>1

```
CCGCCCTCTTCCCATAGAACGGATTAGACATGCACACGCCCACTCGGCAT
CCGCACTTAGGAACTTAGGATCTACCTGCGATTAAAAAAAAGGAAGGCTAT
AATTCAACAACTCCACCACGACTTTAGCGCACATGAAATTCCCATGCACT
ACGAAACTTTTTAACGTAAAGTAGTTTTTCAATTACTTAGAAGCTTACAA
```

>2

```
CCGAACGGTCGCCGGAAAACCAGCTTTGATTACTACCTCAAGATTGTTTA
ACTCCCTATACTAAATATCCGGAAAAAAGACTACACTAGTAACATCAAGA
GACCTACAACCACTTTACCTGAGGAGAAATATGGATCTTCGCGAACAAAC
CCGCTAAAACTTCACTGAACCAATTCAACCTAGTTATTGAGGTGTATACA
```

>3

```
ACTCATCCCACCGCGCCTAGAAATCAAGGTCTAGTAAGCAAACCCTAATG
AAAAATACCGCAAACATGCATACCTCTAAGTGATTTTGCGGTCACTCCAA
AACCCCTAAAATAGCTACTCAAATTACCCCTCCCTCCACCTTTCACCTCC
ATAGGCATAAACTCAGAGTCCTAGTTCAGTTGCATGCTCGATCAGTCAAA
```

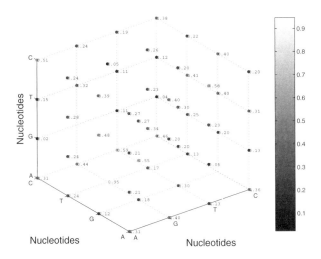

## FIGURE 8.11

Probability of transition in DNA sequences for the 2nd order Markov chains $< \Phi(N_{t-2})\ \Phi(N_{t-1})\ \Phi(N_t) >$.

**Example 8.12** *(Continuation of Example 8.3). The following protein sequences have been generated based on the 2nd order Markov model.*

```
>1
    KPQVAMGWPPGGMTMIGPSGGRMSRGPGPRAYHYSYHQGMGQYNGVLMND
    QHGHGGSGNVTHSGDGVGGWPEGGGRQWAGLGGQATWGGKPLFGQGGQNG
    GTLMAASDVMPKPNGNLGQQWLYGAEESETEGWKTYGNPMDHGSYRPRGP
    KPLSGPHWGFGYPMSYGEVQTYTVGTEYEGNVGPWRPGHQIPGPVVPQMW
>2
    LMPNEPVYMGKDKRGGTYGYAPSPPLLGGGSVKTHHPHYQNHLTKVQGPR
    HGEMCPLWPLAGGGMGMMFMVSQRWMGYHLARGMMKNGNGLGAHTRTTKR
    QPMPAGGPEDNGYTGQQSLEPTTGYNGVLMKYQNARGQMQMQYDGGMGMK
    TSTGEQYRLGNVPTKGGTPMDGPGNPGQWSKQKGRVGGWGACPSQGQEYI
>3
    YGRGPGEQPPSPTPTGPTVWVRVFQHGYGPIYRTQKVRHTIQEQPSKKRS
    VYKLGGIPGTGFCLGRPYLGDSPPGMYGPKILGDGGHGGKPWNGWHWDTN
    AKNVPGWMMAQQGMQVGQTHYAVYVRAHVLYMVNGLYNPAMMNPVGQTTW
    WPHPGKYQWQYNGHTNWHFSPVWGDWDQIGWGVGQLGPYTVGTTCNPAGE
```

## 8.5 Postprocessing: pairwise alignments

During the validation stage, the generated sequence is compared to the existing database of sequences in order to evaluate error rates. If the validation stage returns inappropriate results in terms of error rates, the generated sequence is discarded and not included in the final solution.

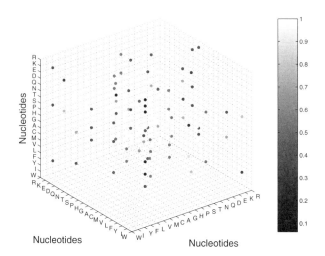

**FIGURE 8.12**
Probability of transition in protein sequences for the 2nd order Markov chains $< \Phi(N_{t-2}) \ \Phi(N_{t-1}) \ \Phi(N_t) >$.

An *alignment* between two sequences is simply a pairwise match between the characters of each sequence. As was pointed out earlier, three kinds of changes can occur at any given position within a sequence:

▶ A mutation that replaces one character with another;

▶ An insertion that adds one or more positions; or

▶ A deletion that eliminates one or more positions. Insertions and deletions have been found to occur in nature at a significantly lower frequency than mutations.

Since there are no common ancestors for inserted or deleted nucleotides while comparing sequences, *gaps* in alignments are commonly added to reflect the occurrence of this type of change.

When validating synthetic sequences, the scoring function is determined by an amount of credit an alignment receives for each aligned pair for

▶ Identical entities (the *match* score),

▶ A penalty for aligned pairs of non-identical entities (the *mismatch score*), and

▶ The *gap penalty* for the alignment with gaps.

A simple alignment score between two sequences *seq1* and *seq2* can be computed as follows:

$$\sum_{i=1}^{n} \begin{cases} gap \ penalty; \ if \ seq1_i =' -' \ or \ seq2_i =' -', \\ match \ score; \ if \ no \ gaps \ and \ seq1_i = seq2_i, \\ mismatch \ score; \ if \ no \ gaps \ and \ seq1_i \neq seq2_i. \end{cases}$$

Two frequently used scoring matrices for DNA sequences are given below:

*Identity matrix*

|   | A | T | C | G |
|---|---|---|---|---|
| A | 1 | 0 | 0 | 0 |
| T | 0 | 1 | 0 | 0 |
| C | 0 | 0 | 1 | 0 |
| G | 0 | 0 | 0 | 1 |

*BLAST matrix*

|   | A | T | C | G |
|---|---|---|---|---|
| A | 5 | -4 | -4 | -4 |
| T | -4 | 5 | -4 | -4 |
| C | -4 | -4 | 5 | -4 |
| G | -4 | -4 | -4 | 5 |

Since an exhaustive search for all possible alignments is an NP-complete problem, it is impossible to compute any real sequences in a reasonable amount of time. We can overcome this problem by applying *dynamic programming* principles (see works by Gusfield [10], and by Krane and Raymer [13]). The alignment problem is broken down into sub-problems, each with reasonable size. Assuming uniform mismatch and gap penalties, there are three possibilities for each position in our alignment:

▶ Place a gap in the first sequence (not likely in the case where the first sequence is longer than the second one);

▶ Place a gap in the second sequence;

▶ Place a gap in neither sequence.

**Example 8.13** *Let us consider two sequences ACAGTAG and ACTCG, where the gap penalty is -1, the match score is +1, and the mismatch score is 0. The alignment of the two sequences is equivalent to a path from the upper left corner of the table to the lower right. A horizontal move in the table represents a gap in the sequence along the left axis. A vertical move represents a gap in the sequence along the top axis, and a diagonal move represents an alignment of nucleotides from each sequence. To fill in the table, we take the maximum value of the described three choices (Figure 8.13).*

The described algorithm performs a *global* alignment. Two sequences are compared entirely. This is not always the most desirable way to align two sequences. When searching for the best alignment between a short sequence and a longer sequence, we might avoid penalizing for gaps that appear at one or both ends of the sequence. This approach is sometimes referred as a *semiglobal* alignment. A few changes need to be made to support semiglobal alignment in our dynamic programming approach. To allow gaps at the end of the sequence without penalty, we need to change the interpretation of the table slightly. We initialize the first column and the first row of the table to all zeros.

|   | A | C | T | C | G |
|---|---|---|---|---|---|
| **0** | -1 | -2 | -3 | -4 | -5 |
| A -1 | 1 |  |  |  |  |
| C -2 |  |  |  |  |  |
| A -3 |  |  |  |  |  |
| G -4 |  |  |  |  |  |
| T -5 |  |  |  |  |  |
| A -6 |  |  |  |  |  |
| G -7 |  |  |  |  |  |

*Step 1: Fill in the first row and the first column assuming that gaps are inserted.*

|   | A | C | T | C | G |
|---|---|---|---|---|---|
| 0 | -1 | -2 | -3 | -4 | -5 |
| A -1 | 1 | 0 | -1 | -2 | -3 |
| C -2 | 0 | 2 | 1 | 0 | -1 |
| A -3 | -1 | 1 | 2 | 1 | 0 |
| G -4 | -2 | 0 | 1 | 2 | 2 |
| T -5 | -3 | -1 | 1 | 1 | 2 |
| A -6 | -4 | -2 | 0 | 1 | 1 |
| G -7 | -5 | -3 | -1 | 0 | 2 |

*Step 2: Fill in the rest of the alignment table by sliding the $2 \times 2$ window from the upper left corner to the lower right. For example, the first iteration with the $2 \times 2$ window analyzes the result of placing gaps (-A in both cases) and the match between A and A. Three moves result in: -2 (vertical), -2 (horizontal) and 1 (diagonal). The maximum value corresponds to the diagonal move and equal 1.*

|   | A | C | T | C | G |
|---|---|---|---|---|---|
| **0** | -1 | -2 | -3 | -4 | -5 |
| A -1 | **1** | 0 | -1 | -2 | -3 |
| C -2 | 0 | **2** | 1 | 0 | -1 |
| A -3 | -1 | **1** | 2 | 1 | 0 |
| G -4 | -2 | **0** | 1 | 2 | 2 |
| T -5 | -3 | -1 | **1** | 1 | 2 |
| A -6 | -4 | -2 | 0 | **1** | 1 |
| G -7 | -5 | -3 | -1 | 0 | **2** |

*Step 3: Trace back from the lower right corner to upper left in order to obtain aligned sequences.*

**FIGURE 8.13**
Algorithm for Example 8.13.

**Example 8.14** *Let us consider the following alignment: ACACTGATCG and ACACTG.*

|   | A | C | A | C | T | G | A | T | C | G |
|---|---|---|---|---|---|---|---|---|---|---|
| **0** | 0 | 0 | 0 | 0 | 0 | 0 | 0 | 0 | 0 | 0 |
| A 0 | **1** | 0 | 1 | 0 | 0 | 0 | 1 | 0 | 0 | 0 |
| C 0 | 0 | **2** | 1 | 2 | 1 | 0 | 0 | 1 | 1 | 0 |
| A 0 | 1 | 1 | **3** | 2 | 2 | 1 | 1 | 0 | 1 | 1 |
| C 0 | 0 | 2 | 2 | **4** | 3 | 2 | 1 | 1 | 1 | 1 |
| T 0 | 0 | 1 | 2 | 3 | **5** | 4 | 3 | 2 | 1 | 1 |
| G 0 | 0 | 0 | 1 | 2 | 3 | **6** | **6** | **6** | **6** | **6** |

*By initializing the first row and column with zero values, and by allowing free horizontal and vertical moves in the last row and column of the table, we modified the dynamic algorithm to search for semiglobal alignments.*

Sometimes even semiglobal alignments do not afford the flexibility needed in a sequence search such as a search for common subsequences. For this sort of comparison, a semiglobal alignment will not suffice, since each alignment will be penalized for every nonmatching position. The appropriate tool for this sort of search is a *local* alignment, which will find the best matching subsequences within the two search sequences.

> **Example 8.15** *Let us consider the following two sequences: AACCTATAGCT and GCGATATA. Using the semiglobal alignment algorithm, we obtain the following alignment:*
>
> ```
> AAC-CTATAGCT
> -GCGATATA---
> ```
>
> *This is a very poor alignment. It does not reveal that there is a matching region in the center of the two sequences: the subsequence TATA. To perform local alignment, we modify the global alignment algorithm by allowing a fourth option when filling in the partial scores table. We place a zero in any position in the table if all of the other cases result in scores less than zero.*

|   | A | A | C | C | T | A | T | A | G | C | T |
|---|---|---|---|---|---|---|---|---|---|---|---|
|   | 0 | 0 | 0 | 0 | 0 | 0 | 0 | 0 | 0 | 0 | 0 |
| G | 0 | 0 | 0 | 0 | 0 | 0 | 0 | 0 | 1 | 0 | 0 |
| C | 0 | 0 | 0 | 1 | 1 | 0 | 0 | 0 | 0 | 2 | 1 |
| G | 0 | 0 | 0 | 0 | 1 | 1 | 0 | 0 | 1 | 1 | 2 |
| A | 0 | 1 | 1 | 0 | 0 | 1 | 2 | 1 | 1 | 0 | 1 | 2 |
| T | 0 | 0 | 1 | 1 | 0 | 1 | 1 | 3 | 2 | 1 | 0 | 2 |
| A | 0 | 1 | 1 | 1 | 1 | 0 | 2 | 2 | 4 | 3 | 2 | 2 |
| T | 0 | 0 | 1 | 1 | 1 | 2 | 1 | 3 | 3 | 4 | 3 | 3 |
| A | 0 | 1 | 1 | 1 | 1 | 1 | 3 | 3 | 4 | 4 | 4 | 4 |

## 8.6 Algorithm for sequence generation

The final algorithm consists of the following steps:

*Step 1:* Select a sample set of sequences from the database.

*Step 2:* Generate statistical data from the sample set.

*Step 3:* Applying a random number generator and considering various distributions from the sample set, determine nucleotides and the parameters of the sequence.

*Step 4:* Perform the necessary format transformations to comply with the format of database storage.

*Step 5:* Perform postprocessing validation of the sequence.

Exploiting the basic principles of sequence generation, the following sequence generation techniques were discussed:

*Generator 1:* The generator is based on the mutation technique.

*Generator 2:* The generator relies on Markov models for sequence generation.

*Generator 3:* The generator is based on sequence processing and generating nucleotides based on pre-existing sequences and their recombination.

## 8.7   Summary

DNA is considered to be a future application of biometrics not affected by environmental and physical disturbances. Historically, DNA signatures used in forensics have proven that people can be identified with high accuracy. The main problem with DNA is the excessive time and resources required for sequencing and processing (alignment, matching, storing, etc.). Another problem is a privacy concern which promotes the work on synthetic DNA instead of actual sequences taken from humans.

1. DNA and protein sequences for identification and verification may hold promise for future application, but the technology and legal precedent, and public willingness to conduct this type of testing for whole populations is lacking.

2. In the inverse biometrics of DNA, it is essential to comply with statistical characteristics, but also to consider genetic attributes. The sequential nature of DNA characteristics allows us to design a model which incorporates the knowledge of all possible combinations of values and their transition from one to another accommodating genetic relevance in nucleotide generation.

3. In our study, we used Markov systems to simulate the sequential behavior of DNA/protein sequences. Models based on Markov chains are proved to be more efficient in terms of storage and run-time for sequence synthesis. In addition to the sequence generator based on Markov chains, we proposed sequence generators based on the mutation technique and sequence recombination.

4. The functionality of the synthesis algorithm is enabled through the integration of three main components:

   ▶ Learning modules which obtain the information from the pre-existing database;
   ▶ A sequence processing toolbox with sequence synthesis tools;
   ▶ An evolving database of DNA and protein sequences.

## 8.8 Problems

**Problem 8.1** Determine the alignment score for the two sequences ACAGTCGAACG and ACCGTCCG (match score = +1, mismatch score = 0, gap penalty = -1). What is the optimal alignment between two sequences? *Hint:* use the global alignment algorithm based on dynamic programming principles (fill in the alignment table where the alignment of the two sequences is equivalent to a path from the upper corner of the table to the lower right).

**Problem 8.2** Determine the semiglobal alignment score for the two sequences ACACGAAGTGAACG and ACGTCG (match score = +1, mismatch score = 0, gap penalty = -1). What is the optimal semiglobal alignment between two sequences? *Hint:* there are no penalties for gaps appearing at one or both ends of the sequence (the interpretation of the alignment table is changed).

**Problem 8.3** Determine the local alignment for the two sequences ACGTATCGCGTATA and GATGCTCTCGGAAA (match score = +1, mismatch score = 0, gap penalty = -1). What is the best matching subsequence? *Hint:* zero is placed in any position in the alignment table if all other cases result in scores less than zero.

**Problem 8.4** Design a generator of random numbers in order to select randomly positions in a given sequence of nucleotides. *Hint:* the generator is supposed to identify a single position for other operations

**Problem 8.5** (Continuation of Problem 8.4) Randomly generating the position of a nucleotide, apply the mutation technique to perform nucleotide mutation for the sequence GATGCTCTCGTAAGTTGGAAA. Perform post-processing comparison with the original sequence in order to determine the alignment score. Repeat the mutation technique for two, three and four mutating positions. Draw a plot to visualize the correspondence between the alignment score and the number of mutating positions. *Hint:* analyze the nucleotide currently located at the randomly generated position, randomly select different nucleotide.

**Problem 8.6** (Continuation of Problem 8.4) Randomly generating the position of a nucleotide, apply the mutation technique to perform nucleotide insertion for the sequence GATGCTCTCGTAAGTTGGAAA. Perform post-processing comparison with the original sequence in order to determine the alignment score. Repeat the operation inserting two, three and four nucleotides. Draw a plot to visualize the correspondence between the

alignment score and the number of inserted nucleotides.
*Hint:* insert a nucleotide which is randomly selected from four different bases A, C, G, and T.

**Problem 8.7** (Continuation of Problem 8.4) Randomly generating the position of a nucleotide, apply the mutation technique to perform nucleotide deletion for the sequence GATGCTCTCGTAAGTTGGAAA. Perform post-processing comparison with the original sequence in order to determine the alignment score. Repeat the operation deleting two, three and four nucleotides. Draw a plot to visualize the correspondence between the alignment score and the number of deleted nucleotides.
*Hint:* post-processing comparison is based on global alignment (use the following characteristics: match score = +1, mismatch score = 0, gap penalty = -1).

**Problem 8.8** (Continuation of Problem 8.4) Randomly generating the position of a nucleotide in the sequence GATGCTCTCGTAAGTTGGAAA, apply the recombination technique replacing the current nucleotide by a nucleotide taken randomly from the second sequence ATCCGTAGGTTGAC. Perform post-processing comparison with the original sequence in order to determine the alignment score.
*Hint:* compare the new sequence with the first sequence using the following characteristics: match score = +1, mismatch score = 0, gap penalty = -1.

## Advanced problems and hands-on projects

**Problem 8.9** Retrieve a mitochondrial DNA (mtDNA) from GenBank with the accession number X93334. Save the retrieved sequence in a FASTA formatted file.

**Problem 8.10** (Continuation of Problem 8.9) Using the FASTA formatted file with the accession number X93334, determine the probabilities for the 0th order Markov model. Plot the probabilities of nucleotides.

**Problem 8.11** (Continuation of Problem 8.9) Using the FASTA formatted file with the accession number X93334, determine the probabilities for the 1st order Markov model. Plot the probabilities of nucleotides.

**Problem 8.12** (Continuation of Problem 8.9) Using the FASTA formatted file with the accession number X93334, determine the probabilities for the 2nd order Markov model. Plot the probabilities of nucleotides.

**Problem 8.13** (Continuation of Problem 8.9) Design a generator of random numbers in order to synthesize a sequence of nucleotides complying with probabilities of the 0th/1st/2nd order Markov models.

## Supporting information for problems

The MATLAB Bioinformatics Toolbox includes many functions to help with biological sequence analysis and synthesis. The sequence databases currently supported by this toolbox are GenBank (functions `getgenbank` and `genbankread`), GenPept (functions `getgenpept` and `genpeptread`), European Molecular Biology Laboratory EMBL (functions `getembl` and `emblread`), Protein Sequence database PIR-PSD (functions `getpir` and `pirread`), and Protein Data Bank PDB (functions `getpdb` and `pdbread`). Here is an overview of some other useful MATLAB functions.

### MATLAB function `fastaread`
Reads a FASTA formatted file into the following fields: `Header` and `Sequence`. For example,
```
[Header, Sequence] = fastaread('fasta.file');
```

### MATLAB function `nwalign`
Implements the Needleman-Wunsch algorithm for global alignment of two sequences. The function aligns both amino acids (type 'AA') and nucleotides (type 'NT'). The function returns the alignment score in bits for the optimal alignment. The scale factor used to calculate the score is provided by the scoring matrix information. If this is not defined, then `nwalign` returns the raw score. The syntax is
```
[Score, Alignment] = nwalign(Sequence1, Sequence2);
```

### MATLAB function `swalign`
Implements the Smith-Waterman algorithm for local alignment of two sequences. The function aligns both amino acids (type 'AA') and nucleotides (type 'NT'). The function returns the alignment score in bits for the optimal alignment. The syntax is
```
[Score, Alignment] = swalign(Sequence1, Sequence2);
```

### MATLAB function `seqdotplot`
Creates the dot plot of two sequences. The dot plot helps to analyze alignments by the visual comparison of graphical dot plots. One sequence is displayed along the horizontal axis while the other is displayed along the vertical axis. Similarity between two sequences results in match scores along the diagonal in the dot matrix. Deviations between the sequences result in off-line dots in the plot. The syntax is
```
seqdotplot(Sequence1, Sequence2);
```

### MATLAB function `showalignment`
Displays a sequence alignment with color. The function accepts the output from either the function `swalign` or `nwalign`. The function displays an

alignment string with matches and similar residues highlighted with color. The syntax is

```
[Score, Alignment] = swalign(Sequence1, Sequence2);
showalignment(Alignment);
```

## 8.9    Further reading

**DNA fingerprinting.** Current DNA fingerprinting technology can perform personal identification with DNA recovered from a variety of human tissues. DNA evidence can give numerous clues to crime investigation and can even be used to trace human migration patterns (some important results are reported by Gill, Jeffreys, and Werrett [9], and by Krawczak and Schmidtke [14]). Current theories say that modern humans evolved in Africa between 100,000 and 200,000 years ago, and originally consisted of a small group of 10,000 individuals as was discussed by Gibbons [8]. Then humans dispersed out of Africa about 100,000 years ago, reaching Asia about 60,000 years ago and migrating into New Guinea and Australia during the ice ages 40,000 years ago. Indeed, DNA studies of mitochondrial DNA (mtDNA) proves close relationship between different ethnic groups (extended experimental results are reported by Bohlmeyer et al. [5]).

A high percentage (about 99%) for DNA similarity between two humans makes the DNA comparison challenging and DNA fingerprinting more sophisticated. The following techniques are used for DNA fingerprinting:

*Satellite DNA.* Some parts of human DNA are made from subsequences, called "satellite DNA," repeated several times in a row. The number of repetitions differs from person to person (see the paper by Benson [2]). The chance that two humans have the same number of repetitions at all areas of satellite DNA is extremely small.

*Enzymes.* The special purpose enzymes can cut DNA sequences at defined positions. For example, the EcoR1 enzyme cuts DNA whenever it sees the letters GAATTC. So, DNA sequences from different humans will contain different distributions of cut positions, which allows one to make decisions about person's identity (see the paper by Jones Ougham and Thomas [11]).

DNA technology continues to improve. For example, a new fingerprinting technique is outlined by Vos in [20]. Other examples include portable credit-card-sized chips, so-called "labs on a chip," which will allow crime scene investigators to process DNA samples at the crime scene and get results within a shorter amount of time. The entire DNA analysis will happen in a portable unit, which contains everything needed to cut, amplify, tag, and analyze the DNA. Another relationship to processors and signal processing techniques is presented by Vaidyanathan in [17].

**Biological sequence processing by HMMs.** A Hidden Markov Model (HMM) is a graph of connected states, where each state possesses a variety of properties. Processes described by HMMs evolve in some dimension, often parameterized by time stamps. HMMs work assuming that the current state is not influenced by the entire history of states, instead only one previous state is considered. Models are interpreted by analyzing possible paths in the structure of states. The structure of model states is built by learning from the data, in which parameters are estimated by applying Bayesian classifiers. Bayesian classifiers provide a framework for obtaining a posteriori probability (the probability of the model, given the observed sequence) that includes the ability to integrate prior knowledge about the sequences as, for example, described in papers by Avery and Henderson [1], Blaisdell [4], and Churchill [6].

For the majority of biological sequences, the time parameterization is replaced by the position in the sequence. HMMs can naturally accommodate variable-length models including operations of mutation, insertion and deletion (see some properties outlined by Durbin et al. [7]). This is generally achieved by having a state which has a transition back to itself. Because most biological data has variable-length properties, machine learning techniques which require a fixed-length input, such as neural networks or support vector machines, are less successful in biological sequence analysis (see the review by Birney [3]).

Future research in HMMs for biological sequences will include the following:

*Biological relevance.* HMMs will accommodate biological meaning and DNA context to improve the accuracy of analysis and prediction.

*Structural relevance.* The structural information will be integrated into HMMs allowing better parameterization of biological sequences.

*Modeling relevance.* Modeling architectures of HMMs will coincide with meaning and parameters obtained from biological sequences.

**Regulatory sequence analysis tools.** Despite the essential role played by non-coding sequences in transcriptional regulation, genome annotations usually focus on identifying the genes and predicting their function through sequence similarity searches. The services offered by most genome centers are restricted to the analysis of coding sequences.

The web resource Regulatory Sequence Analysis Tools (RSAT) (http://rsat.ulb.ac.be/rsat) developed by van Helden [18, 19] is dedicated to the analysis of the other part of the genomes: the non-coding sequences. It offers a collection of software tools dedicated to the prediction of regulatory sites in non-coding DNA sequences. These tools include sequence retrieval, pattern discovery, pattern matching, genome-scale pattern matching, feature-map drawing, random sequence generation and other utilities. Alternative formats are supported for the representation of regulatory motifs (strings or position-specific scoring matrices) and several algorithms are proposed for pattern discovery.

**Available database searches.** The following public DNA and protein repositories have been built throughout past decades to systematize the knowledge about biological sequences in searchable databases:

*GenBank.* GenBank is the NIH (National Institutes of Health, U.S.A.)[†] database maintained and distributed by NCBI (National Center for Biotechnology Information) that stores all known public DNA sequences.[‡] Sequence data are submitted to GenBank by individual scientists from around the world. In addition to GenBank, NCBI supports and distributes a variety of databases for the medical and scientific communities.

*EMBL.* The EMBL (European Molecular Biology Laboratory) Nucleotide Sequence database constitutes Europe's primary nucleotide sequence resource.[§] Main sources for DNA and RNA sequences are direct submissions from individual researchers, genome sequencing projects and patent applications.

*DDBJ.* DDBJ (DNA Data Bank of Japan) functions as the international nucleotide sequence database in collaboration with EBI/EMBL and NCBI/GenBank.[¶]

*mtDNA online.* A human mitochondrial genome database provides comprehensive data on mitochondrial and human nuclear encoded proteins involved in mitochondrial biogenesis.[‖]

There are some successfully adopted formats for DNA and protein sequence representation and their retrieval from databases sources: EMBL, FASTA, GCG, GenBank and others. Without losing generality, we represent sequences using the FASTA format. A sequence file in the FASTA format can contain several sequences, and a FASTA formatted file consists of three parts:

▶ A title line beginning with one or more '>' symbols, which do not form the title.

---

[†]http://www.ncbi.nlm.nih.gov/

[‡]GenBank, http://www.ncbi.nlm.nih.gov/

[§]http://www.ebi.ac.uk/embl/

[¶]http://www.ddbj.nig.ac.jp/

[‖]http://www.mitomap.org/

- ▶ Optional annotation lines, beginning with ';'.
- ▶ The sequence itself, containing one or multiple lines and continuing until the end of file or the next '>' is reached.

All sequences in our experiments are given in the FASTA format.

## 8.10   References

[1] Avery P, and Henderson D. Fitting Markov chain models to discrete state series such as DNA sequences. *Journal of the Royal Statistical Society (Series C): Applied Statistics*, 48(1):53–61, 1999.

[2] Benson G. Tandem repeats finder: a program to analyze DNA sequences. *Nucleic Acids Research*, 27:573–580, 1999.

[3] Birney E. Hidden Markov models in biological sequence analysis. *IBM Journal of Research and Development*, 45(3/4):449–454, 2001.

[4] Blaisdell BE. Markov chain analysis finds a significant influence of neighboring bases on the occurrence of a base in eucariotic nuclear DNA sequences both protein-coding and noncoding. *Journal of Molecular Evolution, Springer, NY*, 21:278–288, 1985.

[5] Bohlmeyer J, Harp B, Heptig B, Kinkade K, Rayson C, Strohm J, Wacker C, Fraga R, and Popel D. Phylogenetic lineage of humans and primates. In *Proceedings of the Midwest Instruction and Computing Symposium*, Morris, MN, April 2004.

[6] Churchill GA. Hidden Markov chains and the analysis of genome structure. *Computers and Chemistry*, 16:107–115, 1992.

[7] Durbin R, Eddy S, Krogh A, and Mitchison G. *Biological Sequence Analysis: Probabilistic Models of Proteins and Nucleic Acids*. Cambridge University Press, Cambridge, 1998.

[8] Gibbons A. Calibrating the mitochondrial clock. *Science*, 279:28–29, 1998.

[9] Gill P, Jeffreys A, and Werrett D. Forensic application of DNA fingerprints. *Nature*, 318(6046):577–579, 1985.

[10] Gusfield D. *Algorithms on Strings, Trees, and Sequences. Computer Science and Computational Biology*. Cambridge University Press, Cambridge, 1997.

[11] Jones N, Ougham H, and Thomas T. Markers and mapping: we are all geneticists now. *New Phytologist*, 137:165–177, 1997.

[12] Khudyakov Yu and Fields H. *Artificial DNA: Methods and Applications.* CRC Press, Boca Raton, FL, 2002.

[13] Krane DE and Raymer ML. *Fundamental Concepts of Bioinformatics.* Benjamin Cummings, San Francisco, CA, 2003.

[14] Krawczak M and Schmidtke J. *DNA Fingerprinting.* BIOS Scientific Publishers, Abington, Oxon, UK, 1994.

[15] Pevzner P. *Computational Molecular Biology. An Algorithmic Approach.* The MIT Press, Cambridge, MA, 2000.

[16] Rudin N, Inman K, Stolvitzky G, and Rigoutsos I. DNA based identification. In Jain AK, Bolle R, and Pankanti S, Eds., *Personal Identification in Networked Society*, pp. 287–325, Kluwer, Dordrecht, 1998.

[17] Vaidyanathan PP. Genomic and proteomics: a signal processor's tour. *IEEE Circuits and Systems Magazine*, 4(4):6–29, 2004.

[18] Van Helden J. Prediction of transcriptional regulation by analysis of the non-coding genome. *Current Genomics*, 4:217–224, 2003.

[19] Van Helden J. Regulatory sequence analysis tools. *Nucleic Acids Research*, 31(13):3593–3596, 2003.

[20] Vos P, Hogers R, Bleeker M, Reijans M, Van de Lee M, Hornes M, Frijters A, Pot A, Peleman J, and Kuiper M. AFLP: a new technique for DNA fingerprinting. *Nucleic Acids Research*, 23:4407–4014, 1995.

# Index

## 3D

curve, 82
face, 239
facial image, 48
image, 345
model, 48, 239
scanner, 247
space, 82, 84
thermogram, 247
visualization, 139

## A

Access to a system, 3, 22
Accuracy, 34, 90, 108, 131, 245, 246, 349
Acoustic signal, 22, 239
Acquisition, 106, 108, 109, 140, 262, 276, 352
Action unit (AU), 206, 239, 242
Aging, 215
Algorithm
Needleman-Wunsch, 373
Smith-Waterman, 373
Alignment, 366
Amino acids, 349, 352, 353
Analysis, 1, 2, 11, 13, 37, 67, 68
graphological, 250
voice stress, 250
Analysis-by-synthesis, 38, 57, 62, 89, 135, 140, 198, 224, 239, 263, 292
Approximation, 15, 80, 82, 84, 89, 158, 160, 292, 327
Arc length, 110
Arch, 153
Architecture, 2, 30

Artificial intelligence, 19, 30–32, 41, 98, 107
Assembling, 17, 32, 63, 66, 115, 154, 233, 336
Attack, 2–4
active, 3, 4
massive, 4
passive, 4
scenario, 4
Authentication, 6, 35, 112, 293
Automated
system, 31, 225, 234
technique, 7, 109
tools, 30, 108, 226, 242
Autonomic nervous system, 200, 201

## B

Background, 178, 269
Band
far-infrared, 9
mid-infrared, 9
spectrum, 246
Banking, 42, 140
Bayesian classifier, 375
Benchmark, 12, 46
Bernstein polynomial, 211
Bezier curve, 108, 159, 278
Bifurcation, 153
Bio-inspired approach, 47
Biomedicine, 43
Biometric
data, 2, 3, 14–16, 48
acceptable, 16, 62, 90, 136
assembling, 17, 63, 64
classification, 14

# U

Unauthorized
    access, 6, 22
    parties, 36
Uncertainty, 32, 34, 46, 154
Unit tangent vector, 110

# V

Validation, 129, 365, 369
Valley, 152
Vein pattern, 23
Verification, 7
Virtual examiner, 249
Voice
    identification, 15
    intonation, 208, 238
Voronoi
    cell, 338
    data structure, 298, 302, 319,
        320, 326
    diagram, 297–300, 326
    region, 302, 303, 305, 314
    site, 299
    transform, 299, 308, 310, 336

# W

Warping, 18, 76, 78, 217, 272
Watermarking, 16, 30, 42, 147,
        180, 346
    robust, 48
Weiner filter, 156
Whorl, 152, 153